States of Matter

" The 19 Recognized States of Matter "

(as of 2016)

Edited by Paul F. Kisak

Contents

9 Liquid crystal   80

Chapter 1

State of matter

Not to be confused with Phase (matter).

In physics, a **state of matter** is one of the distinct forms that matter takes on. Four states of matter are observable in everyday life: solid, liquid, gas, and plasma. Many other states are known to exist only in extreme situations, such as Bose–Einstein condensates, neutron-degenerate matter and quark-gluon plasma, which occur in situations of extreme cold, extreme density and extremely high-energy color-charged matter respectively. Some other states are believed to be possible but remain theoretical for now. For a complete list of all exotic states of matter, see the list of states of matter.

Historically, the distinction is made based on qualitative differences in properties. Matter in the solid state maintains a fixed volume and shape, with component particles (atoms, molecules or ions) close together and fixed into place. Matter in the liquid state maintains a fixed volume, but has a variable shape that adapts to fit its container. Its particles are still close together but move freely. Matter in the gaseous state has both variable volume and shape, adapting both to fit its container. Its particles are neither close together nor fixed in place. Matter in the plasma state has variable volume and shape, but as well as neutral atoms, it contains a significant number of ions and electrons, both of which can move around freely. Plasma is the most common form of visible matter in the universe.[1]

The term phase is sometimes used as a synonym for state of matter, but a system can contain several immiscible phases of the same state of matter (see Phase (matter) for further discussion of the difference between the two terms).

1.1 The four fundamental states

1.1.1 Solid

Main article: Solid

In a solid the particles (ions, atoms or molecules) are closely packed together. The forces between particles are strong so that the particles cannot move freely but can only vibrate. As a result, a solid has a stable, definite shape, and a definite volume. Solids can only change their shape by force, as when broken or cut.

In crystalline solids, the particles (atoms, molecules, or ions) are packed in a regularly ordered, repeating pattern. There are various different crystal structures, and the same substance can have more than one structure (or solid phase). For example, iron has a body-centred cubic structure at temperatures below 912 °C, and a face-centred cubic structure between 912 and 1394 °C. Ice has fifteen known crystal structures, or fifteen solid phases, which exist at various temperatures and pressures.[2]

Glasses and other non-crystalline, amorphous solids without long-range order are not thermal equilibrium ground states; therefore they are described below as nonclassical states of matter.

The four fundamental states of matter. Clockwise from top left, they are solid, liquid, plasma, and gas, represented by an ice sculpture, a drop of water, electrical arcing from a tesla coil, and the air around clouds, respectively.

A crystalline solid: atomic resolution image of strontium titanate. Brighter atoms are Sr and darker ones are Ti.

Solids can be transformed into liquids by melting, and liquids can be transformed into solids by freezing. Solids can also change directly into gases through the process of sublimation, and gases can likewise change directly into solids through deposition.

1.1.2 Liquid

Structure of a classical monatomic liquid. Atoms have many nearest neighbors in contact, yet no long-range order is present.

Main article: Liquid

A liquid is a nearly incompressible fluid that conforms to the shape of its container but retains a (nearly) constant volume independent of pressure. The volume is definite if the temperature and pressure are constant. When a solid is heated above its melting point, it becomes liquid, given that the pressure is higher than the triple point of the substance. Intermolecular (or interatomic or interionic) forces are still important, but the molecules have enough energy to move relative to each other and the structure is mobile. This means that the shape of a liquid is not definite but is determined by its container. The volume is usually greater than that of the corresponding solid, the best known exception being water, H_2O. The highest temperature at which a given liquid can exist is its critical temperature.[3]

1.1.3 Gas

Main article: Gas

A gas is a compressible fluid. Not only will a gas conform to the shape of its container but it will also expand to fill the container.

In a gas, the molecules have enough kinetic energy so that the effect of intermolecular forces is small (or zero for an ideal gas), and the typical distance between neighboring molecules is much greater than the molecular size. A gas has no definite shape or volume, but occupies the entire container in which it is confined. A liquid may be converted to a gas by heating at constant pressure to the boiling point, or else by reducing the pressure at constant temperature.

At temperatures below its critical temperature, a gas is also called a vapor, and can be liquefied by compression alone

The spaces between gas molecules are very big. Gas molecules have very weak or no bonds at all. The molecules in "gas" can move freely and fast.

without cooling. A vapor can exist in equilibrium with a liquid (or solid), in which case the gas pressure equals the vapor pressure of the liquid (or solid).

A supercritical fluid (SCF) is a gas whose temperature and pressure are above the critical temperature and critical pressure respectively. In this state, the distinction between liquid and gas disappears. A supercritical fluid has the physical properties of a gas, but its high density confers solvent properties in some cases, which leads to useful applications. For example, supercritical carbon dioxide is used to extract caffeine in the manufacture of decaffeinated coffee.[4]

1.1.4 Plasma

Main article: Plasma (physics)

Like a gas, plasma does not have definite shape or volume. Unlike gases, plasmas are electrically conductive, produce magnetic fields and electric currents, and respond strongly to electromagnetic forces. Positively charged nuclei swim in a "sea" of freely-moving disassociated electrons, similar to the way such charges exist in conductive metal. In fact it is this electron "sea" that allows matter in the plasma state to conduct electricity.

The plasma state is often misunderstood, but it is actually quite common on Earth, and the majority of people observe it on a regular basis without even realizing it. Lightning, electric sparks, fluorescent lights, neon lights, plasma televisions, some types of flame and the stars are all examples of illuminated matter in the plasma state.

A gas is usually converted to a plasma in one of two ways, either from a huge voltage difference between two points, or by exposing it to extremely high temperatures.

Heating matter to high temperatures causes electrons to leave the atoms, resulting in the presence of free electrons. At very high temperatures, such as those present in stars, it is assumed that essentially all electrons are "free", and that a very high-energy plasma is essentially bare nuclei swimming in a sea of electrons.

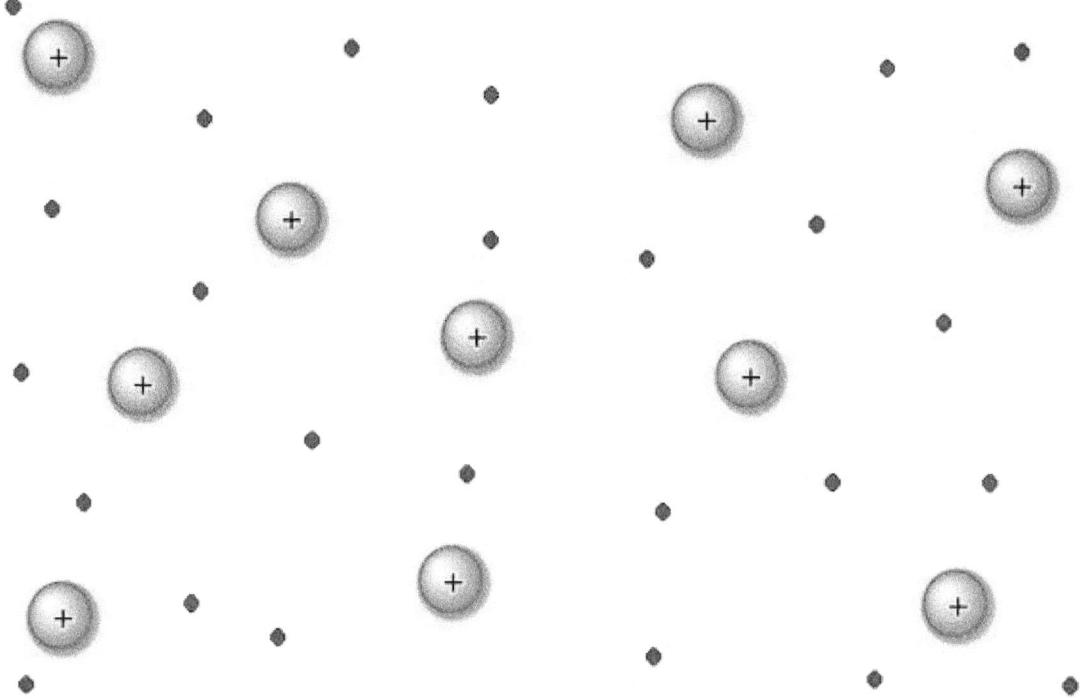

In a plasma, electrons are ripped away from their nuclei, forming an electron "sea". This gives it the ability to conduct electricity.

1.2 Phase transitions

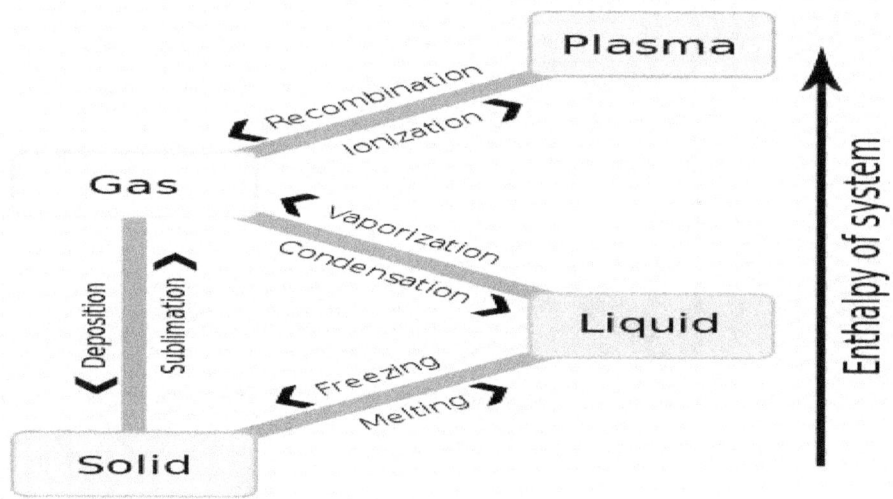

This diagram illustrates transitions between the four fundamental states of matter.

A state of matter is also characterized by phase transitions. A phase transition indicates a change in structure and can be recognized by an abrupt change in properties. A distinct state of matter can be defined as any set of states distinguished from any other set of states by a phase transition. Water can be said to have several distinct solid states.[5] The appearance of superconductivity is associated with a phase transition, so there are superconductive states. Likewise, ferromagnetic states are demarcated by phase transitions and have distinctive properties. When the change of state occurs in stages the intermediate steps are called mesophases. Such phases have been exploited by the introduction of liquid crystal technology. [6][7]

The state or *phase* of a given set of matter can change depending on pressure and temperature conditions, transitioning to other phases as these conditions change to favor their existence; for example, solid transitions to liquid with an increase in temperature. Near absolute zero, a substance exists as a solid. As heat is added to this substance it melts into a liquid at its melting point, boils into a gas at its boiling point, and if heated high enough would enter a plasma state in which the electrons are so energized that they leave their parent atoms.

Forms of matter that are not composed of molecules and are organized by different forces can also be considered different states of matter. Superfluids (like Fermionic condensate) and the quark–gluon plasma are examples.

In a chemical equation, the state of matter of the chemicals may be shown as (s) for solid, (l) for liquid, and (g) for gas. An aqueous solution is denoted (aq). Matter in the plasma state is seldom used (if at all) in chemical equations, so there is no standard symbol to denote it. In the rare equations that plasma is used in plasma is symbolized as (p).

1.3 Non-classical states

1.3.1 Glass

Main article: Glass

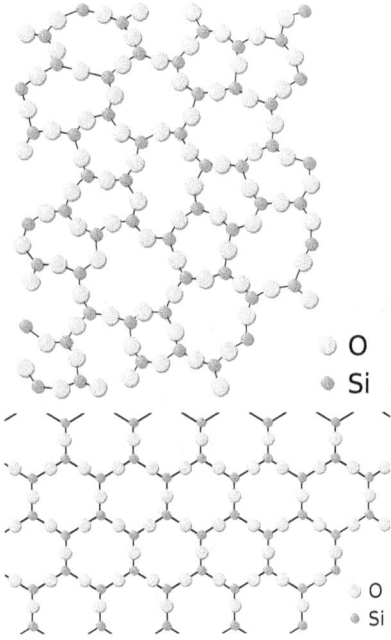

O

Si

O

Si

Schematic representation of a random-network glassy form (left) and ordered crystalline lattice (right) of identical chemical composition.

Glass is a non-crystalline or amorphous solid material that exhibits a glass transition when heated towards the liquid state. Glasses can be made of quite different classes of materials: inorganic networks (such as window glass, made of silicate plus additives), metallic alloys, ionic melts, aqueous solutions, molecular liquids, and polymers. Thermodynamically, a glass is in a metastable state with respect to its crystalline counterpart. The conversion rate, however, is practically zero.

1.3.2 Crystals with some degree of disorder

A plastic crystal is a molecular solid with long-range positional order but with constituent molecules retaining rotational freedom; in an orientational glass this degree of freedom is frozen in a quenched disordered state.

Similarly, in a spin glass magnetic disorder is frozen.

1.3.3 Liquid crystal states

Main article: Liquid crystal

Liquid crystal states have properties intermediate between mobile liquids and ordered solids. Generally, they are able to flow like a liquid, but exhibiting long-range order. For example, the nematic phase consists of long rod-like molecules such as para-azoxyanisole, which is nematic in the temperature range 118–136 °C.[8] In this state the molecules flow as in a liquid, but they all point in the same direction (within each domain) and cannot rotate freely.

Other types of liquid crystals are described in the main article on these states. Several types have technological importance, for example, in liquid crystal displays.

1.3.4 Magnetically ordered

Transition metal atoms often have magnetic moments due to the net spin of electrons that remain unpaired and do not form chemical bonds. In some solids the magnetic moments on different atoms are ordered and can form a ferromagnet, an antiferromagnet or a ferrimagnet.

In a ferromagnet—for instance, solid iron—the magnetic moment on each atom is aligned in the same direction (within a magnetic domain). If the domains are also aligned, the solid is a permanent magnet, which is magnetic even in the absence of an external magnetic field. The magnetization disappears when the magnet is heated to the Curie point, which for iron is 768 °C.

An antiferromagnet has two networks of equal and opposite magnetic moments, which cancel each other out so that the net magnetization is zero. For example, in nickel(II) oxide (NiO), half the nickel atoms have moments aligned in one direction and half in the opposite direction.

In a ferrimagnet, the two networks of magnetic moments are opposite but unequal, so that cancellation is incomplete and there is a non-zero net magnetization. An example is magnetite (Fe_3O_4), which contains Fe^{2+} and Fe^{3+} ions with different magnetic moments.

1.3.5 Microphase-separated

Main article: Copolymer
 Copolymers can undergo microphase separation to form a diverse array of periodic nanostructures, as shown in the example of the styrene-butadiene-styrene block copolymer shown at right. Microphase separation can be understood by analogy to the phase separation between oil and water. Due to chemical incompatibility between the blocks, block copolymers undergo a similar phase separation. However, because the blocks are covalently bonded to each other, they cannot demix macroscopically as water and oil can, and so instead the blocks form nanometer-sized structures. Depending on the relative lengths of each block and the overall block topology of the polymer, many morphologies can be obtained, each its own phase of matter.

1.3.6 Quantum spin liquid

Main article: Quantum spin liquid

A disordered state in a system of interacting quantum spins which preserves its disorder to very low temperatures, unlike other disordered states.

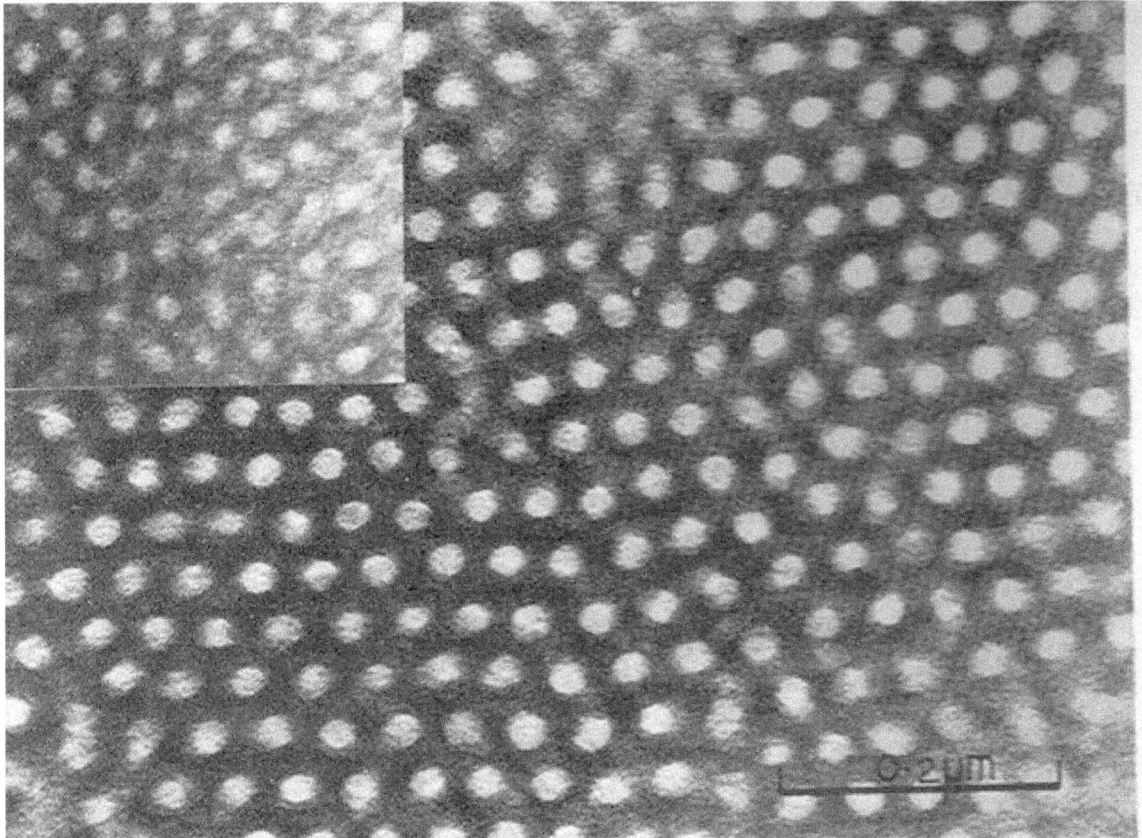

SBS block copolymer in TEM

1.4   Low-temperature states

1.4.1   Superfluid

Main article: Superfluid

Close to absolute zero, some liquids form a second liquid state described as **superfluid** because it has zero viscosity (or infinite fluidity; i.e., flowing without friction). This was discovered in 1937 for helium, which forms a superfluid below the lambda temperature of 2.17 K. In this state it will attempt to "climb" out of its container.[9] It also has infinite thermal conductivity so that no temperature gradient can form in a superfluid. Placing a superfluid in a spinning container will result in quantized vortices.

These properties are explained by the theory that the common isotope helium-4 forms a Bose–Einstein condensate (see next section) in the superfluid state. More recently, Fermionic condensate superfluids have been formed at even lower temperatures by the rare isotope helium-3 and by lithium-6.[10]

1.4.2   Bose–Einstein condensate

Main article: Bose–Einstein condensate

In 1924, Albert Einstein and Satyendra Nath Bose predicted the "Bose–Einstein condensate" (BEC), sometimes referred to as the fifth state of matter. In a BEC, matter stops behaving as independent particles, and collapses into a single quantum state that can be described with a single, uniform wavefunction.

Liquid helium in a superfluid phase creeps up on the walls of the cup in a Rollin film, eventually dripping out from the cup.

In the gas phase, the Bose–Einstein condensate remained an unverified theoretical prediction for many years. In 1995, the research groups of Eric Cornell and Carl Wieman, of JILA at the University of Colorado at Boulder, produced the first such condensate experimentally. A Bose–Einstein condensate is "colder" than a solid. It may occur when atoms have very similar (or the same) quantum levels, at temperatures very close to absolute zero (−273.15 °C).

1.4.3 Fermionic condensate

Main article: Fermionic condensate

A *fermionic condensate* is similar to the Bose–Einstein condensate but composed of fermions. The Pauli exclusion principle prevents fermions from entering the same quantum state, but a pair of fermions can behave as a boson, and multiple such pairs can then enter the same quantum state without restriction.

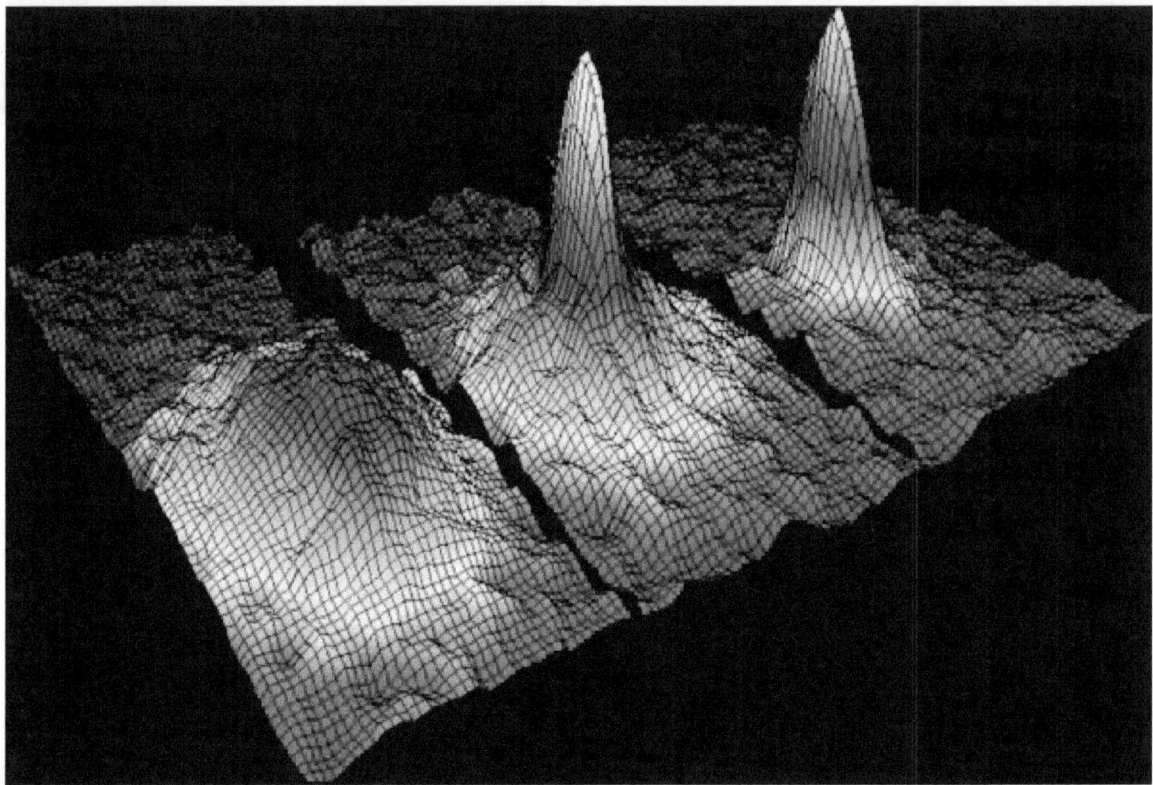

Velocity in a gas of rubidium as it is cooled: the starting material is on the left, and Bose–Einstein condensate is on the right.

1.4.4 Rydberg molecule

One of the metastable states of strongly non-ideal plasma is Rydberg matter, which forms upon condensation of excited atoms. These atoms can also turn into ions and electrons if they reach a certain temperature. In April 2009, *Nature* reported the creation of Rydberg molecules from a Rydberg atom and a ground state atom,[11] confirming that such a state of matter could exist.[12] The experiment was performed using ultracold rubidium atoms.

1.4.5 Quantum Hall state

Main article: Quantum Hall effect

A *quantum Hall state* gives rise to quantized Hall voltage measured in the direction perpendicular to the current flow. A *quantum spin Hall state* is a theoretical phase that may pave the way for the development of electronic devices that dissipate less energy and generate less heat. This is a derivation of the Quantum Hall state of matter.

1.4.6 Strange matter

Main article: Strange matter

Strange matter is a type of quark matter that may exist inside some neutron stars close to the Tolman–Oppenheimer–Volkoff limit (approximately 2–3 solar masses). It may be stable at lower energy states once formed.

1.4.7 Photonic matter

Main article: Photonic matter

In photonic matter, photons behave as if they had mass, and can interact with each other, even forming photonic "molecules". This is in contrast to the usual properties of photons, which have no rest mass, and cannot interact.

1.4.8 Dropleton

Main article: Dropleton

A "quantum fog" of electrons and holes that flow around each other and even ripple like a liquid, rather than existing as discrete pairs.[13]

1.5 High-energy states

1.5.1 Degenerate matter

Main article: Degenerate matter

Under extremely high pressure, ordinary matter undergoes a transition to a series of exotic states of matter collectively known as degenerate matter. In these conditions, the structure of matter is supported by the Pauli exclusion principle. These are of great interest to astrophysicists, because these high-pressure conditions are believed to exist inside stars that have used up their nuclear fusion "fuel", such as the white dwarfs and neutron stars.

Electron-degenerate matter is found inside white dwarf stars. Electrons remain bound to atoms but are able to transfer to adjacent atoms. Neutron-degenerate matter is found in neutron stars. Vast gravitational pressure compresses atoms so strongly that the electrons are forced to combine with protons via inverse beta-decay, resulting in a superdense conglomeration of neutrons. (Normally free neutrons outside an atomic nucleus will decay with a half life of just under 15 minutes, but in a neutron star, as in the nucleus of an atom, other effects stabilize the neutrons.)

1.5.2 Quark–gluon plasma

Main article: Quark–gluon plasma

Quark–gluon plasma is a phase in which quarks become free and able to move independently (rather than being perpetually bound into particles) in a sea of gluons (subatomic particles that transmit the strong force that binds quarks together); this is similar to splitting molecules into atoms. This state may be briefly attainable in particle accelerators, and allows scientists to observe the properties of individual quarks, and not just theorize. See also Strangeness production.

Quark–gluon plasma was discovered at CERN in 2000.

1.5.3 Color-glass condensate

Main article: Color-glass condensate

Color-glass condensate is a type of matter theorized to exist in atomic nuclei traveling near the speed of light. According to Einstein's theory of relativity, a high-energy nucleus appears length contracted, or compressed, along its direction of motion. As a result, the gluons inside the nucleus appear to a stationary observer as a "gluonic wall" traveling near the speed of light. At very high energies, the density of the gluons in this wall is seen to increase greatly. Unlike the quark–gluon plasma produced in the collision of such walls, the color-glass condensate describes the walls themselves, and is an intrinsic property of the particles that can only be observed under high-energy conditions such as those at RHIC and possibly at the Large Hadron Collider as well.

1.6   Very high energy states

The **gravitational singularity** predicted by general relativity to exist at the center of a black hole is *not* a phase of matter; it is not a material object at all (although the mass-energy of matter contributed to its creation) but rather a property of spacetime at a location. It could be argued, of course, that all particles are properties of spacetime at a location,[14] leaving a half-note of controversy on the subject.

1.7   Other proposed states

1.7.1   Supersolid

Main article: Supersolid

A supersolid is a spatially ordered material (that is, a solid or crystal) with superfluid properties. Similar to a superfluid, a supersolid is able to move without friction but retains a rigid shape. Although a supersolid is a solid, it exhibits so many characteristic properties different from other solids that many argue it is another state of matter.[15]

1.7.2   String-net liquid

Main article: String-net liquid

In a string-net liquid, atoms have apparently unstable arrangement, like a liquid, but are still consistent in overall pattern, like a solid. When in a normal solid state, the atoms of matter align themselves in a grid pattern, so that the spin of any electron is the opposite of the spin of all electrons touching it. But in a string-net liquid, atoms are arranged in some pattern that requires some electrons to have neighbors with the same spin. This gives rise to curious properties, as well as supporting some unusual proposals about the fundamental conditions of the universe itself.

1.7.3   Superglass

Main article: Superglass

A superglass is a phase of matter characterized, at the same time, by superfluidity and a frozen amorphous structure.

1.7.4   Dark matter

Main article: Dark matter

While dark matter is estimated to comprise 83% of the mass of matter in the universe, most of its properties remain a mystery due to the fact that it neither absorbs nor emits electromagnetic radiation, and there are many competing theories regarding what dark matter is actually made of. Thus, while it is hypothesized to exist and comprise the vast majority of matter in the universe, almost all of its properties are unknown and a matter of speculation, because it has only been observed through its gravitational effects.[16][17]

1.7.5 Equilibrium gel

Main article: Equilibrium gel

Equilibrium gel is made from a synthetic clay called Laponite. Unlike other gels, it maintains the same consistency throughout its structure and is stable, which means it does not separate into sections of solid mass and those of more liquid mass. Equilibrium gel filtration liquid chromatography is a technique used for the quantitation of ligand binding.[18]

1.8 See also

- Hidden states of matter
- Classical element
- Condensed matter physics
- Cooling curve
- Phase (matter)
- Supercooling
- Superheating

1.9 Notes and references

[1] It is often stated that more than 99% of the material in the visible universe is plasma. See, for instance, D. A. Gurnett; A. Bhattacharjee (2005). *Introduction to Plasma Physics: With Space and Laboratory Applications.* Cambridge, UK: Cambridge University Press. p. 2. ISBN 0-521-36483-3. and K Scherer; H Fichtner; B Heber (2005). *Space Weather: The Physics Behind a Slogan.* Berlin: Springer. p. 138. ISBN 3-540-22907-8.. Essentially, all of the visible light from space comes from stars, which are plasmas with a temperature such that they radiate strongly at visible wavelengths. Most of the ordinary (or baryonic) matter in the universe, however, is found in the intergalactic medium, which is also a plasma, but much hotter, so that it radiates primarily as X-rays. The current scientific consensus is that about 96% of the total energy density in the universe is not plasma or any other form of ordinary matter, but a combination of cold dark matter and dark energy.

[2] M.A. Wahab (2005). *Solid State Physics: Structure and Properties of Materials.* Alpha Science. pp. 1–3. ISBN 1-84265-218-4.

[3] F. White (2003). *Fluid Mechanics.* McGraw-Hill. p. 4. ISBN 0-07-240217-2.

[4] G. Turrell (1997). *Gas Dynamics: Theory and Applications.* John Wiley & Sons. pp. 3–5. ISBN 0-471-97573-7.

[5] M. Chaplin (20 August 2009). "Water phase Diagram". *Water Structure and Science.* Retrieved 23 February 2010.

[6] D.L. Goodstein (1985). *States of Matter.* Dover Phoenix. ISBN 978-0-486-49506-4.

[7] A.P. Sutton (1993). *Electronic Structure of Materials.* Oxford Science Publications. pp. 10–12. ISBN 978-0-19-851754-2.

[8] Shao, Y.; Zerda, T. W. (1998). "Phase Transitions of Liquid Crystal PAA in Confined Geometries". *Journal of Physical Chemistry B* **102** (18): 3387–3394. doi:10.1021/jp9734437.

[9] J.R. Minkel (20 February 2009). "Strange but True: Superfluid Helium Can Climb Walls". *Scientific American.* Retrieved 23 February 2010.

[10] L. Valigra (22 June 2005). "MIT physicists create new form of matter". MIT News. Retrieved 23 February 2010.

[11] V. Bendkowsky; et al. (2009). "Observation of Ultralong-Range Rydberg Molecules". *Nature* **458** (7241): 1005–8. Bibcode: doi:10.1038/nature07945. PMID 19396141.

[12] V. Gill (23 April 2009). "World First for Strange Molecule". BBC News. Retrieved 23 February 2010.

[13] http://www.iflscience.com/physics/new-state-matter-discovered#3Oe9x65kkHViXABt.99

[14] David Chalmers; David Manley; Ryan Wasserman (2009). *Metametaphysics: New Essays on the Foundations of Ontology*. Oxford University Press. pp. 378–. ISBN 978-0-19-954604-6.

[15] G. Murthy; et al. (1997). "Superfluids and Supersolids on Frustrated Two-Dimensional Lattices". *Physical Review B* **55** (5): 3104. arXiv:cond-mat/9607217. Bibcode:1997PhRvB..55.3104M. doi:10.1103/PhysRevB.55.3104.

[16] Trimble, Virginia (1987). "Existence and nature of dark matter in the universe". *Annual Review of Astronomy and Astrophysics* **25**: 425–472. Bibcode:1987ARA&A..25..425T. doi:10.1146/annurev.aa.25.090187.002233.

[17] Hinshaw, Gary F. (29 January 2010). "What is the universe made of?". *Universe 101*. NASA website. Retrieved 17 March 2010.

[18] Cartlidge, Edwin (12 January 2012). "New State of Matter Seen in Clay". *Technology*. Science Now website. Retrieved 10 September 2013.

1.10   External links

- 2005-06-22, MIT News: MIT physicists create new form of matter Citat: "... They have become the first to create a new type of matter, a gas of atoms that shows high-temperature superfluidity."

- 2003-10-10, Science Daily: Metallic Phase For Bosons Implies New State Of Matter

- 2004-01-15, ScienceDaily: Probable Discovery Of A New, Supersolid, Phase Of Matter Citat: "...We apparently have observed, for the first time, a solid material with the characteristics of a superfluid...but because all its particles are in the identical quantum state, it remains a solid even though its component particles are continually flowing..."

- 2004-01-29, ScienceDaily: NIST/University Of Colorado Scientists Create New Form Of Matter: A Fermionic Condensate

- Short videos demonstrating of States of Matter, solids, liquids and gases by Prof. J M Murrell, University of Sussex

Chapter 2

List of states of matter

Main article: States of matter

Classically, states of matter are distinguished by changes in specific heat capacity, pressure and temperature. States are distinguished by a discontinuity in one of those properties: for example, raising the temperature of ice produces a clear discontinuity at 0 °C as energy goes into phase transition, instead of temperature increase.

In the 20th century, increased understanding of the more exotic properties of matter has resulted in many additional states of matter, none of which are observed in normal conditions.

2.1 Low-energy states

2.1.1 Classical states

- **Solid**: A solid holds a rigid shape without a container.The molecules held very closed to each other.

  - **Amorphous solid**: A solid in which there is no long-range order of the positions of the atoms.
  - **Crystalline solid**: A solid in which the constituent atoms, molecules, or ions are packed in a regularly order.
  - **Plastic crystal**: A molecular solid with long-range positional order but with constituent molecules retaining rotational freedom.
  - **Quasicrystal**: A solid in which the positions of the atoms have long-range order, but is not in a repeating pattern.

- **Liquid**: A mostly non-compressible fluid. Able to conform to the shape of its container but retaining a (nearly) constant volume independent of pressure.

  - **Liquid crystal**: Properties intermediate between liquids and crystals: a phase in which a large number of bosons all inhabit the same quantum state, in effect becoming one single wave/particle. This is a low energy phase that can only be formed in laboratory conditions and in very cold temperatures. It must be close to zero : Similar to the Bose–Einstein condensate but composed of fermions. The Pauli exclusion principle prevents fermions from entering the same quantum state, but a pair of fermions can behave as a boson, and multiple such pairs can then enter the same quantum state without restriction.
  - **Disordered Hyperuniformity**: It behaves like a crystal and a liquid. The density of particles over large distances is the same like a crystal. It is also a liquid in the sense that at smaller distances, the particles display same physical properties in all directions.

- **Gas**: A compressible fluid. Not only will a gas conform to the shape of its container but it will also expand to fill the container.

- **Plasma**: free charged particles, usually in equal numbers, such as ions and electrons. Unlike gases, plasmas may self-generate magnetic fields and electric currents, and respond strongly and collectively to electromagnetic forces.

- **Colloids**: "in betweens", containing two or more forms of matter. Examples are shaving foam and butter.

2.1.2   Modern States

- **Degenerate matter**: matter under very high pressure, supported by the Pauli exclusion principle.

  - **Electron-degenerate matter**: found inside white dwarf stars. Electrons remain bound to atoms but are able to transfer to adjacent atoms.

  - **Neutron-degenerate matter**: found in neutron stars. Vast gravitational pressure compresses atoms so strongly that the electrons are forced to combine with protons via inverse beta-decay, resulting in a superdense conglomeration of neutrons. (Normally free neutrons outside an atomic nucleus will decay with a half life of just under 15 minutes, but in a neutron star, as in the nucleus of an atom, other effects stabilize the neutrons.)

  - **Strange matter**: A type of quark matter that may exist inside some neutron stars close to the Tolman–Oppenheimer–Volkoff limit (approximately 2–3 solar masses). May be stable at lower energy states once formed.

- **Photonic matter**: Inside a quantum nonlinear medium, photons can behave as if they had mass, and can interact with each other, forming photonic "molecules".

- **Quantum Hall state**: A state that gives rise to quantized Hall voltage measured in the direction perpendicular to the current flow.

  - **Quantum spin Hall state**: a theoretical phase that may pave the way for the development of electronic devices that dissipate less energy and generate less heat. This is a derivation of the quantum Hall state of matter.

- **Bose–Einstein condensate**: a phase in which a large number of bosons all inhabit the same quantum state, in effect becoming one single wave/particle. This is a low energy phase that can only be formed in laboratory conditions and in very cold temperatures. It must be close to zero kelvin (absolute zero). Satyendra Bose and Albert Einstein predicted the existence of such a state in the 1920s, but it was not observed until 1995 by Eric Cornell and Carl Wieman.

- **Fermionic condensate**: Similar to the Bose–Einstein condensate but composed of fermions. The Pauli exclusion principle prevents fermions from entering the same quantum state, but a pair of fermions can behave as a boson, and multiple such pairs can then enter the same quantum state without restriction.

- **Superconductivity**: is a phenomenon of exactly zero electrical resistance and expulsion of magnetic fields occurring in certain materials when cooled below a characteristic critical temperature. Superconductivity is the ground state of many elemental metals.

- **Superfluid**: A phase achieved by a few cryogenic liquids at extreme temperature where they become able to flow without friction. A superfluid can flow up the side of an open container and down the outside. Placing a superfluid in a spinning container will result in quantized vortices.

- **Supersolid**: similar to a superfluid, a supersolid is able to move without friction but retains a rigid shape.

- **Quantum spin liquid**: A disordered state in a system of interacting quantum spins which preserves its disorder to very low temperatures, unlike other disordered states.

- **String-net liquid**: Atoms in this state have apparently unstable arrangement, like a liquid, but are still consistent in overall pattern, like a solid.

- **Supercritical fluid**: At sufficiently high temperatures and pressures the distinction between liquid and gas disappears.

- **Dropleton**: An artificial quasiparticle, constituting a collection of electrons and places without them inside a semiconductor. Dropleton is the first known quasiparticle that behaves like a liquid.

- **Jahn–Teller metal**: A solid that exhibits many of the characteristics of an insulator, but acts as a conductor due to a distorted crystalline structure.(experiment was not reproduced and confirmed by other scientists)

2.2   Very high energy states

- **Quark–gluon plasma**: A phase in which quarks become free and able to move independently (rather than being perpetually bound into particles, or bound to each other in a quantum lock where exerting force adds energy and eventually solidifies into another quark.) in a sea of gluons (subatomic particles that transmit the strong force that binds quarks together). May be briefly attainable in particle accelerators.

    - **Weakly symmetric matter**: for up to 10^{-12} seconds after the Big Bang the strong, weak and electromagnetic forces were unified.

    - **Strongly symmetric matter**: for up to 10^{-36} seconds after the Big Bang, the energy density of the universe was so high that the four forces of nature — strong, weak, electromagnetic, and gravitational — are thought to have been unified into one single force. As the universe expanded, the temperature and density dropped and the gravitational force separated, a process called symmetry breaking.

Chapter 3

Solid

For other uses, see Solid (disambiguation).

Single crystalline form of solid insulin.

Solid is one of the four fundamental states of matter (the others being liquid, gas, and plasma). It is characterized by structural rigidity and resistance to changes of shape or volume. Unlike a liquid, a solid object does not flow to take on the shape of its container, nor does it expand to fill the entire volume available to it like a gas does. The atoms in a solid are tightly bound to each other, either in a regular geometric lattice (crystalline solids, which include metals and ordinary ice) or irregularly (an amorphous solid such as common window glass).

The branch of physics that deals with solids is called solid-state physics, and is the main branch of condensed matter physics (which also includes liquids). Materials science is primarily concerned with the physical and chemical properties of solids. Solid-state chemistry is especially concerned with the synthesis of novel materials, as well as the science of identification and chemical composition.

3.1 Microscopic description

Model of closely packed atoms within a crystalline solid.

The atoms, molecules or ions which make up solids may be arranged in an orderly repeating pattern, or irregularly. Materials whose constituents are arranged in a regular pattern are known as crystals. In some cases, the regular ordering can continue unbroken over a large scale, for example diamonds, where each diamond is a single crystal. Solid objects that are large enough to see and handle are rarely composed of a single crystal, but instead are made of a large number of single crystals, known as crystallites, whose size can vary from a few nanometers to several meters. Such materials are called polycrystalline. Almost all common metals, and many ceramics, are polycrystalline.

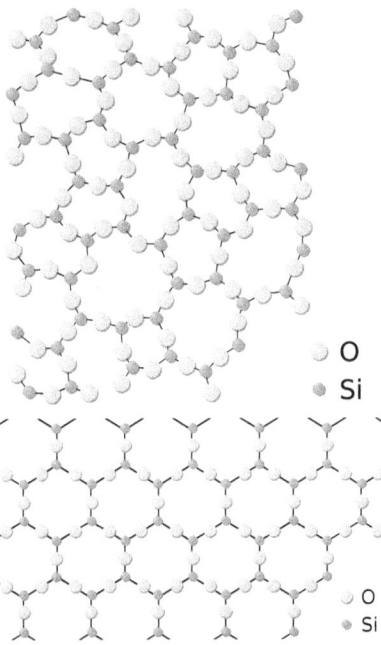

Schematic representation of a random-network glassy form (left) and ordered crystalline lattice (right) of identical chemical composition.

In other materials, there is no long-range order in the position of the atoms. These solids are known as amorphous solids; examples include polystyrene and glass.

Whether a solid is crystalline or amorphous depends on the material involved, and the conditions in which it was formed. Solids which are formed by slow cooling will tend to be crystalline, while solids which are frozen rapidly are more likely to be amorphous. Likewise, the specific crystal structure adopted by a crystalline solid depends on the material involved and on how it was formed.

While many common objects, such as an ice cube or a coin, are chemically identical throughout, many other common materials comprise a number of different substances packed together. For example, a typical rock is an aggregate of several different minerals and mineraloids, with no specific chemical composition. Wood is a natural organic material consisting primarily of cellulose fibers embedded in a matrix of organic lignin. In materials science, composites of more than one constituent material can be designed to have desired properties.

3.2   Classes of solids

Further information: Bonding in solids

The forces between the atoms in a solid can take a variety of forms. For example, a crystal of sodium chloride (common salt) is made up of ionic sodium and chlorine, which are held together by ionic bonds. In diamond or silicon, the atoms share electrons and form covalent bonds. In metals, electrons are shared in metallic bonding. Some solids, particularly most organic compounds, are held together with van der Waals forces resulting from the polarization of the electronic charge cloud on each molecule. The dissimilarities between the types of solid result from the differences between their bonding.

3.2.1   Metals

Main article: Metal

Metals typically are strong, dense, and good conductors of both electricity and heat. The bulk of the elements in the

The pinnacle of New York's Chrysler Building, the world's tallest steel-supported brick building, is clad with stainless steel.

periodic table, those to the left of a diagonal line drawn from boron to polonium, are metals. Mixtures of two or more elements in which the major component is a metal are known as alloys.

People have been using metals for a variety of purposes since prehistoric times. The strength and reliability of metals has led to their widespread use in construction of buildings and other structures, as well as in most vehicles, many appliances and tools, pipes, road signs and railroad tracks. Iron and aluminium are the two most commonly used structural metals, and they are also the most abundant metals in the Earth's crust. Iron is most commonly used in the form of an alloy, steel, which contains up to 2.1% carbon, making it much harder than pure iron.

Because metals are good conductors of electricity, they are valuable in electrical appliances and for carrying an electric current over long distances with little energy loss or dissipation. Thus, electrical power grids rely on metal cables to distribute electricity. Home electrical systems, for example, are wired with copper for its good conducting properties and easy machinability. The high thermal conductivity of most metals also makes them useful for stovetop cooking utensils.

The study of metallic elements and their alloys makes up a significant portion of the fields of solid-state chemistry, physics, materials science and engineering.

Metallic solids are held together by a high density of shared, delocalized electrons, known as "metallic bonding". In a metal, atoms readily lose their outermost ("valence") electrons, forming positive ions. The free electrons are spread over the entire solid, which is held together firmly by electrostatic interactions between the ions and the electron cloud.[1] The large number of free electrons gives metals their high values of electrical and thermal conductivity. The free electrons also prevent transmission of visible light, making metals opaque, shiny and lustrous.

More advanced models of metal properties consider the effect of the positive ions cores on the delocalised electrons. As most metals have crystalline structure, those ions are usually arranged into a periodic lattice. Mathematically, the potential of the ion cores can be treated by various models, the simplest being the nearly free electron model.

3.2.2 Minerals

Main article: Minerals

Minerals are naturally occurring solids formed through various geological processes under high pressures. To be classified as a true mineral, a substance must have a crystal structure with uniform physical properties throughout. Minerals range in composition from pure elements and simple salts to very complex silicates with thousands of known forms. In contrast, a rock sample is a random aggregate of minerals and/or mineraloids, and has no specific chemical composition. The vast majority of the rocks of the Earth's crust consist of quartz (crystalline SiO_2), feldspar, mica, chlorite, kaolin, calcite, epidote, olivine, augite, hornblende, magnetite, hematite, limonite and a few other minerals. Some minerals, like quartz, mica or feldspar are common, while others have been found in only a few locations worldwide. The largest group of minerals by far is the silicates (most rocks are ≥95% silicates), which are composed largely of silicon and oxygen, with the addition of ions of aluminium, magnesium, iron, calcium and other metals.

3.2.3 Ceramics

Main article: Ceramic engineering

Ceramic solids are composed of inorganic compounds, usually oxides of chemical elements. They are chemically inert, and often are capable of withstanding chemical erosion that occurs in an acidic or caustic environment. Ceramics generally can withstand high temperatures ranging from 1000 to 1600 °C (1800 to 3000 °F). Exceptions include non-oxide inorganic materials, such as nitrides, borides and carbides.

Traditional ceramic raw materials include clay minerals such as kaolinite, more recent materials include aluminium oxide (alumina). The modern ceramic materials, which are classified as advanced ceramics, include silicon carbide and tungsten carbide. Both are valued for their abrasion resistance, and hence find use in such applications as the wear plates of crushing equipment in mining operations.

Most ceramic materials, such as alumina and its compounds, are formed from fine powders, yielding a fine grained

A collection of various minerals.

polycrystalline microstructure which is filled with light scattering centers comparable to the wavelength of visible light. Thus, they are generally opaque materials, as opposed to transparent materials. Recent nanoscale (e.g. sol-gel) technology has, however, made possible the production of polycrystalline transparent ceramics such as transparent alumina and alumina compounds for such applications as high-power lasers. Advanced ceramics are also used in the medicine, electrical and electronics industries.

Ceramic engineering is the science and technology of creating solid-state ceramic materials, parts and devices. This is done either by the action of heat, or, at lower temperatures, using precipitation reactions from chemical solutions. The term includes the purification of raw materials, the study and production of the chemical compounds concerned, their formation into components, and the study of their structure, composition and properties.

Mechanically speaking, ceramic materials are brittle, hard, strong in compression and weak in shearing and tension. Brittle materials may exhibit significant tensile strength by supporting a static load. Toughness indicates how much energy a material can absorb before mechanical failure, while fracture toughness (denoted KI_c) describes the ability of a material with inherent microstructural flaws to resist fracture via crack growth and propagation. If a material has a large value of fracture toughness, the basic principles of fracture mechanics suggest that it will most likely undergo ductile fracture. Brittle fracture is very characteristic of most ceramic and glass-ceramic materials which typically exhibit low (and inconsistent) values of KI_c.

For an example of applications of ceramics, the extreme hardness of Zirconia is utilized in the manufacture of knife blades, as well as other industrial cutting tools. Ceramics such as alumina, boron carbide and silicon carbide have been used in bulletproof vests to repel large-caliber rifle fire. Silicon nitride parts are used in ceramic ball bearings, where their high hardness makes them wear resistant. In general, ceramics are also chemically resistant and can be used in wet

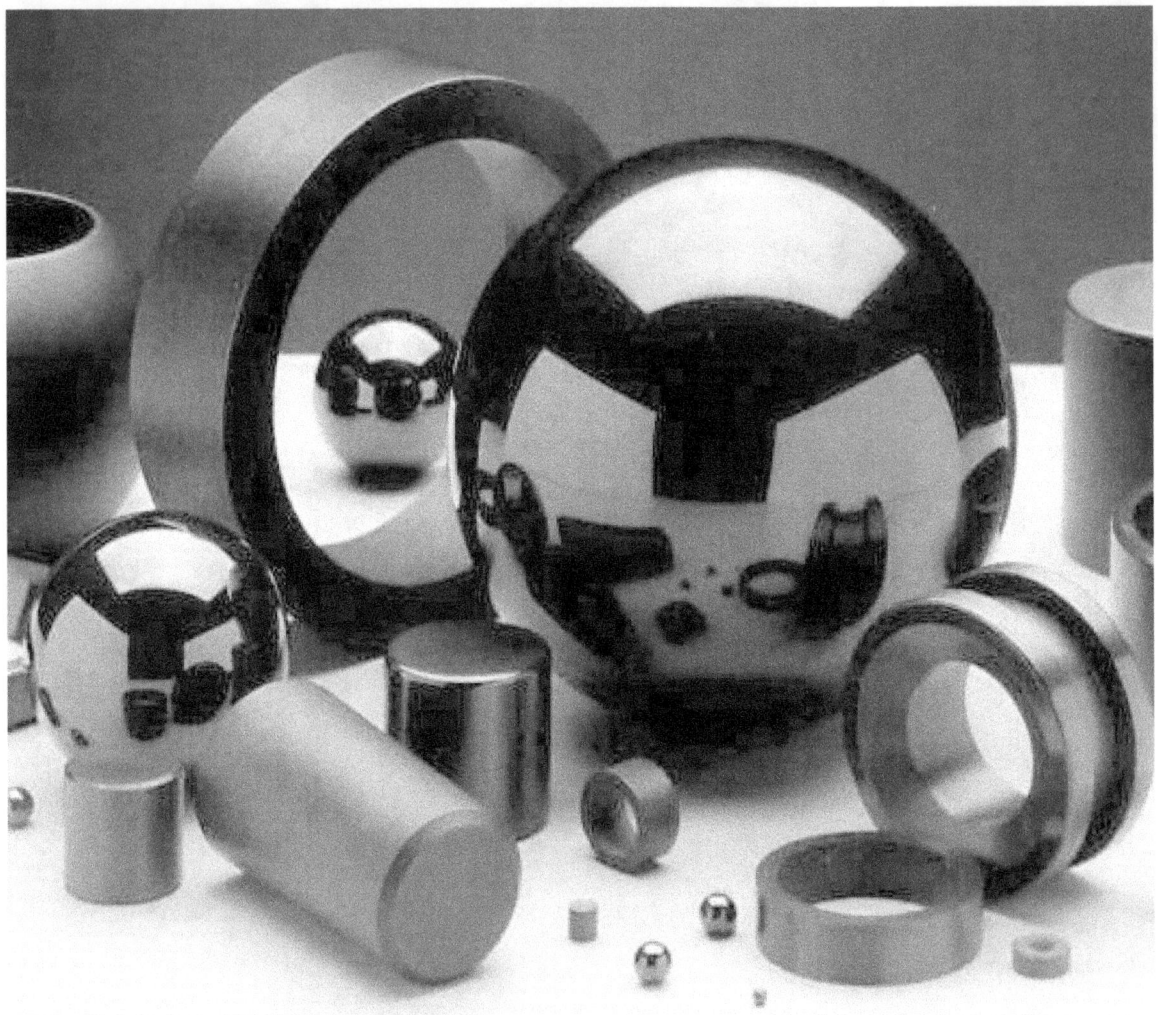

Si₃N₄ ceramic bearing parts

environments where steel bearings would be susceptible to oxidation (or rust).

As another example of ceramic applications, in the early 1980s, Toyota researched production of an adiabatic ceramic engine with an operating temperature of over 6000 °F (3300 °C). Ceramic engines do not require a cooling system and hence allow a major weight reduction and therefore greater fuel efficiency. In a conventional metallic engine, much of the energy released from the fuel must be dissipated as waste heat in order to prevent a meltdown of the metallic parts. Work is also being done in developing ceramic parts for gas turbine engines. Turbine engines made with ceramics could operate more efficiently, giving aircraft greater range and payload for a set amount of fuel. However, such engines are not in production because the manufacturing of ceramic parts in the sufficient precision and durability is difficult and costly. Processing methods often result in a wide distribution of microscopic flaws which frequently play a detrimental role in the sintering process, resulting in the proliferation of cracks, and ultimate mechanical failure.

3.2.4 Glass ceramics

Main article: Glass-ceramic
Glass-ceramic materials share many properties with both non-crystalline glasses and crystalline ceramics. They are formed as a glass, and then partially crystallized by heat treatment, producing both amorphous and crystalline phases so that crystalline grains are embedded within a non-crystalline intergranular phase.

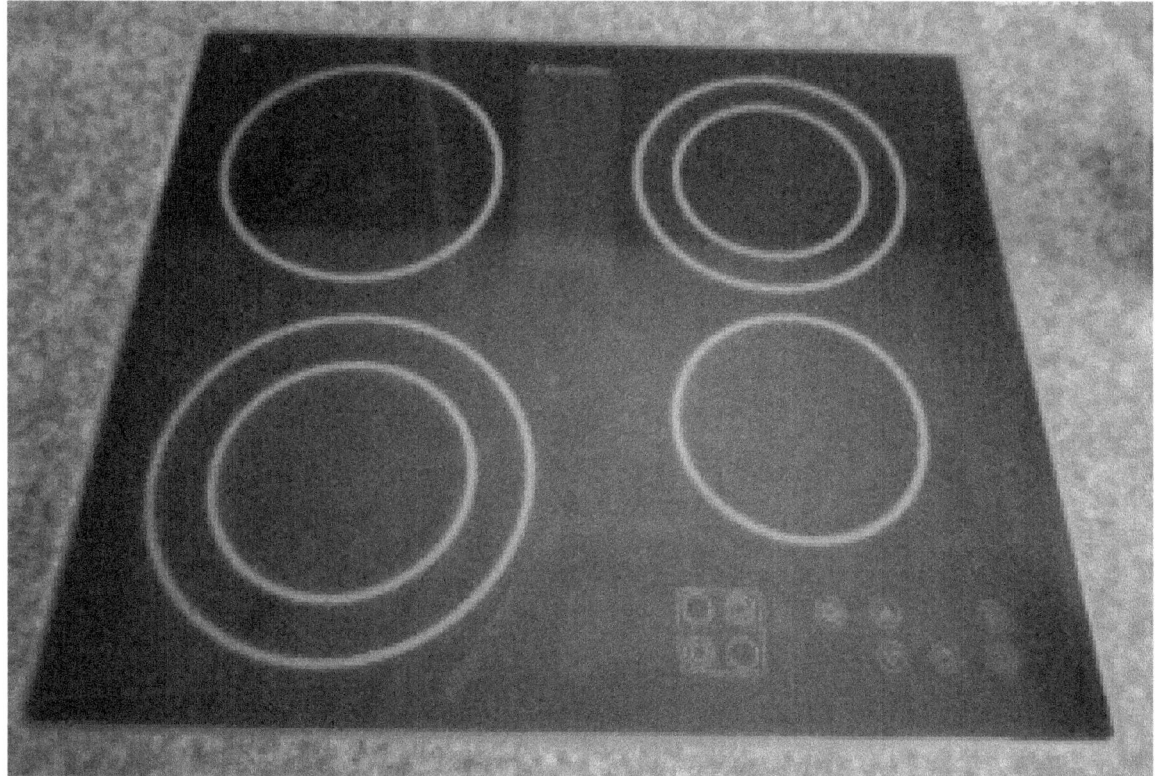

A high strength glass-ceramic cooktop with negligible thermal expansion.

Glass-ceramics are used to make cookware (originally known by the brand name CorningWare) and stovetops which have both high resistance to thermal shock and extremely low permeability to liquids. The negative coefficient of thermal expansion of the crystalline ceramic phase can be balanced with the positive coefficient of the glassy phase. At a certain point (~70% crystalline) the glass-ceramic has a net coefficient of thermal expansion close to zero. This type of glass-ceramic exhibits excellent mechanical properties and can sustain repeated and quick temperature changes up to 1000 °C.

Glass ceramics may also occur naturally when lightning strikes the crystalline (e.g. quartz) grains found in most beach sand. In this case, the extreme and immediate heat of the lightning (~2500 °C) creates hollow, branching rootlike structures called fulgurite via fusion.

3.2.5 Organic solids

Main article: Organic chemistry
 Organic chemistry studies the structure, properties, composition, reactions, and preparation by synthesis (or other means) of chemical compounds of carbon and hydrogen, which may contain any number of other elements such as nitrogen, oxygen and the halogens: fluorine, chlorine, bromine and iodine. Some organic compounds may also contain the elements phosphorus or sulfur. Examples of organic solids include wood, paraffin wax, naphthalene and a wide variety of polymers and plastics.

Wood

Main article: Wood

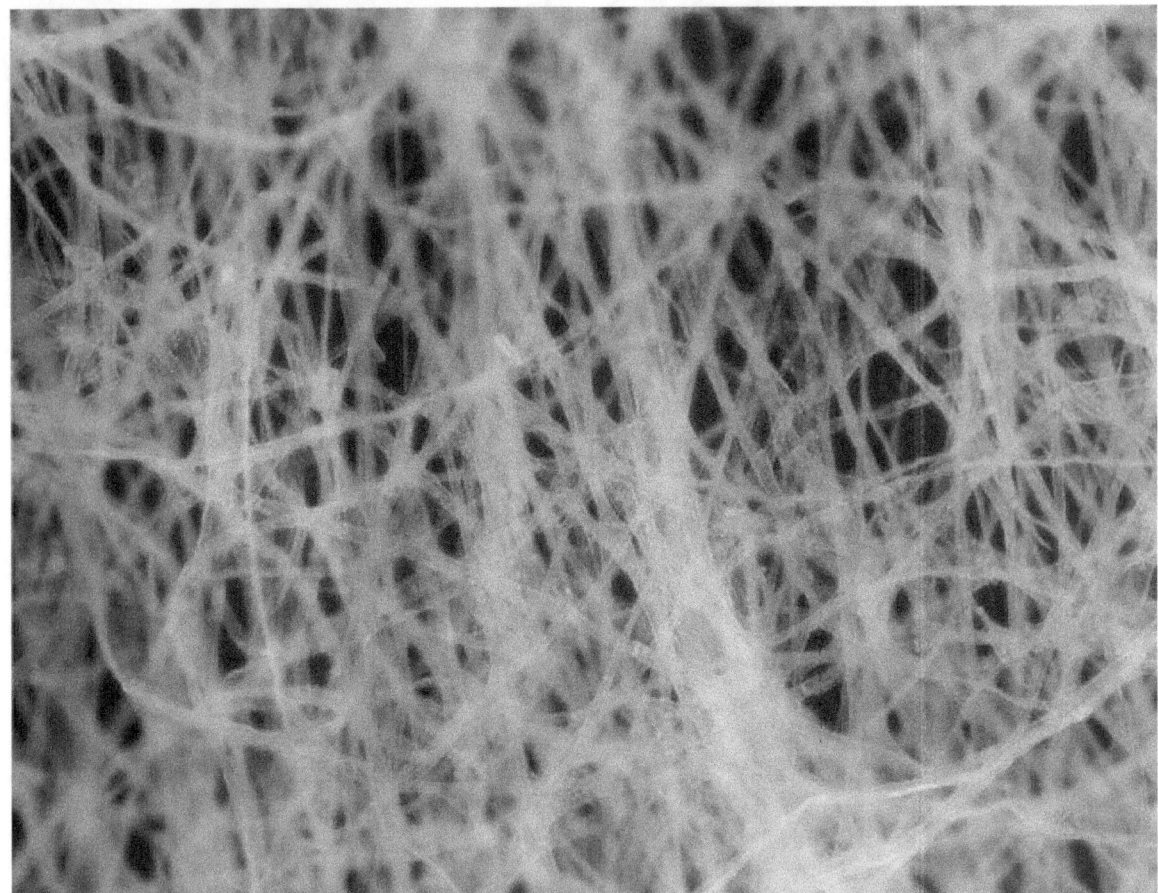

The individual wood pulp fibers in this sample are around 10 μm in diameter.

Wood is a natural organic material consisting primarily of cellulose fibers embedded in a matrix of lignin. Regarding mechanical properties, the fibers are strong in tension, and the lignin matrix resists compression. Thus wood has been an important construction material since humans began building shelters and using boats. Wood to be used for construction work is commonly known as *lumber* or *timber*. In construction, wood is not only a structural material, but is also used to form the mould for concrete.

Wood-based materials are also extensively used for packaging (e.g. cardboard) and paper which are both created from the refined pulp. The chemical pulping processes use a combination of high temperature and alkaline (kraft) or acidic (sulfite) chemicals to break the chemical bonds of the lignin before burning it out.

Polymers

Main article: Polymer

One important property of carbon in organic chemistry is that it can form certain compounds, the individual molecules of which are capable of attaching themselves to one another, thereby forming a chain or a network. The process is called polymerization and the chains or networks polymers, while the source compound is a monomer. Two main groups of polymers exist: those artificially manufactured are referred to as industrial polymers or synthetic polymers (plastics) and those naturally occurring as biopolymers.

Monomers can have various chemical substituents, or functional groups, which can affect the chemical properties of organic compounds, such as solubility and chemical reactivity, as well as the physical properties, such as hardness, density, mechanical or tensile strength, abrasion resistance, heat resistance, transparency, color, etc.. In proteins, these differences

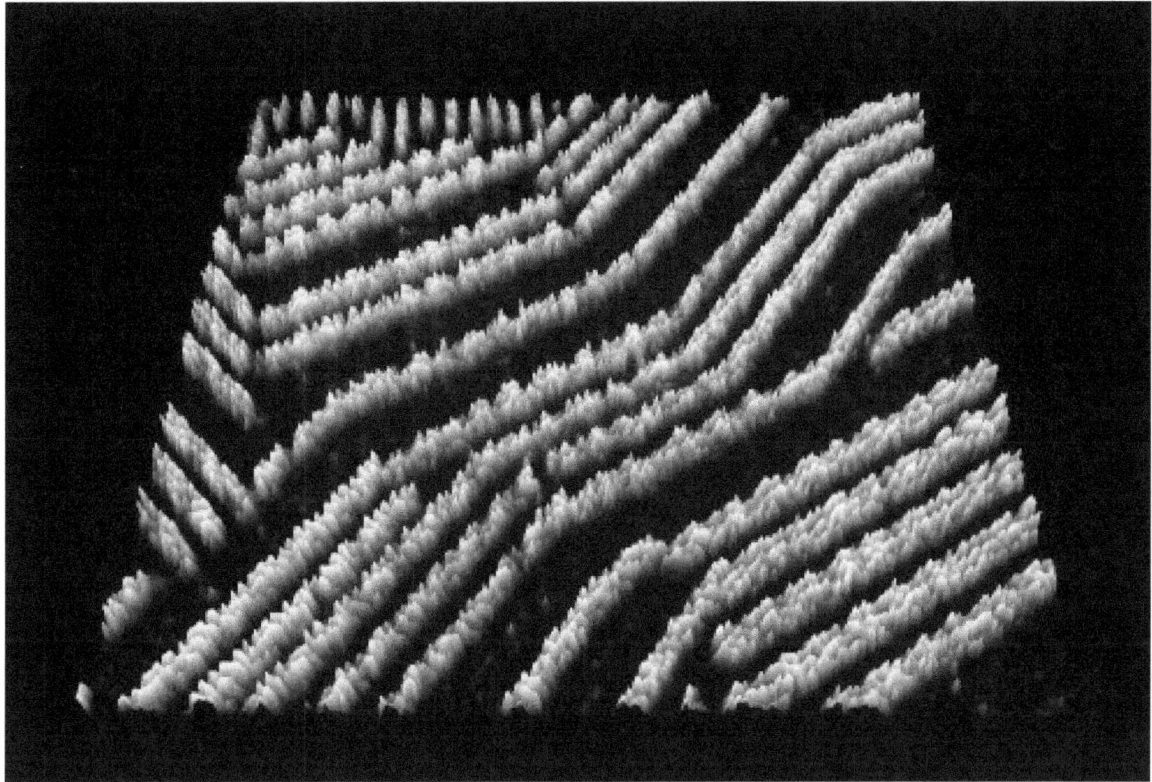

STM image of self-assembled supramolecular chains of the organic semiconductor quinacridone on graphite.

give the polymer the ability to adopt a biologically active conformation in preference to others (see self-assembly).

People have been using natural organic polymers for centuries in the form of waxes and shellac which is classified as a thermoplastic polymer. A plant polymer named cellulose provided the tensile strength for natural fibers and ropes, and by the early 19th century natural rubber was in widespread use. Polymers are the raw materials (the resins) used to make what are commonly called plastics. Plastics are the final product, created after one or more polymers or additives have been added to a resin during processing, which is then shaped into a final form. Polymers which have been around, and which are in current widespread use, include carbon-based polyethylene, polypropylene, polyvinyl chloride, polystyrene, nylons, polyesters, acrylics, polyurethane, and polycarbonates, and silicon-based silicones. Plastics are generally classified as "commodity", "specialty" and "engineering" plastics.

3.2.6   Composite materials

Main article: Composite material

Composite materials contain two or more macroscopic phases, one of which is often ceramic. For example, a continuous matrix, and a dispersed phase of ceramic particles or fibers.

Applications of composite materials range from structural elements such as steel-reinforced concrete, to the thermally insulative tiles which play a key and integral role in NASA's Space Shuttle thermal protection system which is used to protect the surface of the shuttle from the heat of re-entry into the Earth's atmosphere. One example is Reinforced Carbon-Carbon (RCC), the light gray material which withstands reentry temperatures up to 1510 °C (2750 °F) and protects the nose cap and leading edges of Space Shuttle's wings. RCC is a laminated composite material made from graphite rayon cloth and impregnated with a phenolic resin. After curing at high temperature in an autoclave, the laminate is pyrolized to convert the resin to carbon, impregnated with furfural alcohol in a vacuum chamber, and cured/pyrolized to convert the furfural alcohol to carbon. In order to provide oxidation resistance for reuse capability, the outer layers of the RCC

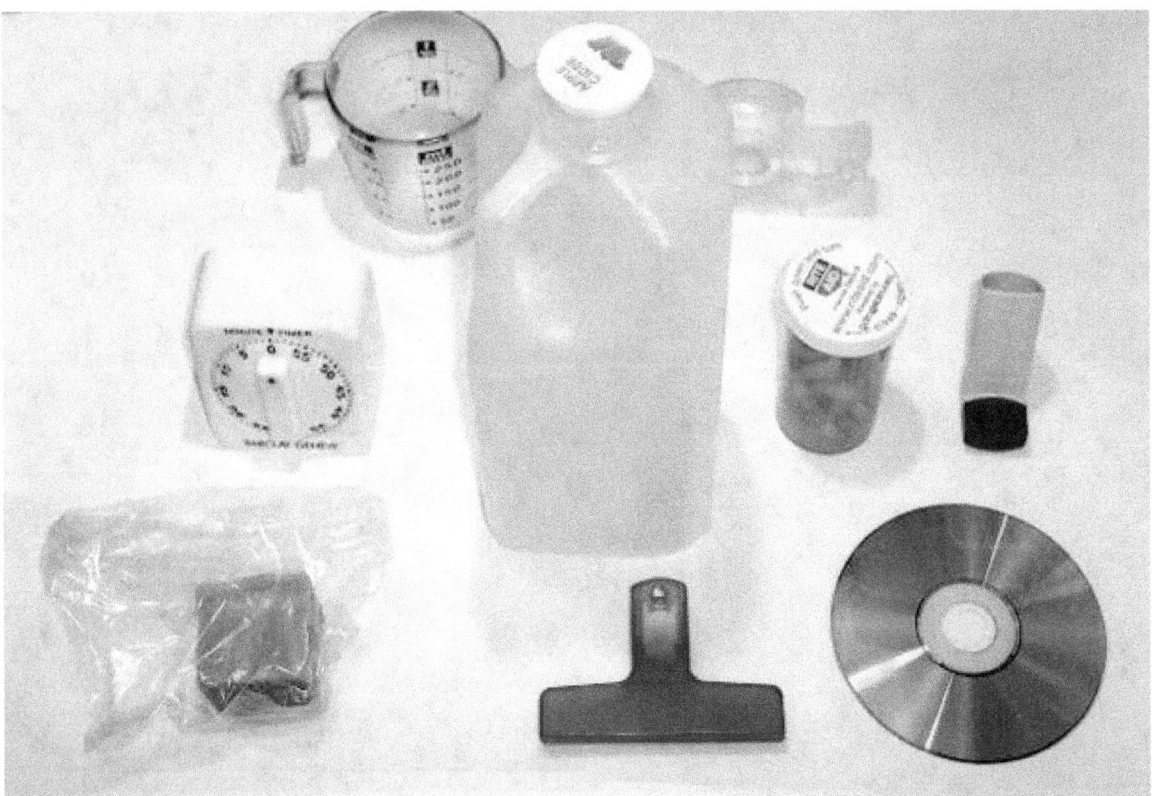

Household items made of various kinds of plastic.

are converted to silicon carbide.

Domestic examples of composites can be seen in the "plastic" casings of television sets, cell-phones and so on. These plastic casings are usually a composite made up of a thermoplastic matrix such as acrylonitrile butadiene styrene (ABS) in which calcium carbonate chalk, talc, glass fibers or carbon fibers have been added for strength, bulk, or electro-static dispersion. These additions may be referred to as reinforcing fibers, or dispersants, depending on their purpose.

Thus, the matrix material surrounds and supports the reinforcement materials by maintaining their relative positions. The reinforcements impart their special mechanical and physical properties to enhance the matrix properties. A synergism produces material properties unavailable from the individual constituent materials, while the wide variety of matrix and strengthening materials provides the designer with the choice of an optimum combination.

3.2.7   Semiconductors

Main article: Semiconductors

Semiconductors are materials that have an electrical resistivity (and conductivity) between that of metallic conductors and non-metallic insulators. They can be found in the periodic table moving diagonally downward right from boron. They separate the electrical conductors (or metals, to the left) from the insulators (to the right).

Devices made from semiconductor materials are the foundation of modern electronics, including radio, computers, telephones, etc. Semiconductor devices include the transistor, solar cells, diodes and integrated circuits. Solar photovoltaic panels are large semiconductor devices that directly convert light into electrical energy.

In a metallic conductor, current is carried by the flow of electrons", but in semiconductors, current can be carried either by electrons or by the positively charged "holes" in the electronic band structure of the material. Common semiconductor materials include silicon, germanium and gallium arsenide.

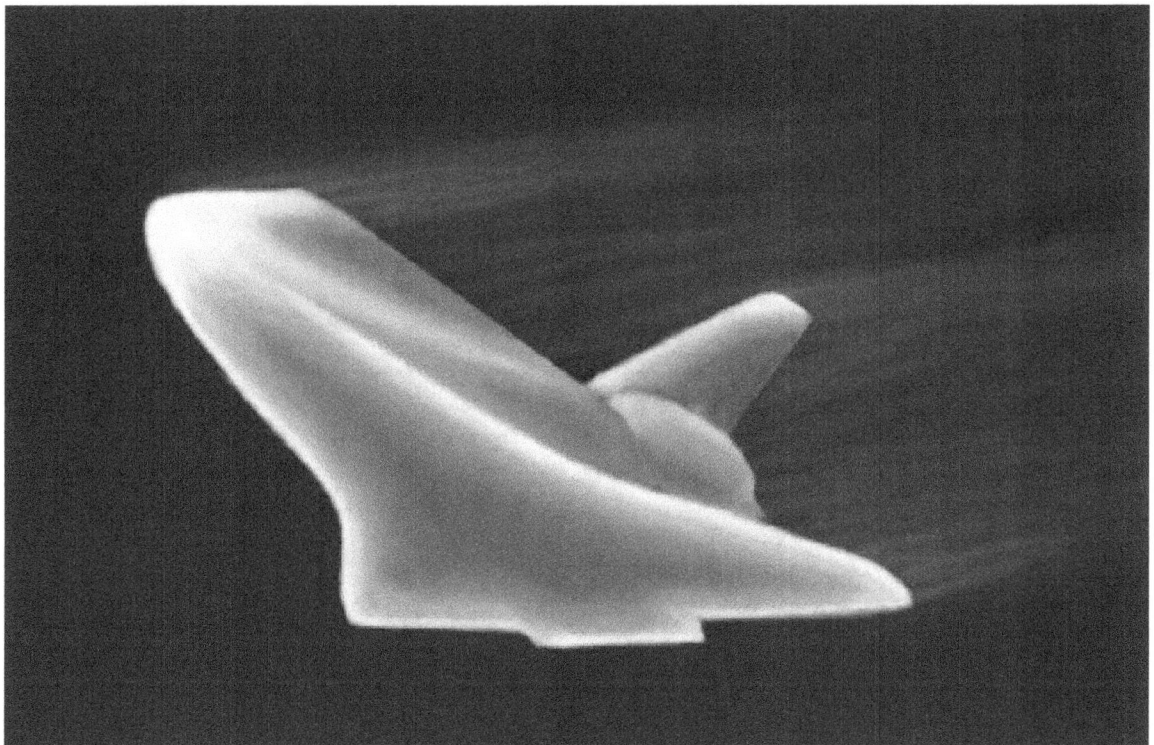

Simulation of the outside of the Space Shuttle as it heats up to over 1500 °C during re-entry

3.2.8 Nanomaterials

Main article: Nanotechnology

Many traditional solids exhibit different properties when they shrink to nanometer sizes. For example, nanoparticles of usually yellow gold and gray silicon are red in color; gold nanoparticles melt at much lower temperatures (~300 °C for 2.5 nm size) than the gold slabs (1064 °C);[2] and metallic nanowires are much stronger than the corresponding bulk metals.[3][4] The high surface area of nanoparticles makes them extremely attractive for certain applications in the field of energy. For example, platinum metals may be provide improvements as automotive fuel catalysts, as well as proton exchange membrane (PEM) fuel cells. Also, ceramic oxides (or cermets) of lanthanum, cerium, manganese and nickel are now being developed as solid oxide fuel cells (SOFC). Lithium, lithium–titanate and tantalum nanoparticles are being applied in lithium ion batteries. Silicon nanoparticles have been shown to dramatically expand the storage capacity of lithium ion batteries during the expansion/contraction cycle. Silicon nanowires cycle without significant degradation and present the potential for use in batteries with greatly expanded storage times. Silicon nanoparticles are also being used in new forms of solar energy cells. Thin film deposition of silicon quantum dots on the polycrystalline silicon substrate of a photovoltaic (solar) cell increases voltage output as much as 60% by fluorescing the incoming light prior to capture. Here again, surface area of the nanoparticles (and thin films) plays a critical role in maximizing the amount of absorbed radiation.

3.2.9 Biomaterials

Main article: Biomaterials

Many natural (or biological) materials are complex composites with remarkable mechanical properties. These complex structures, which have risen from hundreds of million years of evolution, are inspiring materials scientists in the design of novel materials. Their defining characteristics include structural hierarchy, multifunctionality and self-healing capability. Self-organization is also a fundamental feature of many biological materials and the manner by which the structures are assembled from the molecular level up. Thus, self-assembly is emerging as a new strategy in the chemical synthesis of

A cloth of woven carbon fiber filaments, a common element in composite materials

high performance biomaterials.

3.3   Physical properties

Physical properties of elements and compounds which provide conclusive evidence of chemical composition include odor, color, volume, density (mass per unit volume), melting point, boiling point, heat capacity, physical form and shape at room temperature (solid, liquid or gas; cubic, trigonal crystals, etc.), hardness, porosity, index of refraction and many others. This section discusses some physical properties of materials in the solid state.

3.3.1   Mechanical

The mechanical properties of materials describe characteristics such as their strength and resistance to deformation. For example, steel beams are used in construction because of their high strength, meaning that they neither break nor bend significantly under the applied load.

Mechanical properties include elasticity and plasticity, tensile strength, compressive strength, shear strength, fracture toughness, ductility (low in brittle materials), and indentation hardness. Solid mechanics is the study of the behavior of solid matter under external actions such as external forces and temperature changes.

A solid does not exhibit macroscopic flow, as fluids do. Any degree of departure from its original shape is called deformation. The proportion of deformation to original size is called strain. If the applied stress is sufficiently low,

Semiconductor chip on crystalline silicon substrate.

almost all solid materials behave in such a way that the strain is directly proportional to the stress (Hooke's law). The coefficient of the proportion is called the modulus of elasticity or Young's modulus. This region of deformation is known as the linearly elastic region. Three models can describe how a solid responds to an applied stress:

- Elasticity – When an applied stress is removed, the material returns to its undeformed state.

- Viscoelasticity – These are materials that behave elastically, but also have damping. When the applied stress is removed, work has to be done against the damping effects and is converted to heat within the material. This results in a hysteresis loop in the stress–strain curve. This implies that the mechanical response has a time-dependence.

- Plasticity – Materials that behave elastically generally do so when the applied stress is less than a yield value. When the stress is greater than the yield stress, the material behaves plastically and does not return to its previous state. That is, irreversible plastic deformation (or viscous flow) occurs after yield which is permanent.

Many materials become weaker at high temperatures. Materials which retain their strength at high temperatures, called refractory materials, are useful for many purposes. For example, glass-ceramics have become extremely useful for countertop cooking, as they exhibit excellent mechanical properties and can sustain repeated and quick temperature changes up to 1000 °C. In the aerospace industry, high performance materials used in the design of aircraft and/or spacecraft exteriors must have a high resistance to thermal shock. Thus, synthetic fibers spun out of organic polymers and polymer/ceramic/metal composite materials and fiber-reinforced polymers are now being designed with this purpose in mind.

3.3.2 Thermal

Because solids have thermal energy, their atoms vibrate about fixed mean positions within the ordered (or disordered) lattice. The spectrum of lattice vibrations in a crystalline or glassy network provides the foundation for the kinetic theory

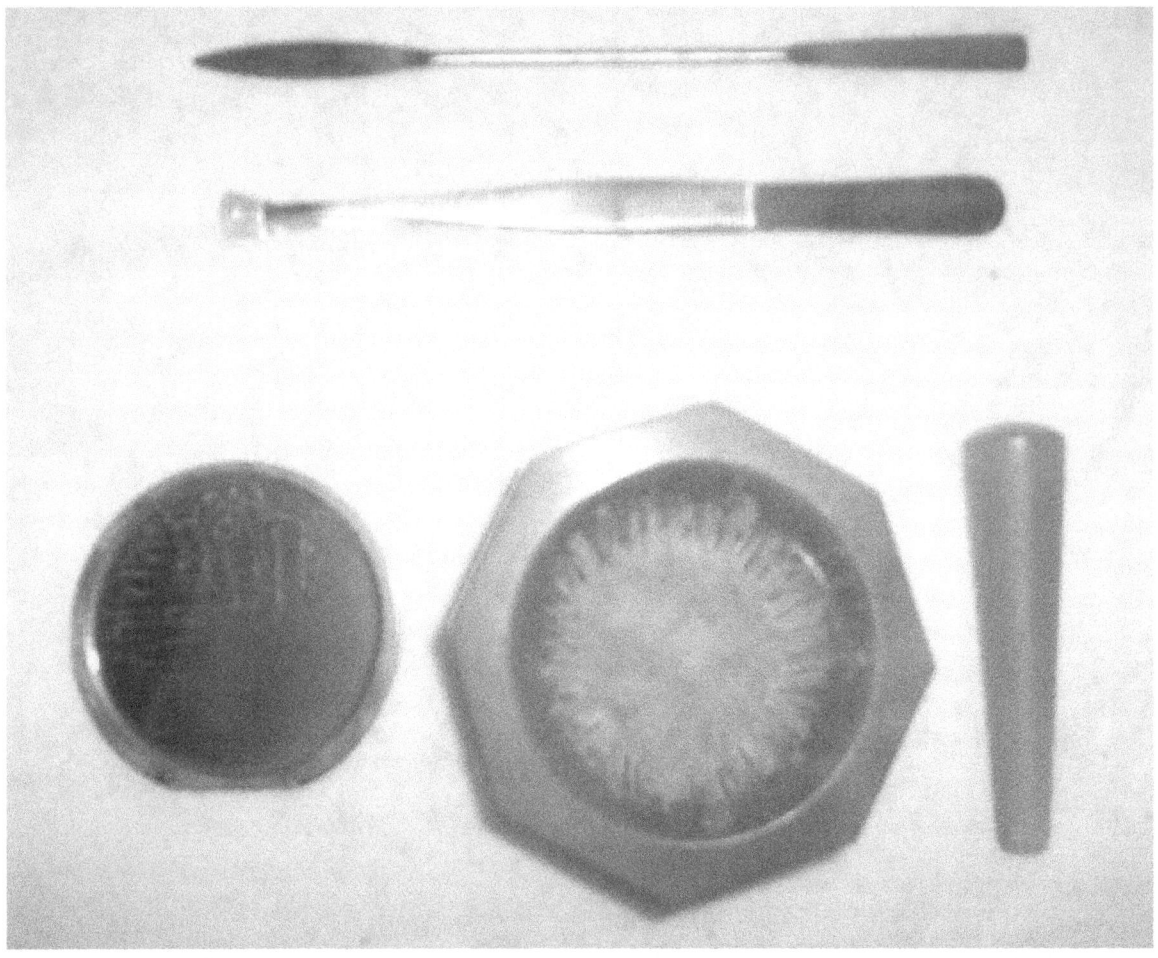

Bulk silicon (left) and silicon nanopowder (right)

of solids. This motion occurs at the atomic level, and thus cannot be observed or detected without highly specialized equipment, such as that used in spectroscopy.

Thermal properties of solids include thermal conductivity, which is the property of a material that indicates its ability to conduct heat. Solids also have a specific heat capacity, which is the capacity of a material to store energy in the form of heat (or thermal lattice vibrations).

3.3.3 Electrical

Electrical properties include conductivity, resistance, impedance and capacitance. Electrical conductors such as metals and alloys are contrasted with electrical insulators such as glasses and ceramics. Semiconductors behave somewhere in between. Whereas conductivity in metals is caused by electrons, both electrons and holes contribute to current in semiconductors. Alternatively, ions support electric current in ionic conductors.

Many materials also exhibit superconductivity at low temperatures; they include metallic elements such as tin and aluminium, various metallic alloys, some heavily doped semiconductors, and certain ceramics. The electrical resistivity of most electrical (metallic) conductors generally decreases gradually as the temperature is lowered, but remains finite. In a superconductor however, the resistance drops abruptly to zero when the material is cooled below its critical temperature. An electric current flowing in a loop of superconducting wire can persist indefinitely with no power source.

A dielectric, or electrical insulator, is a substance that is highly resistant to the flow of electric current. A dielectric, such as plastic, tends to concentrate an applied electric field within itself which property is used in capacitors. A capacitor is

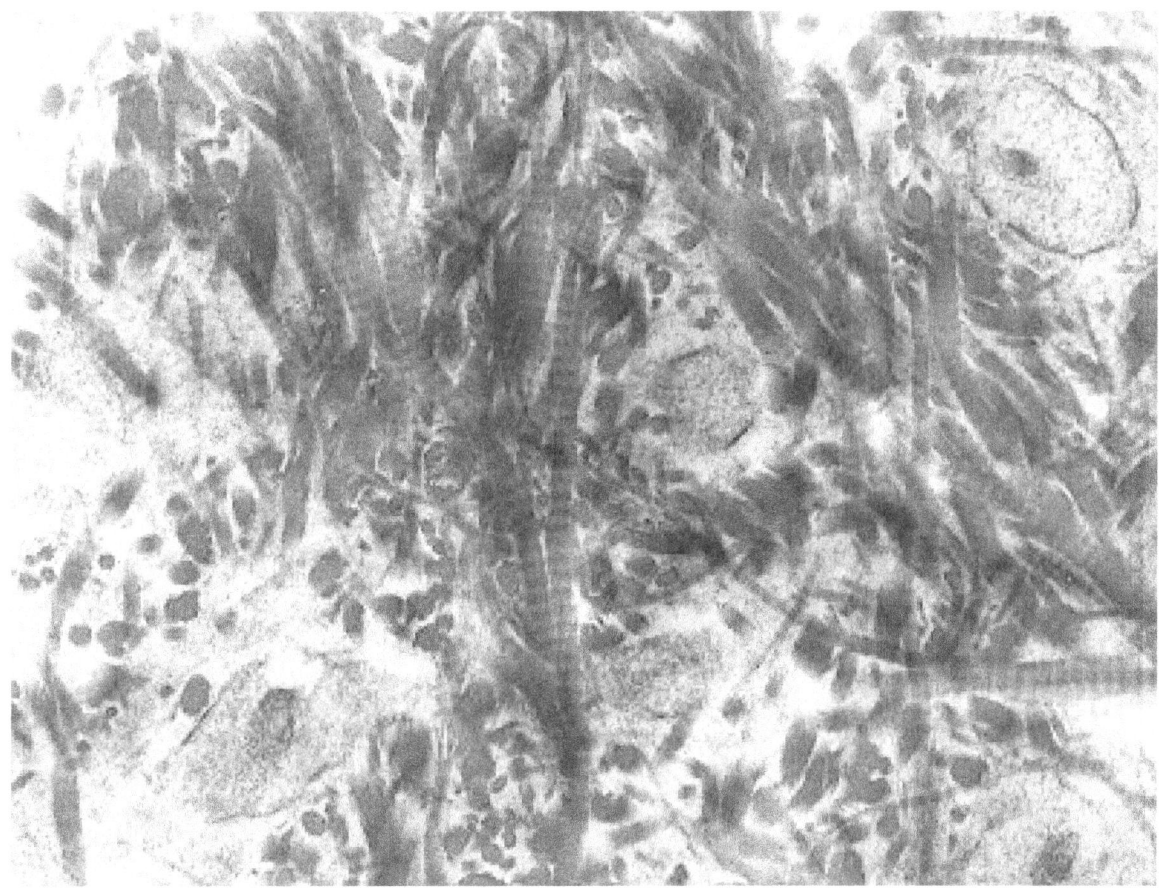

Collagen fibers of woven bone

an electrical device that can store energy in the electric field between a pair of closely spaced conductors (called 'plates'). When voltage is applied to the capacitor, electric charges of equal magnitude, but opposite polarity, build up on each plate. Capacitors are used in electrical circuits as energy-storage devices, as well as in electronic filters to differentiate between high-frequency and low-frequency signals.

Electro-mechanical

Piezoelectricity is the ability of crystals to generate a voltage in response to an applied mechanical stress. The piezoelectric effect is reversible in that piezoelectric crystals, when subjected to an externally applied voltage, can change shape by a small amount. Polymer materials like rubber, wool, hair, wood fiber, and silk often behave as electrets. For example, the polymer polyvinylidene fluoride (PVDF) exhibits a piezoelectric response several times larger than the traditional piezoelectric material quartz (crystalline SiO_2). The deformation (~0.1%) lends itself to useful technical applications such as high-voltage sources, loudspeakers, lasers, as well as chemical, biological, and acousto-optic sensors and/or transducers.

3.3.4 Optical

Materials can transmit (e.g. glass) or reflect (e.g. metals) visible light.

Many materials will transmit some wavelengths while blocking others. For example, window glass is transparent to visible light, but much less so to most of the frequencies of ultraviolet light that cause sunburn. This property is used for frequency-selective optical filters, which can alter the color of incident light.

For some purposes, both the optical and mechanical properties of a material can be of interest. For example, the sensors

Granite rock formation in the Chilean Patagonia. Like most inorganic minerals formed by oxidation in the Earth's atmosphere, granite consists primarily of crystalline silica SiO_2 and alumina Al_2O_3.

on an infrared homing ("heat-seeking") missile must be protected by a cover which is transparent to infrared radiation. The current material of choice for high-speed infrared-guided missile domes is single-crystal sapphire. The optical transmission of sapphire does not actually extend to cover the entire mid-infrared range (3–5 µm), but starts to drop off at wavelengths greater than approximately 4.5 µm at room temperature. While the strength of sapphire is better than that of other available mid-range infrared dome materials at room temperature, it weakens above 600 °C. A long-standing trade-off exists between optical bandpass and mechanical durability; new materials such as transparent ceramics or optical nanocomposites may provide improved performance.

Guided lightwave transmission involves the field of fiber optics and the ability of certain glasses to transmit, simultaneously and with low loss of intensity, a range of frequencies (multi-mode optical waveguides) with little interference between them. Optical waveguides are used as components in integrated optical circuits or as the transmission medium in optical communication systems.

Opto-electronic

Main article: Solar cell

A solar cell or photovoltaic cell is a device that converts light energy into electrical energy. Fundamentally, the device needs to fulfill only two functions: photo-generation of charge carriers (electrons and holes) in a light-absorbing material, and separation of the charge carriers to a conductive contact that will transmit the electricity (simply put, carrying electrons off through a metal contact into an external circuit). This conversion is called the photoelectric effect, and the field of

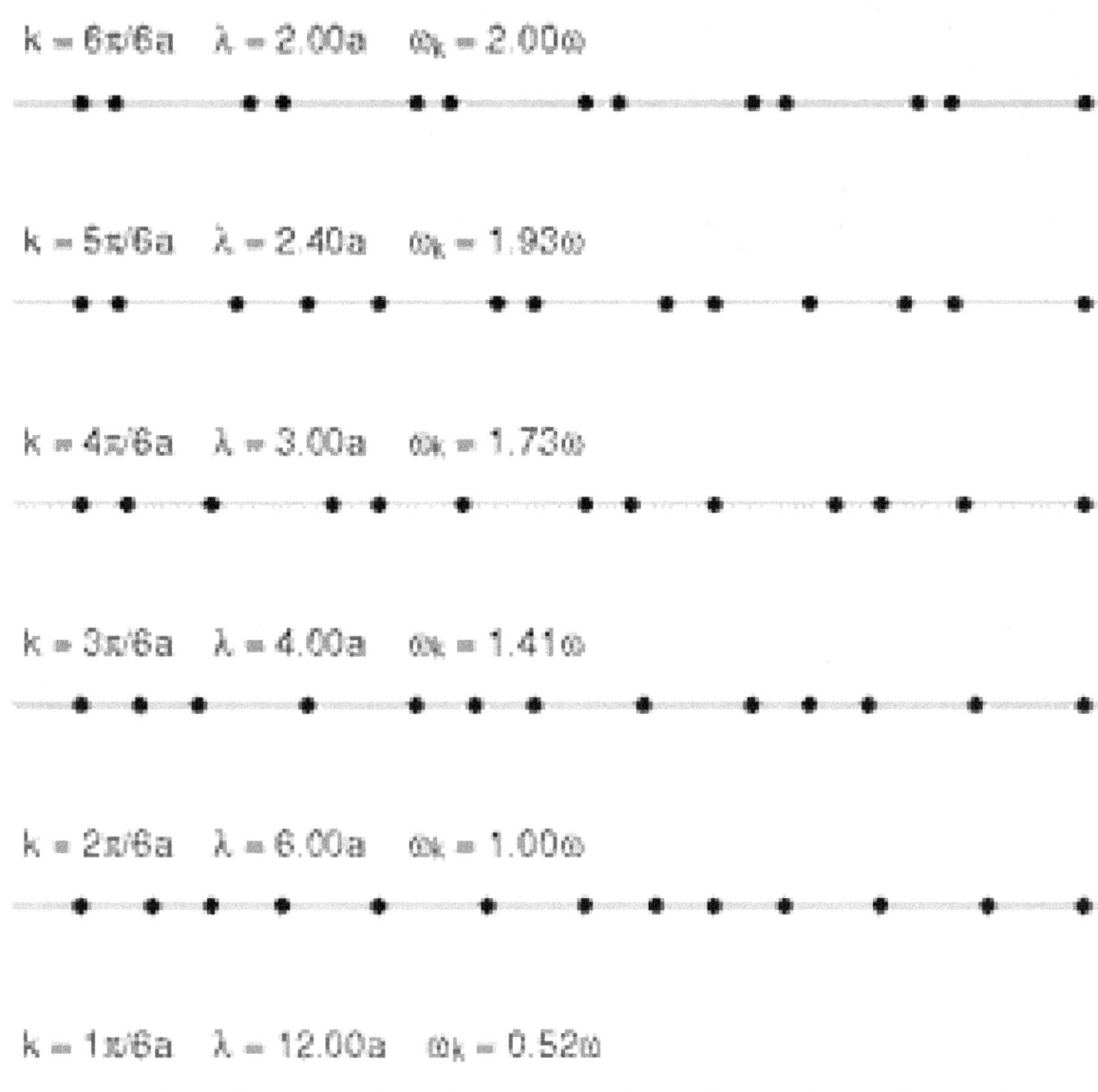

Normal modes of atomic vibration in a crystalline solid.

research related to solar cells is known as photovoltaics.

Solar cells have many applications. They have long been used in situations where electrical power from the grid is unavailable, such as in remote area power systems, Earth-orbiting satellites and space probes, handheld calculators, wrist watches, remote radiotelephones and water pumping applications. More recently, they are starting to be used in assemblies of solar modules (photovoltaic arrays) connected to the electricity grid through an inverter, that is not to act as a sole supply but as an additional electricity source.

All solar cells require a light absorbing material contained within the cell structure to absorb photons and generate electrons via the photovoltaic effect. The materials used in solar cells tend to have the property of preferentially absorbing the wavelengths of solar light that reach the earth surface. However, some solar cells are optimized for light absorption beyond Earth's atmosphere as well.

Video of superconducting levitation of YBCO

3.4   References

[1] Mortimer, Charles E. (1975). *Chemistry: A Conceptual Approach* (3rd ed.). New York:: D. Van Nostrad Company. ISBN 0-442-25545-4.

[2] Buffat, Ph.; Borel, J.-P. (1976). "Size effect on the melting temperature of gold particles". *Physical Review A* **13** (6): 2287. Bibcode:1976PhRvA..13.2287B. doi:10.1103/PhysRevA.13.2287.

[3] Walter H. Kohl (1995). *Handbook of materials and techniques for vacuum devices.* Springer. pp. 164–167. ISBN 1-56396-387-6.

[4] Shpak, Anatoly P; Kotrechko, Sergiy O; Mazilova, Tatjana I; Mikhailovskij, Igor M (2009). "Inherent tensile strength of molybdenum nanocrystals". *Science and Technology of Advanced Materials* **10** (4): 045004. Bibcode:2009STAdM..10d5004S. doi:10.1088/1468-6996/10/4/045004.

3.5   External links

- Wiki on equipment for handling and processing Bulk Solids

Chapter 4

Amorphous solid

"Amorphous" redirects here. For amorphousness in computational systems, see amorphous computing. For amorphousness in science fiction, see amorphous creature. For amorphousness in set theory, see amorphous set.

In condensed matter physics and materials science, an **amorphous** (from the Greek *a*, without, *morphé*, shape, form) or **non-crystalline solid** is a solid that lacks the long-range order characteristic of a crystal. In some older books, the term has been used synonymously with glass. Nowadays, "amorphous solid" is considered to be the overarching concept, and glass the more special case: A glass is an amorphous solid that exhibits a glass transition.[1] Polymers are often amorphous. Other types of amorphous solids include gels, thin films, and nanostructured materials such as glass.

Amorphous materials have an internal structure made of interconnected structural blocks. Whether a material is liquid or solid depends primarily on the connectivity between its elementary building blocks so that solids are characterized by a high degree of connectivity whereas structural blocks in fluids have lower connectivity (see figure on amorphous material states).[2]

4.1   Nano-structured materials

Even amorphous materials have some shortrange order at the atomic length scale due to the nature of chemical bonding (see structure of liquids and glasses for more information on non-crystalline material structure). Furthermore, in very small crystals a large fraction of the atoms are the crystal; relaxation of the surface and interfacial effects distort the atomic positions, decreasing the structural order. Even the most advanced structural characterization techniques, such as x-ray diffraction and transmission electron microscopy, have difficulty in distinguishing between amorphous and crystalline structures on these length scales.

4.2   Amorphous thin films

Amorphous phases are important constituents of thin films, which are solid layers of a few nm to some tens of μm thickness deposited upon a substrate. So-called structure zone models were developed to describe the micro structure and ceramics of thin films as a function of the homologous temperature Th that is the ratio of deposition temperature over melting temperature.[3][4] According to these models, a necessary (but not sufficient) condition for the occurrence of amorphous phases is that Th has to be smaller than 0.3, that is the deposition temperature must be below 30% of the melting temperature. For higher values, the surface diffusion of deposited atomic species would allow for the formation of crystallites with long range atomic order.

Regarding their applications, amorphous metallic layers played an important role in the discussion of a suspected superconductivity in amorphous metals.[5] Today, optical coatings made from TiO_2, SiO_2, Ta_2O_5 etc. and combinations of them in most cases consist of amorphous phases of these compounds. Much research is carried out into thin amorphous films as a gas

Amorphous metals have low toughness, but high strength

separating membrane layer.[6] The technologically most important thin amorphous film is probably represented by few nm thin SiO_2 layers serving as isolator above the conducting channel of a metal-oxide semiconductor field-effect transistor (MOSFET). Also, hydrogenated amorphous silicon, a-Si:H in short, is of technical significance for thin film solar cells. In case of a-Si:H the missing long-range order between silicon atoms is partly induced by the presence by hydrogen in the percent range.

The occurrence of amorphous phases turned out as a phenomenon of particular interest for studying thin film growth.[7] Remarkably, the growth of polycrystalline films is often used and preceded by an initial amorphous layer, the thickness of which may amount to only a few nm. The most investigated example is represented by thin multicrystalline silicon films, where such as the unoriented molecule. An initial amorphous layer was observed in many studies.[8] Wedge-shaped polycrystals were identified by transmission electron microscopy to grow out of the amorphous phase only after the latter has exceeded a certain thickness, the precise value of which depends on deposition temperature, background pressure and various other process parameters. The phenomenon has been interpreted in the framework of Ostwald's rule of stages[9] that predicts the formation of phases to proceed with increasing condensation time towards increasing stability.[5][8] Experimental studies of the phenomenon require a clearly defined state of the substrate surface and its contaminant density etc., upon which the thin film is deposited.

4.3 References

[1] J. Zarzycki: Les verres et l'état vitreux. Paris: Masson 1982. English translation available.

[2] M.I. Ojovan, W.E. Lee. Connectivity and glass transition in disordered oxide systems. J. Non-Cryst. Solids, 356, 2534-2540 (2010). doi:10.1016/j.jnoncrysol.2010.05.012

Ordered structures
Liquid Crystals | Crystals

Disordered structures
Melts | Glasses

Degree of order

Degree of Connectivity

States of crystalline and amorphous materials as a function of connectivity

[3] B. A. Movchan and A. V. Demchishin (1969). "Study of the structure and properties of thick vacuum condensates of nickel, titanium, tungsten, aluminium oxide and zirconium dioxide". *Phys. Met. Metallogr.* **28**: 83–90.

[4] J.A. Thornton (1974). "Influence of apparatus geometry and deposition conditions on the structure and topography of thick sputtered coatings". *Journal of Vacuum Science and Technology* **11** (4): 666–670. Bibcode:1974JVST...11..666T. doi:10.1116/1.1312732.

[5] Buckel, W. (1961). "The influence of crystal bonds on film growth". *Elektrische en Magnetische Eigenschappen van dunne Metallaagies*. Leuven, Belgium.

[6] R.M. de Vos, H. Verweij (1998). "High-Selectivity, High-Flux Silica Membranes for Gas Separation". *Science* **279** (5357): 1710–1. Bibcode:1998Sci...279.1710D. doi:10.1126/science.279.5357.1710. PMID 9497287.

[7] Magnuson et al. Electronic Structure and Chemical Bonding of Amorphous Chromium Carbide Thin Films; J. Phys. - Cond. Mat. 24 , 225004 (2012).

[8] M. Birkholz, B. Selle, W. Fuhs, S. Christiansen, H. P. Strunk, and R. Reich (2001). "Amorphous-crystalline phase transition during the growth of thin films: the case of microcrystalline silicon" (PDF). *Phys. Rev. B* **64** (8): 085402. Bibcode:2001PhRvB..64h5402B. doi:10.1103/PhysRevB.64.085402.

[9] W. Ostwald (1897). "Studien über die Umwandlung fester Körper". *Z. Phys. Chem.* **22**: 289–330.

4.4 Further reading

- R. Zallen (1969). *The Physics of Amorphous Solids*. Wiley Interscience.

- S.R. Elliot (1969). *The Physics of Amorphous Materials* (2nd ed.). Longman.

- N. Cusack (1969). *The Physics of Structurally Disordered Matter: An Introduction*. IOP Publishing.

- N.H. March, R.A. Street, M.P. Tosi, Eds., (1969). *Amorphous Solids and the Liquid State*. Springer.

- D.A. Adler, B.B. Schwartz, M.C. Steele, Eds. (1969). *Physical Properties of Amorphous Materials*. Springer.

- A. Inoue, K. Hasimoto, Eds. (1969). *Amorphous and Nanocrystalline Materials*. Springer.

4.5   External links

- Journal of non-crystalline solids (Elsevier)

Chapter 5

Crystal

"Xtal" redirects here. For other uses, see Xtal (disambiguation).
This article is about crystalline solids. For the type of glass, see Lead glass. For other uses, see Crystal (disambiguation).

A **crystal** or **crystalline solid** is a solid material whose constituents, such as atoms, molecules or ions, are arranged in a

A crystal of amethyst quartz

highly ordered microscopic structure, forming a crystal lattice that extends in all directions. In addition, macroscopic single crystals are usually identifiable by their geometrical shape, consisting of flat faces with specific, characteristic orientations.

The scientific study of crystals and crystal formation is known as crystallography. The process of crystal formation via mechanisms of crystal growth is called crystallization or solidification. The word *crystal* is derived from the Ancient Greek word κρύσταλλος (*krustallos*), meaning both "ice" and "rock crystal",[1] from κρύος (*kruos*), "icy cold, frost".[2][3]

Examples of large crystals include snowflakes, diamonds, and table salt. Most inorganic solids are not crystals but

41

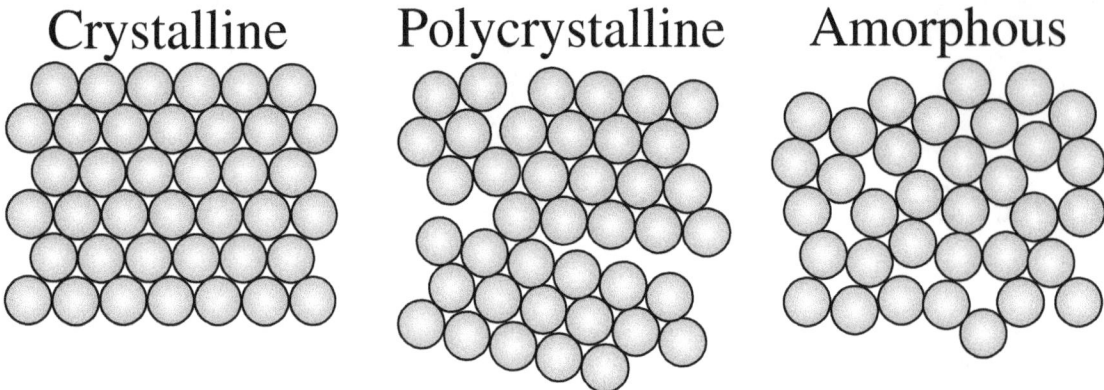

Microscopically, a single crystal has atoms in a near-perfect periodic arrangement; a polycrystal is composed of many microscopic crystals (called "crystallites" or "grains"); and an amorphous solid (such as glass) has no periodic arrangement even microscopically.

polycrystals, i.e. many microscopic crystals fused together into a single solid. Examples of polycrystals include most metals, rocks, ceramics, and ice. A third category of solids is amorphous solids, where the atoms have no periodic structure whatsoever. Examples of amorphous solids include glass, wax, and many plastics.

5.1   Crystal structure (microscopic)

Halite (table salt, NaCl): Microscopic and macroscopic

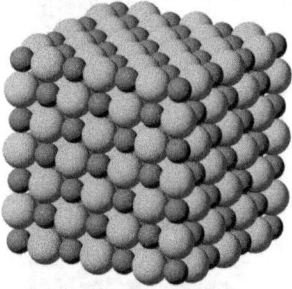

Microscopic structure of a halite crystal. (Purple is sodium ion, green is chlorine ion.) There is cubic symmetry in the atoms' arrangement.

Macroscopic (~16cm) halite crystal. The right-angles between crystal faces are due to the cubic symmetry of the atoms' arrangement.
Main article: Crystal structure

The scientific definition of a "crystal" is based on the microscopic arrangement of atoms inside it, called the crystal structure. A crystal is a solid where the atoms form a periodic arrangement. (Quasicrystals are an exception, see below.)

Not all solids are crystals. For example, when liquid water starts freezing, the phase change begins with small ice crystals that grow until they fuse, forming a *polycrystalline* structure. In the final block of ice, each of the small crystals (called "crystallites" or "grains") is a true crystal with a periodic arrangement of atoms, but the whole polycrystal does *not* have a periodic arrangement of atoms, because the periodic pattern is broken at the grain boundaries. Most macroscopic inorganic solids are polycrystalline, including almost all metals, ceramics, ice, rocks, etc. Solids that are neither crystalline nor polycrystalline, such as glass, are called *amorphous solids*, also called glassy, vitreous, or noncrystalline. These have no periodic order, even microscopically. There are distinct differences between crystalline solids and amorphous solids: most notably, the process of forming a glass does not release the latent heat of fusion, but forming a crystal does.

A crystal structure (an arrangement of atoms in a crystal) is characterized by its *unit cell*, a small imaginary box containing one or more atoms in a specific spatial arrangement. The unit cells are stacked in three-dimensional space to form the crystal.

The symmetry of a crystal is constrained by the requirement that the unit cells stack perfectly with no gaps. There are 219 possible crystal symmetries, called crystallographic space groups. These are grouped into 7 crystal systems, such as cubic crystal system (where the crystals may form cubes or rectangular boxes, such as halite shown at right) or hexagonal crystal system (where the crystals may form hexagons, such as ordinary water ice).

5.2   Crystal faces and shapes

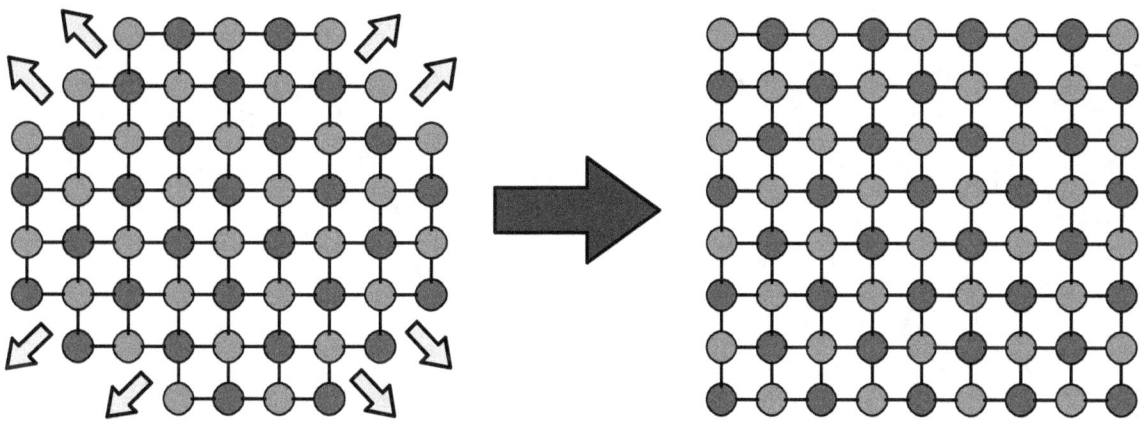

As a halite crystal is growing, new atoms can very easily attach to the parts of the surface with rough atomic-scale structure and many dangling bonds. Therefore, these parts of the crystal grow out very quickly (yellow arrows). Eventually, the whole surface consists of smooth, stable faces, where new atoms cannot as easily attach themselves.

Crystals are commonly recognized by their shape, consisting of flat faces with sharp angles. These shape characteristics are not *necessary* for a crystal—a crystal is scientifically defined by its microscopic atomic arrangement, not its macroscopic shape—but the characteristic macroscopic shape is often present and easy to see.

Euhedral crystals are those with obvious, well-formed flat faces. Anhedral crystals do not, usually because the crystal is one grain in a polycrystalline solid.

The flat faces (also called facets) of a euhedral crystal are oriented in a specific way relative to the underlying atomic arrangement of the crystal: They are planes of relatively low Miller index.[4] This occurs because some surface orientations are more stable than others (lower surface energy). As a crystal grows, new atoms attach easily to the rougher and less stable parts of the surface, but less easily to the flat, stable surfaces. Therefore, the flat surfaces tend to grow larger and smoother, until the whole crystal surface consists of these plane surfaces. (See diagram on right.)

One of the oldest techniques in the science of crystallography consists of measuring the three-dimensional orientations of the faces of a crystal, and using them to infer the underlying crystal symmetry.

A crystal's habit is its visible external shape. This is determined by the crystal structure (which restricts the possible facet orientations), the specific crystal chemistry and bonding (which may favor some facet types over others), and the conditions under which the crystal formed.

5.3   Occurrence in nature

Ice crystals.

5.3.1   Rocks

By volume and weight, the largest concentrations of crystals in the Earth are part of its solid bedrock. Crystals found in rocks typically range in size from a fraction of a millimetre to several centimetres across, although exceptionally large crystals are occasionally found. As of 1999, the world's largest known naturally occurring crystal is a crystal of beryl from Malakialina, Madagascar, 18 m (59 ft) long and 3.5 m (11 ft) in diameter, and weighing 380,000 kg (840,000 lb).[5]

Some crystals have formed by magmatic and metamorphic processes, giving origin to large masses of crystalline rock. The vast majority of igneous rocks are formed from molten magma and the degree of crystallization depends primarily on the conditions under which they solidified. Such rocks as granite, which have cooled very slowly and under great pressures, have completely crystallized; but many kinds of lava were poured out at the surface and cooled very rapidly, and in this latter group a small amount of amorphous or glassy matter is common. Other crystalline rocks, the metamorphic rocks such as marbles, mica-schists and quartzites, are recrystallized. This means that they were at first fragmental rocks like limestone, shale and sandstone and have never been in a molten condition nor entirely in solution, but the high temperature and pressure conditions of metamorphism have acted on them by erasing their original structures and inducing recrystallization in the solid state.[6]

Fossil shell with calcite crystals.

Other rock crystals have formed out of precipitation from fluids, commonly water, to form druses or quartz veins. The evaporites such as halite, gypsum and some limestones have been deposited from aqueous solution, mostly owing to evaporation in arid climates.

5.3.2 Ice

Water-based ice in the form of snow, sea ice and glaciers is a very common manifestation of crystalline or polycrystalline matter on Earth. A single snowflake is typically a single crystal, while an ice cube is a polycrystal.

5.3.3 Organigenic crystals

Many living organisms are able to produce crystals, for example calcite and aragonite in the case of most molluscs or hydroxylapatite in the case of vertebrates.

5.4 Polymorphism and allotropy

Main articles: Polymorphism (materials science) and Allotropy

The same group of atoms can often solidify in many different ways. Polymorphism is the ability of a solid to exist in more than one crystal form. For example, water ice is ordinarily found in the hexagonal form Ice I_h, but can also exist as the cubic Ice I_c, the rhombohedral ice II, and many other forms. The different polymorphs are usually called different *phases*.

In addition, the same atoms may be able to form noncrystalline phases. For example, water can also form amorphous ice, while SiO_2 can form both fused silica (an amorphous glass) and quartz (a crystal). Likewise, if a substance can form crystals, it can also form polycrystals.

For pure chemical elements, polymorphism is known as allotropy. For example, diamond and graphite are two crystalline forms of carbon, while amorphous carbon is a noncrystalline form. Polymorphs, despite having the same atoms, may have wildly different properties. For example, diamond is among the hardest substances known, while graphite is so soft that it is used as a lubricant.

Polyamorphism is a similar phenomenon where the same atoms can exist in more than one amorphous solid form.

5.5   Crystallization

Main articles: Crystallization and Crystal growth
Crystallization is the process of forming a crystalline structure from a fluid or from materials dissolved in a fluid. (More rarely, crystals may be deposited directly from gas; see thin-film deposition and epitaxy.)

Crystallization is a complex and extensively-studied field, because depending on the conditions, a single fluid can solidify into many different possible forms. It can form a single crystal, perhaps with various possible phases, stoichiometries, impurities, defects, and habits. Or, it can form a polycrystal, with various possibilities for the size, arrangement, orientation, and phase of its grains. The final form of the solid is determined by the conditions under which the fluid is being solidified, such as the chemistry of the fluid, the ambient pressure, the temperature, and the speed with which all these parameters are changing.

Specific industrial techniques to produce large single crystals (called *boules*) include the Czochralski process and the Bridgman technique. Other less exotic methods of crystallization may be used, depending on the physical properties of the substance, including hydrothermal synthesis, sublimation, or simply solvent-based crystallization.

Large single crystals can be created by geological processes. For example, selenite crystals in excess of 10 meters are found in the Cave of the Crystals in Naica, Mexico.[7] For more details on geological crystal formation, see above.

Crystals can also be formed by biological processes, see above. Conversely, some organisms have special techniques to *prevent* crystallization from occurring, such as antifreeze proteins.

5.6   Defects, impurities, and twinning

Main articles: Crystallographic defect, Impurity, Crystal twinning and Mosaicity
An *ideal* crystal has every atom in a perfect, exactly repeating pattern. However, in reality, most crystalline materials have a variety of crystallographic defects, places where the crystal's pattern is interrupted. The types and structures of these defects may have a profound effect on the properties of the materials.

A few examples of crystallographic defects include vacancy defects (an empty space where an atom should fit), interstitial defects (an extra atom squeezed in where it does not fit), and dislocations (see figure at right). Dislocations are especially important in materials science, because they help determine the mechanical strength of materials.

Another common type of crystallographic defect is an impurity, meaning that the "wrong" type of atom is present in a crystal. For example, a perfect crystal of diamond would only contain carbon atoms, but a real crystal might perhaps contain a few boron atoms as well. These boron impurities change the diamond's color to slightly blue. Likewise, the only difference between ruby and sapphire is the type of impurities present in a corundum crystal.

In semiconductors, a special type of impurity, called a dopant, drastically changes the crystal's electrical properties. Semiconductor devices, such as transistors, are made possible largely by putting different semiconductor dopants into different places, in specific patterns.

Twinning is a phenomenon somewhere between a crystallographic defect and a grain boundary. Like a grain boundary, a twin boundary has different crystal orientations on its two sides. But unlike a grain boundary, the orientations are not random, but related in a specific, mirror-image way.

Mosaicity is a spread of crystal plane orientations. A mosaic crystal is supposed to consist of smaller crystalline units that are somewhat misaligned with respect to each other.

5.7 Chemical bonds

Crystalline structures occur in all classes of materials, with all types of chemical bonds. Almost all metal exists in a poly-crystalline state; amorphous or single-crystal metals must be produced synthetically, often with great difficulty. Ionically bonded crystals can form upon solidification of salts, either from a molten fluid or upon crystallization from a solution. Covalently bonded crystals are also very common, notable examples being diamond, silica, and graphite. Polymer materials generally will form crystalline regions, but the lengths of the molecules usually prevent complete crystallization. Weak van der Waals forces can also play a role in a crystal structure; for example, this type of bonding loosely holds together the hexagonal-patterned sheets in graphite.

5.7.1 Properties

5.8 Quasicrystals

Main article: Quasicrystal

A quasicrystal consists of arrays of atoms that are ordered but not strictly periodic. They have many attributes in common with ordinary crystals, such as displaying a discrete pattern in x-ray diffraction, and the ability to form shapes with smooth, flat faces.

Quasicrystals are most famous for their ability to show five-fold symmetry, which is impossible for an ordinary periodic crystal (see crystallographic restriction theorem).

The International Union of Crystallography has redefined the term "crystal" to include both ordinary periodic crystals and quasicrystals ("any solid having an essentially discrete diffraction diagram"[8]).

Quasicrystals, first discovered in 1982, are quite rare in practice. Only about 100 solids are known to form quasicrystals, compared to about 400,000 periodic crystals measured to date.[9] The 2011 Nobel Prize in Chemistry was awarded to Dan Shechtman for the discovery of quasicrystals.[10]

5.9 Special properties from anisotropy

See also: Crystal optics

Crystals can have certain special electrical, optical, and mechanical properties that glass and polycrystals normally cannot. These properties are related to the anisotropy of the crystal, i.e. the lack of rotational symmetry in its atomic arrangement. One such property is the piezoelectric effect, where a voltage across the crystal can shrink or stretch it. Another is birefringence, where a double image appears when looking through a crystal. Moreover, various properties of a crystal, including electrical conductivity, electrical permittivity, and Young's modulus, may be different in different directions in a crystal. For example, graphite crystals consist of a stack of sheets, and although each individual sheet is mechanically very strong, the sheets are rather loosely bound to each other. Therefore, the mechanical strength of the material is quite different depending on the direction of stress.

Not all crystals have all of these properties. Conversely, these properties are not quite exclusive to crystals. They can appear in glasses or polycrystals that have been made anisotropic by working or stress—for example, stress-induced birefringence.

5.10 Crystallography

Main article: Crystallography

Crystallography is the science of measuring the crystal structure (in other words, the atomic arrangement) of a crystal. One widely used crystallography technique is X-ray diffraction. Large numbers of known crystal structures are stored in crystallographic databases.

5.11 Gallery

- Insulin crystals grown in earth orbit.

- Hoar frost: A type of ice crystal (picture taken from a distance of about 5 cm).

- Gallium, a metal that easily forms large crystals.

- An apatite crystal sits front and center on cherry-red rhodochroite rhombs, purple fluorite cubes, quartz and a dusting of brass-yellow pyrite cubes.

- Boules of silicon, like this one, are an important type of industrially-produced single crystal.

- A specimen consisting of a bornite-coated chalcopyrite crystal nestled in a bed of clear quartz crystals and lustrous pyrite crystals. The bornite-coated crystal is up to 1.5 cm across.

5.12 See also

- Atomic packing factor

- Anticrystal

- Cocrystal

- Colloidal crystal

- Crystal growth

- Crystal oscillator

- Liquid crystal

5.13 References

[1] κρύσταλλος, Henry George Liddell, Robert Scott, *A Greek-English Lexicon*, on Perseus Digital Library

[2] κρύος, Henry George Liddell, Robert Scott, *A Greek-English Lexicon*, on Perseus Digital Library

[3] "The American Heritage Dictionary of the English Language". Kreus. 2000.

[4] *The surface science of metal oxides*, by Victor E. Henrich, P. A. Cox, page 28, google books link

[5] G. Cressey and I. F. Mercer, (1999) *Crystals*, London, Natural History Museum, page 58

[6] One or more of the preceding sentences incorporates text from a publication now in the public domain: Chisholm, Hugh, ed. (1911). "Petrology". *Encyclopædia Britannica* (11th ed.). Cambridge University Press.

[7] "Cave of Crystal Giants — National Geographic Magazine". *nationalgeographic.com*.

[8] International Union of Crystallography (1992). "Report of the Executive Committee for 1991". *Acta Cryst.* **A48** (6): 922. doi:10.1107/S0108767392008328.

[9] Steurer W. (2004). "Twenty years of structure research on quasicrystals. Part I. Pentagonal, octagonal, decagonal and dodecagonal quasicrystals". *Z. Kristallogr.* **219** (7–2004): 391–446. Bibcode:2004ZK....219..391S. doi:10.1524/zkri.219.7.391.35643.

[10] "The Nobel Prize in Chemistry 2011". Nobelprize.org. Retrieved 2011-12-29.

5.14 Further reading

- Howard, J. Michael; Darcy Howard (Illustrator) (1998). "Introduction to Crystallography and Mineral Crystal Systems". Bob's Rock Shop. Retrieved 2008-04-20. Cite uses deprecated parameter |coauthors= (help)

- Krassmann, Thomas (2005–2008). "The Giant Crystal Project". Krassmann. Retrieved 2008-04-20.

- Various authors (2007). "Teaching Pamphlets". Commission on Crystallographic Teaching. Retrieved 2008-04-20.

- Various authors (2004). "Crystal Lattice Structures:Index by Space Group". U.S. Naval Research Laboratory, Center for Computational Materials Science. Retrieved 2008-04-20.

- Various authors (2010). "Crystallography". Spanish National Research Council, Department of Crystallography. Retrieved 2010-01-08.

Vertical cooling crystallizer in a beet sugar factory.

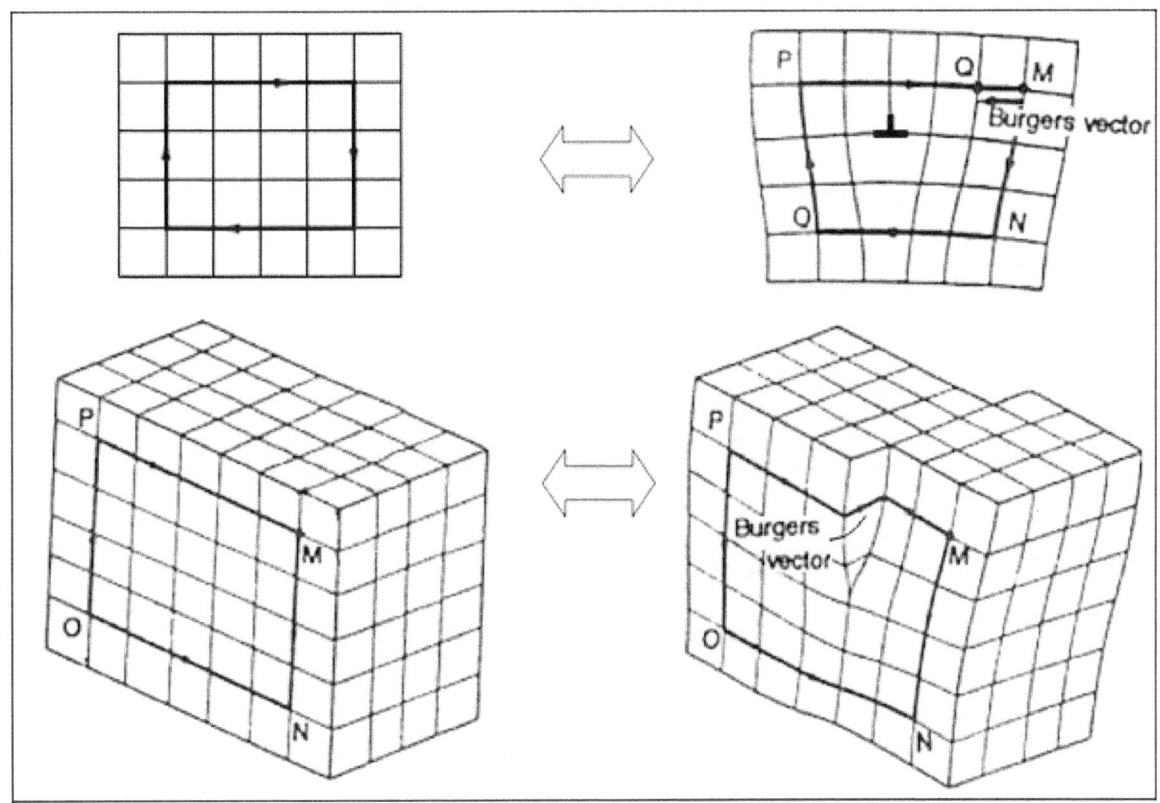

Two types of crystallographic defects. Top right: edge dislocation. Bottom right: screw dislocation.

Twinned pyrite crystal group.

The material Ho-Mg-Zn forms quasicrystals, which can take on the macroscopic shape of a dodecahedron. (Only a quasicrystal, not a normal crystal, can take this shape.) The edges are 2 mm long.

Chapter 6

Plastic crystal

A **plastic crystal** is a crystal composed of weakly interacting molecules that possess some orientational or conformational degree of freedom. The name plastic crystal refers to the mechanical softness of such phases: they resemble waxes and are easily deformed. If the internal degree of freedom is molecular rotation, the name **rotor phase** or **rotatory phase** is also used. Typical examples are the modifications Methane I and Ethane I. In addition to the conventional molecular plastic crystals, there are also emerging ionic plastic crystals, particularly organic ionic plastic crystals (OIPCs) and protic organic ionic plastic crystals (POIPCs).[1][2] POIPCs are solid protic organic salts formed by proton transfer from a Brønsted acid to a Brønsted base and in essence are protic ionic liquids in the molten state, have found to be promising solid-state proton conductors for high temperature proton exchange membrane fuel cells.[1] Examples include 1,2,4-triazolium perfluorobutanesulfonate[1] and imidazolium methanesulfonate.[2]

If the internal degree of freedom freezes in a disordered way, an orientational glass is obtained.

The orientational degree of freedom may be an almost free rotation, or it may be a jump diffusion between a restricted number of possible orientations, as was shown for carbon tetrabromide.[3]

The X-ray diffraction patterns of plastic crystals are characterized by strong diffuse intensity in addition to the sharp Bragg peaks.[1] In a powder pattern this intensity appears to resemble an amorphous background as one would expect for a liquid,[1] but for a single crystal the diffuse contribution reveals itself to be highly structured. The Bragg peaks can be used to determine an average structure but due to the large amount of disorder this is not very insightful. It is the structure of the diffuse scattering that reflects the details of the constrained disorder in the system. Recent advances in two-dimensional detection at synchrotron beam lines facilitate the study of such patterns.

6.1   Plastic crystals versus liquid crystals

Like liquid crystals, plastic crystals can be considered a transitional stage between real solids and real liquids and can be considered *soft matter*. Another common denominator is the simultaneous presence of order and disorder. Both types of phases are usually observed between the true solid and liquid phases on the temperature scale:

true crystal -> plastic crystal -> true liquid

true crystal -> liquid crystal -> true liquid

The difference between liquid and plastic crystals is easily observed in X-ray diffraction. Plastic crystals possess strong long range order and therefore show sharp Bragg reflections.[1] Liquid crystals show no or very broad Bragg peaks because the order is not long range. The molecules that give rise to liquid crystalline behavior often have a strongly elongated or disc like shape. Plastic crystals consist usually of almost spherical objects. In this respect one could see them as opposites.

6.2 References

[1] Jiangshui Luo; Annemette H. Jensen; Neil R. Brooks; Jeroen Sniekers; et al. (2015). "1,2,4-Triazolium perfluorobutanesulfonate as an archetypal pure protic organic ionic plastic crystal electrolyte for all-solid-state fuel cells". *Energy & Environmental Science* **8**. doi:10.1039/C4EE02280G.

[2] Jiangshui Luo; Olaf Conrad & Ivo F. J. Vankelecom (2013). "Imidazolium methanesulfonate as a high temperature proton conductor". *Journal of Materials Chemistry A* **1**. doi:10.1039/C2TA00713D.

[3] Coupled orientational and displacive degrees of freedom in the high-temperature plastic phase of the carbon tetrabromide α-CBr4 Jacob C. W. Folmer, Ray L. Withers, T. R. Welberry, and James D. Martin. Physical Review B 77, 144205 (2008).

Chapter 7

Quasicrystal

Not to be confused with Quasi-crystals (supramolecular).

A quasiperiodic crystal, or **quasicrystal**, is a structure that is ordered but not periodic. A quasicrystalline pattern can continuously fill all available space, but it lacks translational symmetry. While crystals, according to the classical crystallographic restriction theorem, can possess only two, three, four, and six-fold rotational symmetries, the Bragg diffraction pattern of quasicrystals shows sharp peaks with other symmetry orders, for instance five-fold.

Aperiodic tilings were discovered by mathematicians in the early 1960s, and, some twenty years later, they were found to apply to the study of quasicrystals. The discovery of these aperiodic forms in nature has produced a paradigm shift in the fields of crystallography. Quasicrystals had been investigated and observed earlier,[2] but, until the 1980s, they were disregarded in favor of the prevailing views about the atomic structure of matter. In 2009, after a dedicated search, a mineralogical finding, icosahedrite, offered evidence for the existence of natural quasicrystals.[3]

Roughly, an ordering is non-periodic if it lacks translational symmetry, which means that a shifted copy will never match exactly with its original. The more precise mathematical definition is that there is never translational symmetry in more than $n - 1$ linearly independent directions, where n is the dimension of the space filled, e.g., the three-dimensional tiling displayed in a quasicrystal may have translational symmetry in two dimensions. The ability to diffract comes from the existence of an indefinitely large number of elements with a regular spacing, a property loosely described as long-range order. Experimentally, the aperiodicity is revealed in the unusual symmetry of the diffraction pattern, that is, symmetry of orders other than two, three, four, or six. In 1982 materials scientist Dan Shechtman observed that certain aluminium-manganese alloys produced the unusual diffractograms which today are seen as revelatory of quasicrystal structures. Due to fear of the scientific community's reaction, it took him two years to publish the results[4][5] for which he was awarded the Nobel Prize in Chemistry in 2011.[6]

7.1 History

In 1961, Hao Wang asked whether determining if a set of tiles admits a tiling of the plane is an algorithmically unsolvable problem or not. He conjectured that it is solvable, relying on the hypothesis that every set of tiles that can tile the plane can do it *periodically* (hence, it would suffice to try to tile bigger and bigger patterns until obtaining one that tiles periodically). Nevertheless, two years later, his student, Robert Berger, constructed a set of some 20,000 square tiles (now called Wang tiles) that can tile the plane but not in a periodic fashion. As the number of known aperiodic sets of tiles grew, each set seemed to contain even fewer tiles than the previous one. In particular, in 1976, Roger Penrose proposed a set of just two tiles, up to rotation, (referred to as Penrose tiles) that produced only non-periodic tilings of the plane. These tilings displayed instances of fivefold symmetry. One year later, Alan Mackay showed experimentally that the diffraction pattern from the Penrose tiling had a two-dimensional Fourier transform consisting of sharp 'delta' peaks arranged in a fivefold symmetric pattern.[7] Around the same time, Robert Ammann had created a set of aperiodic tiles that produced eightfold symmetry.

Mathematically, quasicrystals have been shown to be derivable from a general method that treats them as projections of a

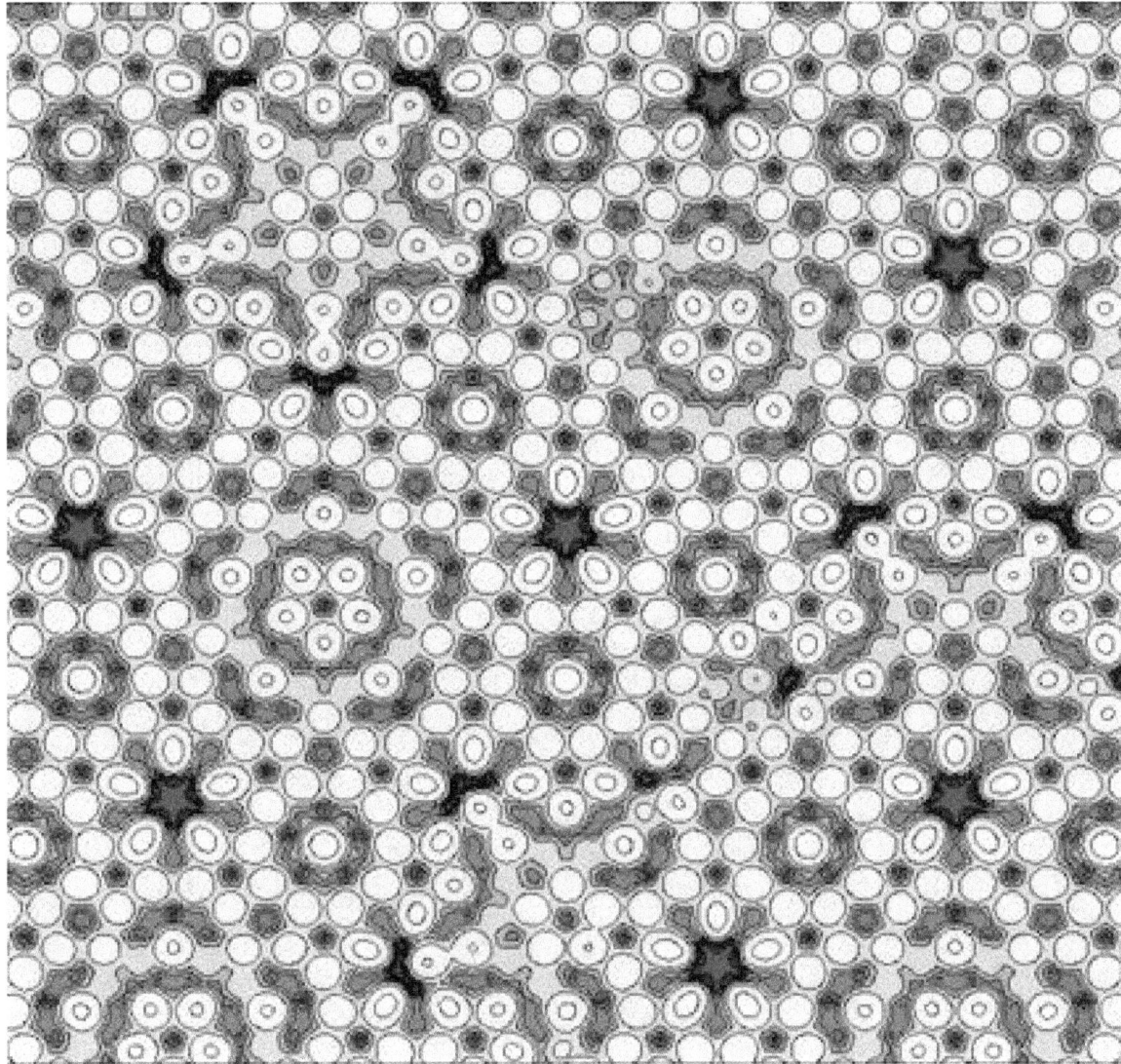

Potential energy surface for silver depositing on an aluminium-palladium-manganese (Al-Pd-Mn) quasicrystal surface. Similar to Fig. 6 in Ref.[11]

higher-dimensional lattice. Just as circles, ellipses, and hyperbolic curves in the plane can be obtained as sections from a three-dimensional double cone, so too various (aperiodic or periodic) arrangements in two and three dimensions can be obtained from postulated hyperlattices with four or more dimensions. Icosahedral quasicrystals in three dimensions were projected from a six-dimensional hypercubic lattice by Peter Kramer and Roberto Neri in 1984.[8] The tiling is formed by two tiles with rhombohedral shape.

Shechtman first observed ten-fold electron diffraction patterns in 1982, as described in his notebook. The observation was made during a routine investigation, by electron microscopy, of a rapidly cooled alloy of aluminium and manganese prepared at the National Bureau of Standards (now NIST).

In the summer of the same year, Shechtman visited Ilan Blech and related his observation to him. Blech responded that such diffractions were seen before.[9][10] Around that time, Shechtman also related his finding to John Cahn of NIST who did not offer any explanation and challenged him to solve the observation. Shechtman quoted Cahn as saying: "Danny, this material is telling us something and I challenge you to find out what it is".

The observation of the ten-fold diffraction pattern lay unexplained by Shechtman and others for two years until the spring of 1984, when Blech asked Shechtman to show him his results again. A quick study of Shechtman's results showed that

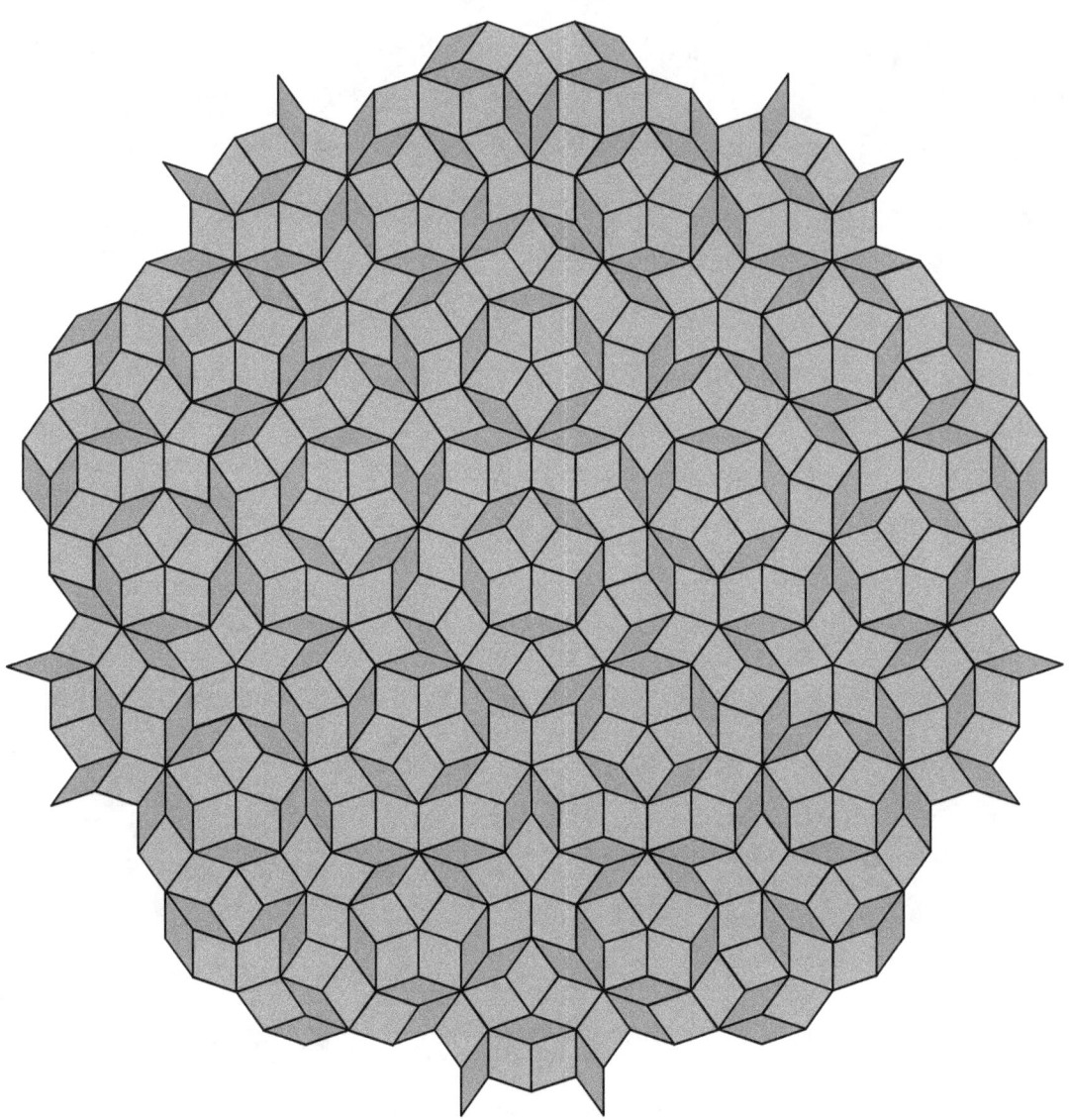

A Penrose tiling

the common explanation for a ten-fold symmetrical diffraction pattern, namely the existence of twins, was ruled out by his experiments. Since periodicity and twins were ruled out, Blech, unaware of the two-dimensional tiling work, was looking for another possibility: a completely new structure containing cells connected to each other by defined angles and distances but without translational periodicity. Blech decided to use a computer simulation to calculate the diffraction intensity from a cluster of such a material without long-range translational order but still not random. He termed this new structure multiple polyhedral.

The idea of a new structure was the necessary paradigm shift to break the impasse. The "Eureka moment" came when the computer simulation showed sharp ten-fold diffraction patterns, similar to the observed ones, emanating from the three-dimensional structure devoid of periodicity. The multiple polyhedral structure was termed later by many researchers as icosahedral glass but in effect it embraces *any arrangement of polyhedra connected with definite angles and distances* (this general definition includes tiling, for example).

Shechtman accepted Blech's discovery of a new type of material and it gave him the courage to publish his experimental observation. Shechtman and Blech jointly wrote a paper entitled "The Microstructure of Rapidly Solidified Al_6Mn" [11] and sent it for publication around June 1984 to the Journal of Applied Physics (JAP). The JAP editor promptly rejected the paper as being better fit for a metallurgical readership. As a result, the same paper was re-submitted for publication to the Metallurgical Transactions A, where it was accepted. Although not noted in the body of the published text, the published paper was slightly revised prior to publication.

Meanwhile, on seeing the draft of the Shechtman-Blech paper in the summer of 1984, John Cahn suggested that Shechtman's experimental results merit a fast publication in a more appropriate scientific journal. Shechtman agreed and, in hindsight, called this fast publication - "a winning move". This paper, published in the Physical Review Letters",[5] repeated Shechtman's observation and used the same illustrations as the original Shechtman-Blech paper in the *Metallurgical Transactions A*. Naturally, being the first paper to appear in print, the *Physical Review Letters* paper caused considerable excitement in the scientific community.

Next year, Ishimasa *et al.* reported twelvefold symmetry in Ni-Cr particles.[12] Soon, eightfold diffraction patterns were recorded in V-Ni-Si and Cr-Ni-Si alloys.[13] Over the years, hundreds of quasicrystals with various compositions and different symmetries have been discovered. The first quasicrystalline materials were thermodynamically unstable—when heated, they formed regular crystals. However, in 1987, the first of many stable quasicrystals were discovered, making it possible to produce large samples for study and opening the door to potential applications. In 2009, following a 10-year systematic search, scientists reported the first natural quasicrystal, a mineral found in the Khatyrka River in eastern Russia.[3] This natural quasicrystal exhibits high crystalline quality, equalling the best artificial examples.[14] The natural quasicrystal phase, with a composition of $Al_{63}Cu_{24}Fe_{13}$, was named icosahedrite and it was approved by the International Mineralogical Association in 2010. Furthermore, analysis indicates it may be meteoritic in origin, possibly delivered from a carbonaceous chondrite asteroid.[15]

A further study of Khatyrka meteorites revealed micron-sized grains of another natural quasicrystal, which has a ten-fold symmetry and a chemical formula of $Al_{71}Ni_{24}Fe_5$. This quasicrystal is stable in a narrow temperature range, from 1120 to 1200 K at ambient pressure, which suggests that natural quasicrystals are formed by rapid quenching of a meteorite heated during an impact-induced shock.[16]

In 1972, de Wolf and van Aalst[17] reported that the diffraction pattern produced by a crystal of sodium carbonate cannot be labeled with three indices but needed one more, which implied that the underlying structure had four dimensions in reciprocal space. Other puzzling cases have been reported,[18] but until the concept of quasicrystal came to be established, they were explained away or denied.[19][20] However, at the end of the 1980s, the idea became acceptable, and in 1992 the International Union of Crystallography altered its definition of a crystal, broadening it as a result of Shechtman's findings, reducing it to the ability to produce a clear-cut diffraction pattern and acknowledging the possibility of the ordering to be either periodic or aperiodic.[4][notes 1] Now, the symmetries compatible with translations are defined as "crystallographic", leaving room for other "non-crystallographic" symmetries. Therefore, aperiodic or quasiperiodic structures can be divided into two main classes: those with crystallographic point-group symmetry, to which the incommensurately modulated structures and composite structures belong, and those with non-crystallographic point-group symmetry, to which quasicrystal structures belong.

Originally, the new form of matter was dubbed "Shechtmanite".[21] The term "quasicrystal" was first used in print by Steinhardt and Levine[22] shortly after Shechtman's paper was published. The adjective *quasicrystalline* has been already in use but now it came to be applied to any pattern with unusual symmetry.[notes 2] 'Quasiperiodical' structures were claimed to be observed in some decorative tilings devised by medieval Islamic architects.[23][24] For example, Girih tiles in a medieval Islamic mosque in Isfahan, Iran, are arranged in a two-dimensional quasicrystalline pattern.[25] These claims have, however, been under some debate.[26]

Shechtman was awarded the Nobel Prize in Chemistry in 2011 for his work on quasicrystals. "His discovery of quasicrystals revealed a new principle for packing of atoms and molecules," stated the Nobel Committee and pointed that "this led to a paradigm shift within chemistry." [4][27]

7.2 Mathematics

There are several ways to mathematically define quasicrystalline patterns. One definition, the "cut and project" construction, is based on the work of Harald Bohr. The concept of an almost periodic function (also called a quasiperiodic

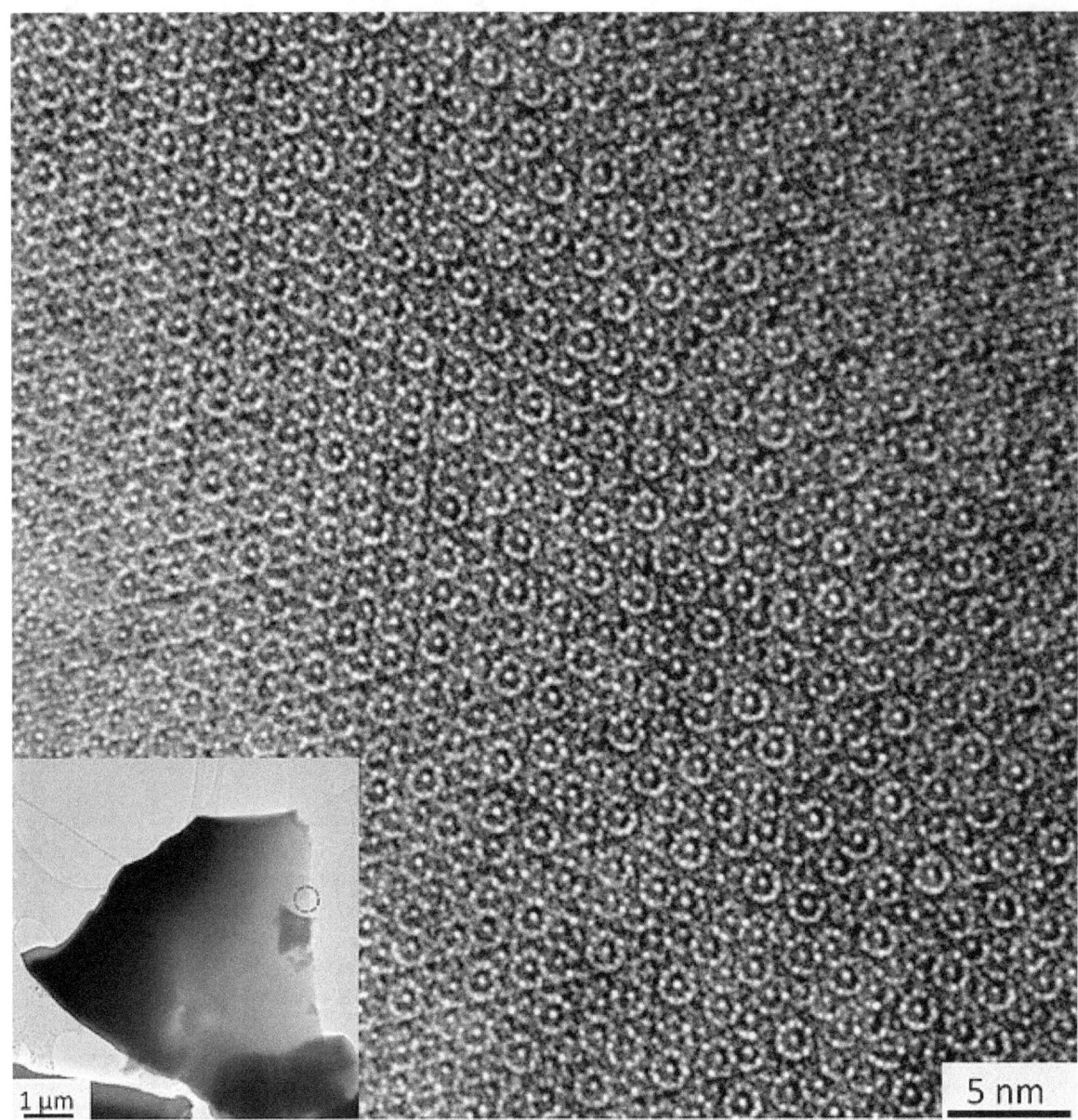

Atomic image of a micron-sized grain of the natural $Al_{71}Ni_{24}Fe_5$ quasicrystal (shown in the inset) from a Khatyrka meteorite. The corresponding diffraction patterns reveal a ten-fold symmetry.[16]

function) was studied by Bohr, including work of Bohl and Escanglon.[28] He introduced the notion of a superspace. Bohr showed that quasiperiodic functions arise as restrictions of high-dimensional periodic functions to an irrational slice (an intersection with one or more hyperplanes), and discussed their Fourier point spectrum. These functions are not exactly periodic, but they are arbitrarily close in some sense, as well as being a projection of an exactly periodic function.

In order that the quasicrystal itself be aperiodic, this slice must avoid any lattice plane of the higher-dimensional lattice. De Bruijn showed that Penrose tilings can be viewed as two-dimensional slices of five-dimensional hypercubic structures.[29] Equivalently, the Fourier transform of such a quasicrystal is nonzero only at a dense set of points spanned by integer multiples of a finite set of basis vectors (the projections of the primitive reciprocal lattice vectors of the higher-dimensional lattice).[30] The intuitive considerations obtained from simple model aperiodic tilings are formally expressed in the concepts of Meyer and Delone sets. The mathematical counterpart of physical diffraction is the Fourier transform and the qualitative description of a diffraction picture as 'clear cut' or 'sharp' means that singularities are present in the Fourier spectrum. There are different methods to construct model quasicrystals. These are the same methods that produce aperi-

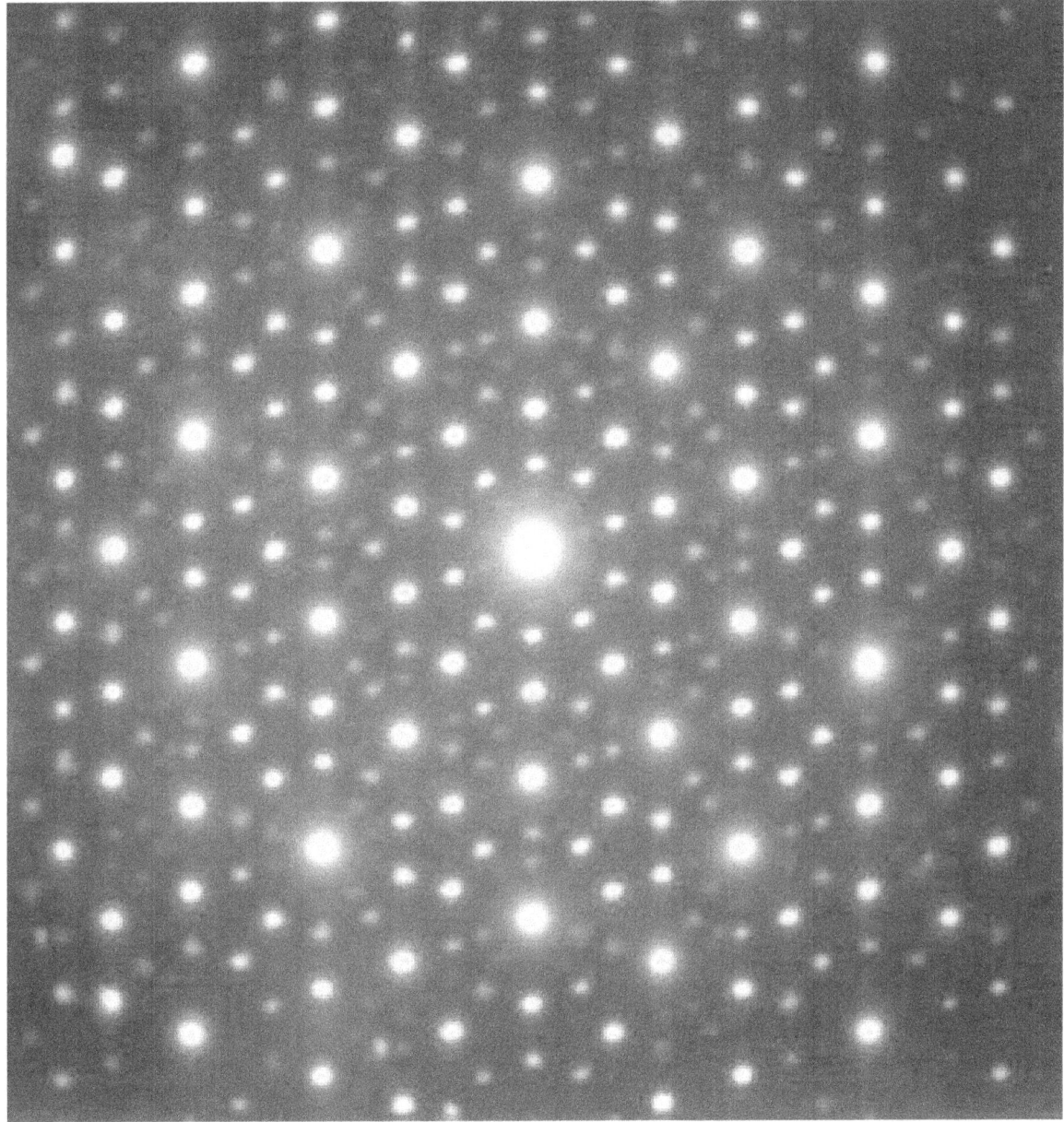

Electron diffraction pattern of an icosahedral Ho-Mg-Zn quasicrystal

odic tilings with the additional constraint for the diffractive property. Thus, for a substitution tiling the eigenvalues of the substitution matrix should be Pisot numbers. The aperiodic structures obtained by the cut-and-project method are made diffractive by choosing a suitable orientation for the construction; this is a geometric approach that has also a great appeal for physicists.

Classical theory of crystals reduces crystals to point lattices where each point is the center of mass of one of the identical units of the crystal. The structure of crystals can be analyzed by defining an associated group. Quasicrystals, on the other hand, are composed of more than one type of unit, so, instead of lattices, quasilattices must be used. Instead of groups, groupoids, the mathematical generalization of groups in category theory, is the appropriate tool for studying quasicrystals.[31]

Using mathematics for construction and analysis of quasicrystal structures is a difficult task for most experimentalists. Computer modeling, based on the existing theories of quasicrystals, however, greatly facilitated this task. Advanced pro-

A penteract (5-cube) pattern using 5D orthographic projection to 2D using Petrie polygon basis vectors overlaid on the diffractogram from an Icosahedral Ho-Mg-Zn quasicrystal

grams have been developed[32] allowing one to construct, visualize and analyze quasicrystal structures and their diffraction patterns.

Interacting spins were also analyzed in quasicrystals: AKLT Model and 8-vertex model were solved in quasicrystals analytically [33]

7.3   Materials science of quasicrystals

Since the original discovery by Dan Shechtman, hundreds of quasicrystals have been reported and confirmed. Undoubtedly, the quasicrystals are no longer a unique form of solid; they exist universally in many metallic alloys and some polymers. Quasicrystals are found most often in aluminium alloys (Al-Li-Cu, Al-Mn-Si, Al-Ni-Co, Al-Pd-Mn, Al-Cu-Fe,

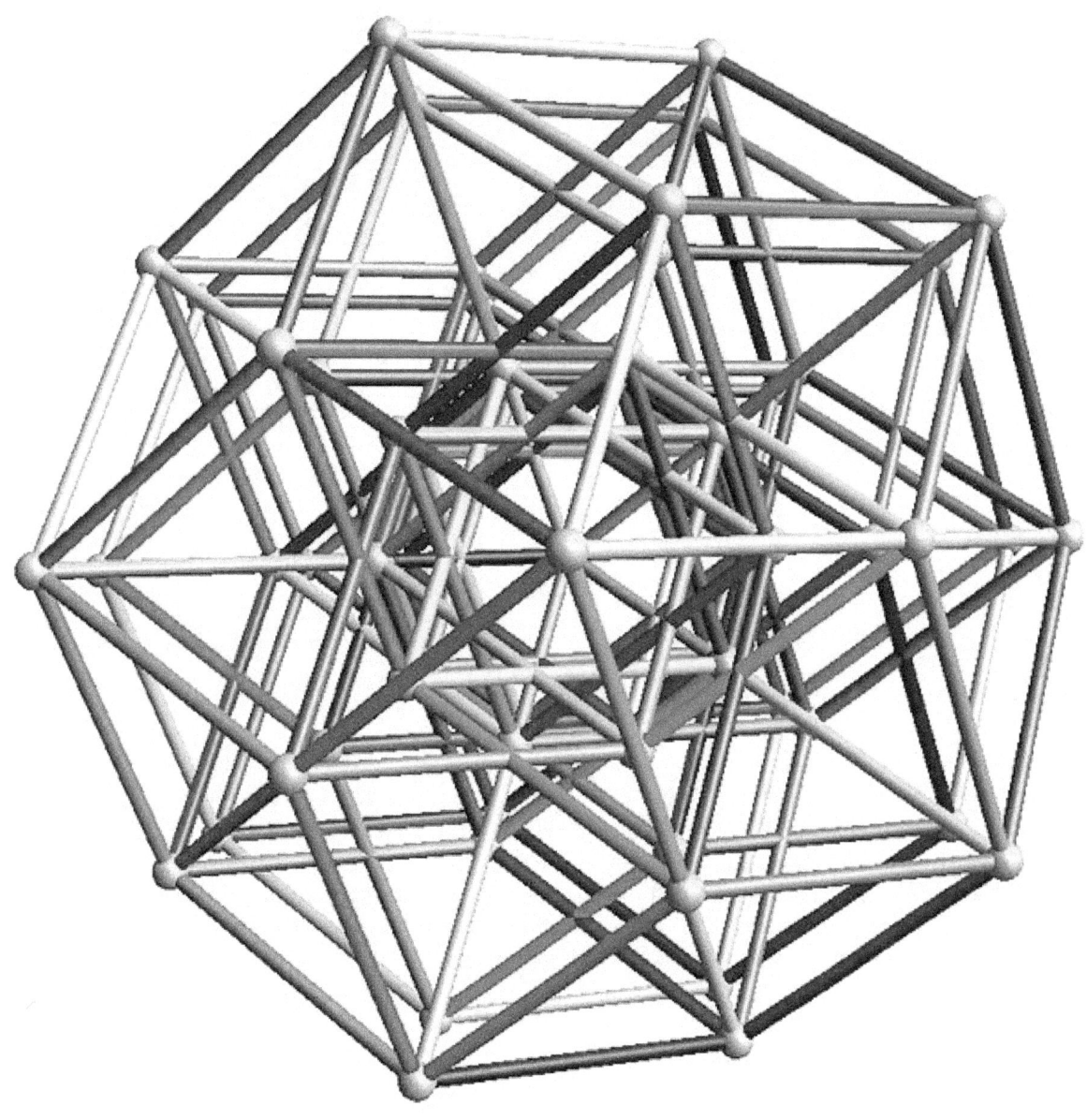

A hexeract (6-cube) pattern using 6D orthographic projection to a 3D Perspective (visual) object (the Rhombic triacontahedron) using the Golden ratio in the basis vectors. This is used to understand the aperiodic Icosahedral structure of Quasicrystals.

Al-Cu-V, etc.), but numerous other compositions are also known (Cd-Yb, Ti-Zr-Ni, Zn-Mg-Ho, Zn-Mg-Sc, In-Ag-Yb, Pd-U-Si, etc.).[34]

There are two types of known quasicrystals.[32] The first type, polygonal (dihedral) quasicrystals, have an axis of eight, ten, or 12-fold local symmetry (octagonal, decagonal, or dodecagonal quasicrystals, respectively). They are periodic along this axis and quasiperiodic in planes normal to it. The second type, icosahedral quasicrystals, are aperiodic in all directions.

Regarding thermal stability, three types of quasicrystals are distinguished:[35]

- Stable quasicrystals grown by slow cooling or casting with subsequent annealing,

- Metastable quasicrystals prepared by melt spinning, and

- Metastable quasicrystals formed by the crystallization of the amorphous phase.

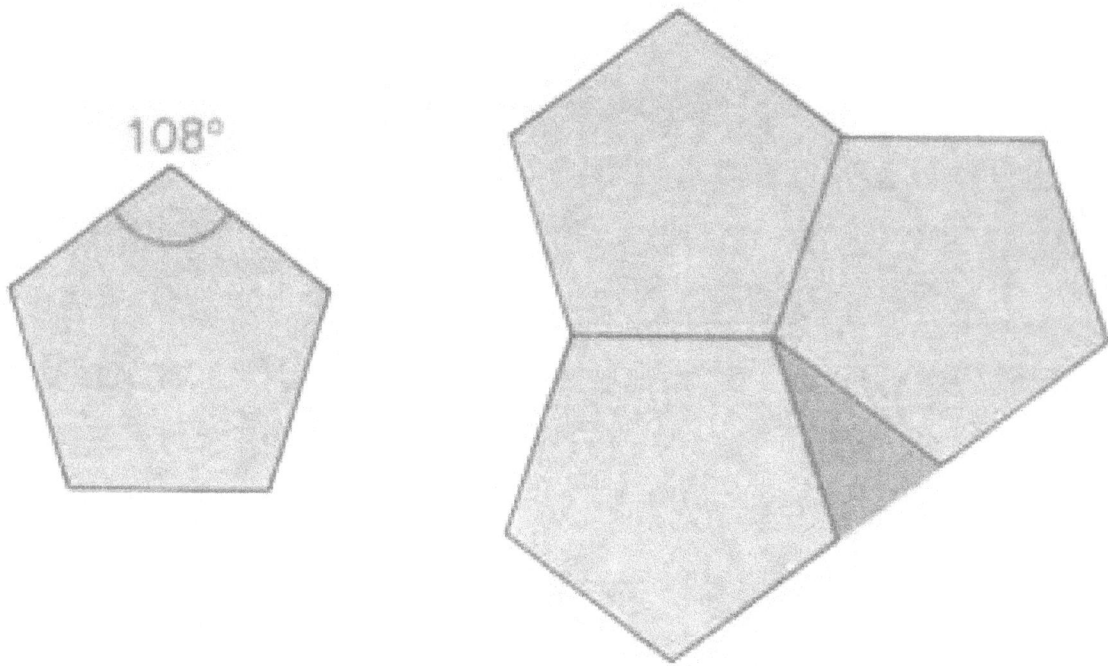

Tiling of a plane by regular pentagons is impossible but can be realized on a sphere in the form of pentagonal dodecahedron.

Except for the Al–Li–Cu system, all the stable quasicrystals are almost free of defects and disorder, as evidenced by x-ray and electron diffraction revealing peak widths as sharp as those of perfect crystals such as Si. Diffraction patterns exhibit fivefold, threefold, and twofold symmetries, and reflections are arranged quasiperiodically in three dimensions.

The origin of the stabilization mechanism is different for the stable and metastable quasicrystals. Nevertheless, there is a common feature observed in most quasicrystal-forming liquid alloys or their undercooled liquids: a local icosahedral order. The icosahedral order is in equilibrium in the *liquid state* for the stable quasicrystals, whereas the icosahedral order prevails in the *undercooled liquid state* for the metastable quasicrystals.

A nanoscale icosahedral phase was formed in Zr-, Cu- and Hf-based bulk metallic glasses alloyed with noble metals.[36]

7.4 See also

- Archimedean solid
- Fibonacci quasicrystal
- Phason
- Tessellation
- Icosahedral twins

7.5 Notes

[1] The concept of *aperiodic crystal* was coined by Erwin Schrödinger in another context with a somewhat different meaning. In his popular book *What is life?* in 1944, Schrödinger sought to explain how hereditary information is stored: molecules were deemed too small, amorphous solids were plainly chaotic, so it had to be a kind of crystal; as a periodic structure could encode very little information, it had to be aperiodic. DNA was later discovered, and, although not crystalline, it possesses properties predicted by Schrödinger—it is a regular but aperiodic molecule.

A Ho-Mg-Zn icosahedral quasicrystal formed as a pentagonal dodecahedron, the dual of the icosahedron. Unlike the similar pyritohedron shape of some cubic-system crystals such as pyrite, the quasicrystal has faces that are true regular pentagons

[2] The use of the adjective 'quasicrystalline' for qualifying a structure can be traced back to the mid-1940-50s, e.g. in Kratky, O; Porod, G (1949). "Diffuse small-angle scattering of x-rays in colloid systems". *Journal of Colloid Science* **4** (1): 35–70. doi:10.1016/0095-8522(49)90032-X. PMID 18110601.; Gunn, R (1955). "The statistical electrification of aerosols by ionic diffusion". *Journal of Colloid Science* **10**: 107–119. doi:10.1016/0095-8522(55)90081-7.

7.6 References

[1] Ünal, B; V. Fournée; K.J. Schnitzenbaumer; C. Ghosh; C.J. Jenks; A.R. Ross; T.A. Lograsso; J.W. Evans; P.A. Thiel (2007). "Nucleation and growth of Ag islands on fivefold Al-Pd-Mn quasicrystal surfaces: Dependence of island density on temperature and flux". *Physical Review B* **75** (6): 064205. Bibcode:2007PhRvB..75f4205U. doi:10.1103/PhysRevB.75.064205.

[2] Steurer W. (2004). "Twenty years of structure research on quasicrystals. Part I. Pentagonal, octagonal, decagonal and dodecagonal quasicrystals". *Z. Kristallogr.* **219** (7–2004): 391–446. Bibcode:2004ZK....219..391S. doi:10.1524/zkri.219.7.391.35643.

[3]Bindi, L.; Steinhardt, P. J.; Yao, N.; Lu, P. J. (2009). "Natural Quasicrystals".*Science***324**(5932): 1306–9.Bibcode:2009Sci...3 doi:10.1126/science.1170827. PMID 19498165.

[4] Andrea Gerlin (5 October 2011). "Technion's Shechtman Wins Nobel in Chemistry for Quasicrystals Discovery". *Bloomberg.*

TiMn quasicrystal approximant lattice.

[5] Shechtman, D.; Blech, I.; Gratias, D.; Cahn, J. (1984). "Metallic Phase with Long-Range Orientational Order and No Transla-
     tional Symmetry".*Physical Review Letters***53**(20): 1951–1953.Bibcode:1984PhRvL..53.1951S.doi:10.1103/PhysRevLett.53.

[6] "The Nobel Prize in Chemistry 2011". Nobelprize.org. Retrieved 2011-10-06.

[7] Mackay, A.L. (1982). "Crystallography and the Penrose Pattern". *Physica A* **114**: 609–613. Bibcode:1982PhyA..114..609M.
     doi:10.1016/0378-4371(82)90359-4.

[8] Kramer, P.; Neri, R. (1984). "On periodic and non-periodic space fillings of E^m obtained by projection". *Acta Crystallographica*
     A40 (5): 580–587. doi:10.1107/S0108767384001203.

[9]Yang, C. Y. (1979). "Crystallography of decahedral and icosahedral particles".*J. Cryst. Growth***47**(2): 274–282.Bibcode:1979
     doi:10.1016/0022-0248(79)90252-5.

[10] Yang, C. Y.; M. J. Yacaman, K, Heinemann (1979). "Crystallography of decahedral and icosahedral particles". *J. Cryst. Growth*

47 (2): 283–290. Bibcode:1979JCrGr..47..283Y. doi:10.1016/0022-0248(79)90253-7. Cite uses deprecated parameter |coauthors= (help)

[11] Shechtman, Dan; I. A. Blech (1985). "The Microstructure of Rapidly Solidified Al$_6$Mn". *Met. Trans. A* **16A** (6): 1005–1012. Bibcode:1985MTA....16.1005S. doi:10.1007/BF02811670.

[12] Ishimasa, T.; Nissen, H.-U.; Fukano, Y. (1985). "New ordered state between crystalline and amorphous in Ni-Cr particles". *Physical Review Letters* **55** (5): 511–513. Bibcode:1985PhRvL..55..511I. doi:10.1103/PhysRevLett.55.511. PMID 10032372.

[13] Wang, N.; Chen, H.; Kuo, K. (1987). "Two-dimensional quasicrystal with eightfold rotational symmetry". *Physical Review Letters* **59** (9): 1010–1013. Bibcode:1987PhRvL..59.1010W. doi:10.1103/PhysRevLett.59.1010. PMID 10035936.

[14] Steinhardt, Paul; Bindi, Luca (2010). "Once upon a time in Kamchatka: the search for natural quasicrystals". *Philosophical Magazine* **91** (19–21): 1. Bibcode:2011PMag...91.2421S. doi:10.1080/14786435.2010.510457.

[15] Bindi, Luca; John M. Eiler; Yunbin Guan; Lincoln S. Hollister; Glenn MacPherson; Paul J. Steinhardt; Nan Yao (2012-01-03). "Evidence for the extraterrestrial origin of a natural quasicrystal". *Proceedings of the National Academy of Sciences* **109** (5): 1396–1401. Bibcode:2012PNAS..109.1396B. doi:10.1073/pnas.1111115109. Retrieved 2012-01-04.

[16] Bindi, L.; Yao, N.; Lin, C.; Hollister, L. S.; Andronicos, C. L.; Distler, V. V.; Eddy, M. P.; Kostin, A.; Kryachko, V.; MacPherson, G. J.; Steinhardt, W. M.; Yudovskaya, M.; Steinhardt, P. J. (2015). "Natural quasicrystal with decagonal symmetry". *Scientific Reports* **5**: 9111. Bibcode:2015NatSR...5E9111B. doi:10.1038/srep09111. PMC 4357871. PMID 25765857.

[17] de Wolf, R.M. and van Aalst, W. (1972). "The four dimensional group of γ-Na$_2$CO$_3$". *Acta Cryst. A* **28**: S111.

[18] Kleinert H. and Maki K. (1981). "Lattice Textures in Cholesteric Liquid Crystals" (PDF). *Fortschritte der Physik* **29** (5): 219–259. Bibcode:1981ForPh..29..219K. doi:10.1002/prop.19810290503.

[19] Pauling, L (1987-01-26). "So-called icosahedral and decagonal quasicrystals are twins of an 820-atom cubic crystal. Pauling L". *Physical Review Letters* **58** (4): 365–368. Bibcode:1987PhRvL..58..365P. doi:10.1103/PhysRevLett.58.365. PMID 10034915.

[20] Kenneth Chang (October 5, 2011). "Israeli Scientist Wins Nobel Prize for Chemistry". NY Times.

[21] Browne, Malcolm W. (1989-09-05). "Impossible' Form of Matter Takes Spotlight In Study of Solids". New York Times.

[22] Levine, Dov; Steinhardt, Paul (1984). "Quasicrystals: A New Class of Ordered Structures". *Physical Review Letters* **53** (26): 2477–2480. Bibcode:1984PhRvL..53.2477L. doi:10.1103/PhysRevLett.53.2477.

[23] Makovicky, E. (1992), 800-year-old pentagonal tiling from Maragha, Iran, and the new varieties of aperiodic tiling it inspired. In: I. Hargittai, editor: Fivefold Symmetry, pp. 67–86. World Scientific, Singapore-London

[24] Peter J. Lu and Paul J. Steinhardt (2007). "Decagonal and Quasi-crystalline Tilings in Medieval Islamic Architecture" (PDF). *Science* **315** (5815): 1106–1110. Bibcode:2007Sci...315.1106L. doi:10.1126/science.1135491. PMID 17322056.

[25] Lu, P. J.; Steinhardt, P. J. (2007). "Decagonal and Quasi-Crystalline Tilings in Medieval Islamic Architecture". *Science* **315** (5815): 1106–1110. Bibcode:2007Sci...315.1106L. doi:10.1126/science.1135491. PMID 17322056.

[26] Makovicky, Emil (2007). "Comment on "Decagonal and Quasi-Crystalline Tilings in Medieval Islamic Architecture"". *Science* **318** (5855): 1383–1383. Bibcode:2007Sci...318.1383M. doi:10.1126/science.1146262.

[27] "Nobel win for crystal discovery". *BBC News*. 2011-10-05. Retrieved 2011-10-05.

[28] Bohr, H. (1925). "Zur Theorie fastperiodischer Funktionen I". *Acta Mathematicae* **45**: 580. doi:10.1007/BF02395468.

[29] de Bruijn, N. (1981). "Algebraic theory of Penrose's non-periodic tilings of the plane". *Nederl. Akad. Wetensch. Proc* **A84**: 39.

[30] Suck, Jens-Boie; Schreiber, M.; Häussler, Peter (2002). *Quasicrystals: An Introduction to Structure, Physical Properties and Applications*. Springer Science & Business Media. pp. 1–. ISBN 978-3-540-64224-4.

[31] Paterson, Alan L. T. (1999). *Groupoids, inverse semigroups, and their operator algebras*. Springer. p. 164. ISBN 0-8176-4051-7.

[32] Yamamoto, Akiji (2008). "Software package for structure analysis of quasicrystals". *Science and Technology of Advanced Materials* (free-download review) **9**: 013001. Bibcode:2008STAdM...9a3001Y. doi:10.1088/1468-6996/9/3/013001.

[33] Korepin, V.E. *Completely integrable models in quasicrystals*. Comm. Math. Phys. Volume 110, Number 1 (1987), 157–171.

[34] Maclá, Enrique (2006). "The role of aperiodic order in science and technology". *Reports on Progress in Physics* **69** (2): 397–441. Bibcode:2006RPPh...69..397M. doi:10.1088/0034-4885/69/2/R03.

[35] Tsai, An Pang (2008). "Icosahedral clusters, icosaheral order and stability of quasicrystals—a view of metallurgy". *Science and Technology of Advanced Materials* (free-download review) **9**: 013008. Bibcode:2008STAdM...9a3008T. doi:10.1088/1468-6996/9/1/013008.

[36] Louzguine-Luzgin, D. V.; Inoue, A. (2008). "Formation and Properties of Quasicrystals". *Annual Review of Materials Research* **38**: 403–423. doi:10.1146/annurev.matsci.38.060407.130318.

7.7   Further reading

- V.I. Arnold, *Huygens and Barrow, Newton and Hooke: Pioneers in mathematical analysis and catastrophe theory from evolvents to quasicrystals*, Eric J.F. Primrose translator, Birkhäuser Verlag (1990) ISBN 3-7643-2383-3 .

- Barber, Enrique Macia (2010). *Aperiodic Structures in Condensed Matter: Fundamentals and Applications*. Taylor & Francis. ISBN 978-1-4200-6827-6.

- Baake, Michael; Moody, Robert V., eds. (2000). *Directions in mathematical quasicrystals*. CRM Monograph Series **13**. Providence, RI: American Mathematical Society. ISBN 0-8218-2629-8. Zbl 0955.00025.

- Jean-Marie Dubois, *Useful quasicrystals*, World Scientific, Singapore 2005.

- Christian Janot, *Quasicrystals – a primer*, 2nd ed. Oxford UP 1997.

- Peter Kramer and Zorka Papadopolos (editors), *Coverings of discrete quasiperiodic sets: theory and applications to quasicrystals*, Springer. Berlin 2003.

- Ron Lifshitz, Dan Shechtman, Shelomo I. Ben-Abraham (editors), *Quasicrystals: The Silver Jubilee*, Philosophical Magazine Special Issue 88/13-15 (2008).

- Pampaloni, E.; Ramazza, P. L.; Residori, S.; Arecchi, F. T. (1995). "Two-Dimensional Crystals and Quasicrystals in Nonlinear Optics". *Phys. Rev. Lett* **74**: 258–261. doi:10.1103/physrevlett.74.258.

- Marjorie Senechal, *Quasicrystals and geometry*, Cambridge UP 1995.

- Walter Steurer, Sofia Deloudi, *Crystallography of quasicrystals*, Springer, Heidelberg 2009.

- Hans-Rainer Trebin (editor), *Quasicrystals*, Wiley-VCH. Weinheim 2003.

7.8   External links

- A Partial Bibliography of Literature on Quasicrystals (1996–2008).

- BBC webpage showing pictures of Quasicrystals

- What is... a Quasicrystal?, *Notices of the AMS* 2006, Volume 53, Number 8

- Gateways towards quasicrystals: a short history by P. Kramer

- Quasicrystals: an introduction by R. Lifshitz

- Quasicrystals: an introduction by S. Weber

- Steinhardt's proposal

- Quasicrystal Research – Documentary 2011 on the research of the University of Stuttgart

- •Thiel, P.A. (2008). "Quasicrystal Surfaces". *Annual Review of Physical Chemistry* **59**: 129–152. Bibcode:2008AR doi:10.1146/annurev.physchem.59.032607.093736. PMID 17988201.

- Foundations of Crystallography.

- Pentagon tile by Alexander Braun based on quasicrystals.

Chapter 8

Liquid

For other uses, see Liquid (disambiguation).
"Liquid State" redirects here. For the song by Muse, see Liquid State (song).
 A **liquid** is a nearly incompressible fluid that conforms to the shape of its container but retains a (nearly) constant volume

The formation of a spherical droplet of liquid water minimizes the surface area, which is the natural result of surface tension in liquids.

independent of pressure. As such, it is one of the four fundamental states of matter (the others being solid, gas, and plasma), and is the only state with a definite volume but no fixed shape. A liquid is made up of tiny vibrating particles of matter, such as atoms, held together by intermolecular bonds. Water is, by far, the most common liquid on Earth. Like a gas, a liquid is able to flow and take the shape of a container. Most liquids resist compression, although others can be compressed. Unlike a gas, a liquid does not disperse to fill every space of a container, and maintains a fairly constant density. A distinctive property of the liquid state is surface tension, leading to wetting phenomena.

The density of a liquid is usually close to that of a solid, and much higher than in a gas. Therefore, liquid and solid are both termed condensed matter. On the other hand, as liquids and gases share the ability to flow, they are both called fluids. Although liquid water is abundant on Earth, this state of matter is actually the least common in the known universe, because liquids require a relatively narrow temperature/pressure range to exist. Most known matter in the universe is in gaseous form (with traces of detectable solid matter) as interstellar clouds or in plasma form within stars.

8.1 Introduction

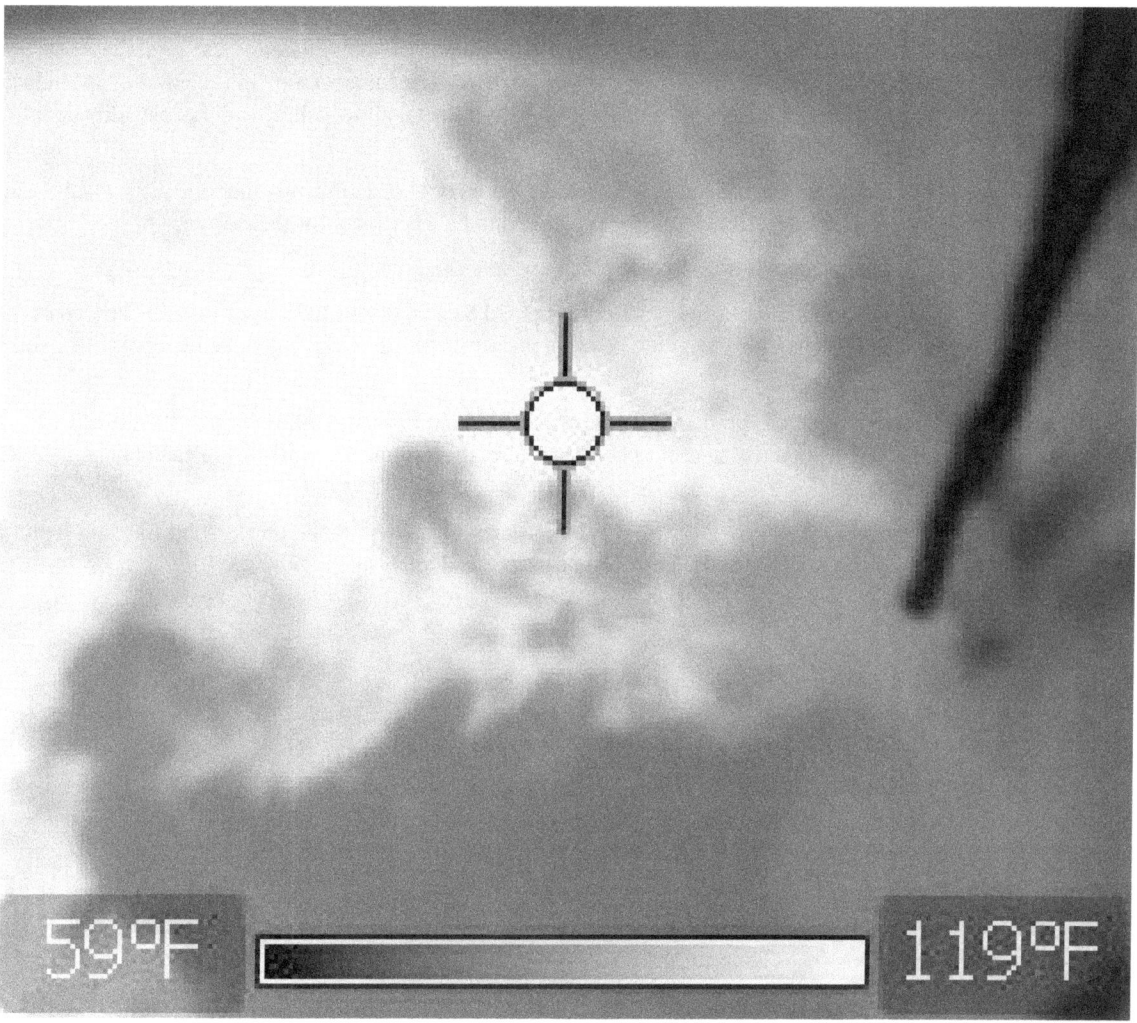

Thermal image of a sink full of hot water with cold water being added, showing how the hot and the cold water flow into each other.

Liquid is one of the four primary states of matter, with the others being solid, gas and plasma. A liquid is a fluid. Unlike a solid, the molecules in a liquid have a much greater freedom to move. The forces that bind the molecules together in a solid are only temporary in a liquid, allowing a liquid to flow while a solid remains rigid.

A liquid, like a gas, displays the properties of a fluid. A liquid can flow, assume the shape of a container, and, if placed in a sealed container, will distribute applied pressure evenly to every surface in the container. If you place the liquid in a bag, you can squeeze it into any shape you want. Unlike a gas, a liquid may not always mix readily with another liquid, will not always fill every space in the container, forming its own surface, and will not compress significantly, except under extremely high pressures. These properties make a liquid suitable for applications such as hydraulics.

Liquid particles are bound firmly but not rigidly. They are able to move around one another freely, resulting in a limited degree of particle mobility. As the temperature increases, the increased vibrations of the molecules causes distances between the molecules to increase. When a liquid reaches its boiling point, the cohesive forces that bind the molecules closely together break, and the liquid changes to its gaseous state (unless superheating occurs). If the temperature is decreased, the distances between the molecules become smaller. When the liquid reaches its freezing point the molecules will usually lock into a very specific order, called crystallizing, and the bonds between them become more rigid, changing the liquid into its solid state (unless supercooling occurs).

8.2 Examples

Only two elements are liquid at standard conditions for temperature and pressure: mercury and bromine. Four more elements have melting points slightly above room temperature: francium, caesium, gallium and rubidium.[1] Metal alloys that are liquid at room temperature include NaK, a sodium-potassium metal alloy, galinstan, a fusible alloy liquid, and some amalgams (alloys involving mercury).

Pure substances that are liquid under normal conditions include water, ethanol and many other organic solvents. Liquid water is of vital importance in chemistry and biology; it is believed to be a necessity for the existence of life.

Inorganic liquids include water, magma, inorganic nonaqueous solvents and many acids.

Important everyday liquids include aqueous solutions like household bleach, other mixtures of different substances such as mineral oil and gasoline, emulsions like vinaigrette or mayonnaise, suspensions like blood, and colloids like paint and milk.

Many gases can be liquefied by cooling, producing liquids such as liquid oxygen, liquid nitrogen, liquid hydrogen and liquid helium. Not all gases can be liquified at atmospheric pressure, for example carbon dioxide can only be liquified at pressures above 5.1 atm.

Some materials cannot be classified within the classical three states of matter; they possess solid-like and liquid-like properties. Examples include liquid crystals, used in LCD displays, and biological membranes.

8.3 Applications

Liquids have a variety of uses, as lubricants, solvents, and coolants. In hydraulic systems, liquid is used to transmit power.

In tribology, liquids are studied for their properties as lubricants. Lubricants such as oil are chosen for viscosity and flow characteristics that are suitable throughout the operating temperature range of the component. Oils are often used in engines, gear boxes, metalworking, and hydraulic systems for their good lubrication properties.[2]

Many liquids are used as solvents, to dissolve other liquids or solids. Solutions are found in a wide variety of applications, including paints, sealants, and adhesives. Naphtha and acetone are used frequently in industry to clean oil, grease, and tar from parts and machinery. Body fluids are water based solutions.

Surfactants are commonly found in soaps and detergents. Solvents like alcohol are often used as antimicrobials. They are found in cosmetics, inks, and liquid dye lasers. They are used in the food industry, in processes such as the extraction of vegetable oil.[3]

Liquids tend to have better thermal conductivity than gases, and the ability to flow makes a liquid suitable for removing excess heat from mechanical components. The heat can be removed by channeling the liquid through a heat exchanger, such as a radiator, or the heat can be removed with the liquid during evaporation.[4] Water or glycol coolants are used to keep engines from overheating.[5] The coolants used in nuclear reactors include water or liquid metals, such as sodium or bismuth.[6] Liquid propellant films are used to cool the thrust chambers of rockets.[7] In machining, water and oils are used to remove the excess heat generated, which can quickly ruin both the work piece and the tooling. During perspiration, sweat removes heat from the human body by evaporating. In the heating, ventilation, and air-conditioning industry (HVAC), liquids such as water are used to transfer heat from one area to another.[8]

Liquid is the primary component of hydraulic systems, which take advantage of Pascal's law to provide fluid power. Devices such as pumps and waterwheels have been used to change liquid motion into mechanical work since ancient times. Oils are forced through hydraulic pumps, which transmit this force to hydraulic cylinders. Hydraulics can be found in many applications, such as automotive brakes and transmissions, heavy equipment, and airplane control systems. Various hydraulic presses are used extensively in repair and manufacturing, for lifting, pressing, clamping and forming.[9]

Liquids are sometimes used in measuring devices. A thermometer often uses the thermal expansion of liquids, such as mercury, combined with their ability to flow to indicate temperature. A manometer uses the weight of the liquid to indicate air pressure.[10]

8.4 Mechanical properties

8.4.1 Volume

Quantities of liquids are commonly measured in units of volume. These include the SI unit cubic metre (m^3) and its divisions, in particular the cubic decimeter, more commonly called the litre (1 dm^3 = 1 L = 0.001 m^3), and the cubic centimetre, also called millilitre (1 cm^3 = 1 mL = 0.001 L = 10^{-6} m^3).

The volume of a quantity of liquid is fixed by its temperature and pressure. Liquids generally expand when heated, and contract when cooled. Water between 0 °C and 4 °C is a notable exception. Liquids have little compressibility. Water, for example, will compress by only 46.4 parts per million for every unit increase in atmospheric pressure (bar).[11] At around 4000 bar (58,000 psi) of pressure, at room temperature, water only experiences an 11% decrease in volume.[12] In the study of fluid dynamics, liquids are often treated as incompressible, especially when studying incompressible flow. This incompressible nature makes a liquid suitable for transmitting hydraulic power, because very little of the energy is lost in the form of compression.[12] However, the very slight compressibility does lead to other phenomena. The banging of pipes, called water hammer, occurs when a valve is suddenly closed, creating a huge pressure-spike at the valve that travels backward through the system. Another phenomenon caused by liquid's incompressibility is cavitation, where liquid in an area of low pressure vaporizes and forms bubbles, which then collapse as they enter high pressure areas. This causes liquid to fill the cavity left by the bubble with tremendous, localized force, eroding any adjacent solid surface.[13]

8.4.2 Pressure and buoyancy

Main article: fluid statics

In a gravitational field, liquids exert pressure on the sides of a container as well as on anything within the liquid itself. This pressure is transmitted in all directions and increases with depth. If a liquid is at rest in a uniform gravitational field, the pressure, p, at any depth, z, is given by

$$p = \rho g z$$

where:

ρ is the density of the liquid (assumed constant)

g is the gravitational acceleration.

Note that this formula assumes that the pressure *at* the free surface is zero, and that surface tension effects may be neglected.

Objects immersed in liquids are subject to the phenomenon of buoyancy. (Buoyancy is also observed in other fluids, but is especially strong in liquids due to their high density.)

8.4.3 Surfaces

Main article: surface science
 Unless the volume of a liquid exactly matches the volume of its container, one or more surfaces are observed. The

Surface waves in water

surface of a liquid behaves like an elastic membrane in which surface tension appears, allowing the formation of drops and bubbles. Surface waves, capillary action, wetting, and ripples are other consequences of surface tension. In a confined liquid, defined by geometric constraints on a nanoscopic scale, most molecules sense some surface effects, which can result in physical properties grossly deviating from those of the bulk liquid.

Free surface

Main article: Free surface

A **free surface** is the surface of a fluid that is subject to both zero perpendicular normal stress and parallel shear stress, such as the boundary between, e.g., liquid water and the air in the Earth's atmosphere.

Level

The **liquid level** (as in, e.g., water level) is the height associated with the liquid free surface, especially when it's the top-most surface. It may be measured with a level sensor.

8.4.4 Flow

Main articles: fluid mechanics and fluid dynamics

Viscosity measures the resistance of a liquid which is being deformed by either shear stress or extensional stress.

When a liquid is supercooled towards the glass transition, the viscosity increases dramatically. The liquid then becomes a viscoelastic medium that shows both the elasticity of a solid and the fluidity of a liquid, depending on the time scale of observation or on the frequency of perturbation.

8.4.5 Sound propagation

Main article: speed of sound § Speed of sound in liquids

Hence the speed of sound in a fluid is given by $c = \sqrt{K/\rho}$ where K is the bulk modulus of the fluid, and ϱ the density. To give a typical value, in fresh water c=1497 m/s at 25 °C.

8.5 Thermodynamics

8.5.1 Phase transitions

Main articles: boiling, boiling point, melting and melting point
At a temperature below the boiling point, any matter in liquid form will evaporate until the condensation of gas above reach an equilibrium. At this point the gas will condense at the same rate as the liquid evaporates. Thus, a liquid cannot exist permanently if the evaporated liquid is continually removed. A liquid at its boiling point will evaporate more quickly than the gas can condense at the current pressure. A liquid at or above its boiling point will normally boil, though superheating can prevent this in certain circumstances.

At a temperature below the freezing point, a liquid will tend to crystallize, changing to its solid form. Unlike the transition to gas, there is no equilibrium at this transition under constant pressure, so unless supercooling occurs, the liquid will eventually completely crystallize. Note that this is only true under constant pressure, so e.g. water and ice in a closed, strong container might reach an equilibrium where both phases coexist. For the opposite transition from solid to liquid, see melting.

8.5.2 Liquids in space

The phase diagram explains why liquids do not exist in space or any other vacuum. Since the pressure is zero (except on surfaces or interiors of planets and moons) water and other liquids exposed to space will either immediately boil or freeze depending on the temperature. In regions of space near the earth, water will freeze if the sun is not shining directly on it and vapourize (sublime) as soon as it is in sunlight. If water exists as ice on the moon, it can only exist in shadowed holes where the sun never shines and where the surrounding rock doesn't heat it up too much. At some point near the orbit of Saturn, the light from the sun is too faint to sublime ice to water vapour. This is evident from the longevity of the ice that composes Saturn's rings.

8.5.3 Solutions

Main article: solution

Liquids can display immiscibility. The most familiar mixture of two immiscible liquids in everyday life is the vegetable oil and water in Italian salad dressing. A familiar set of miscible liquids is water and alcohol. Liquid components in a

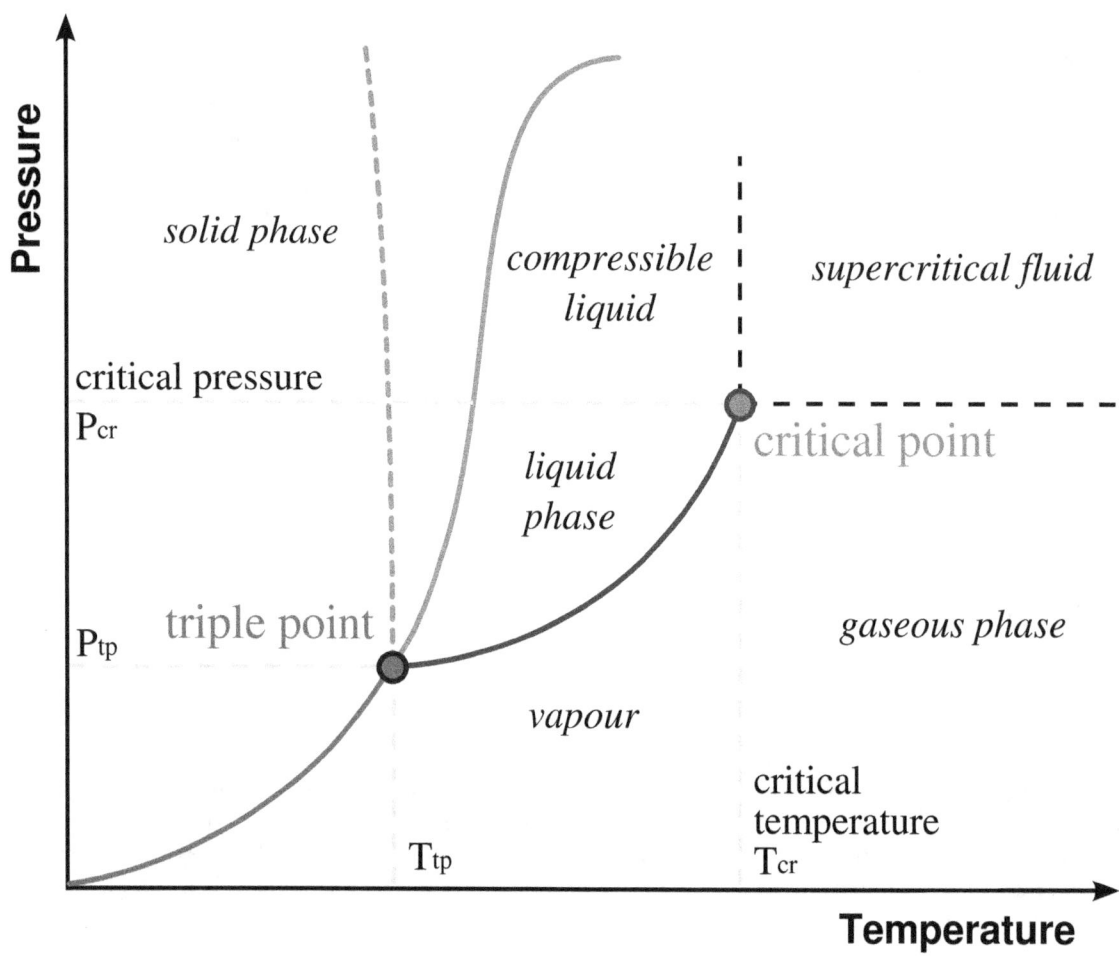

A typical phase diagram. The dotted line gives the anomalous behaviour of water. The green lines show how the freezing point can vary with pressure, and the blue line shows how the boiling point can vary with pressure. The red line shows the boundary where sublimation or deposition can occur.

mixture can often be separated from one another via fractional distillation.

8.6   Microscopic properties

8.6.1   Static structure factor

Main article: structure of liquids and glasses

In a liquid, atoms do not form a crystalline lattice, nor do they show any other form of long-range order. This is evidenced by the absence of Bragg peaks in X-ray and neutron diffraction. Under normal conditions, the diffraction pattern has circular symmetry, expressing the isotropy of the liquid. In radial direction, the diffraction intensity smoothly oscillates. This is usually described by the static structure factor $S(q)$, with wavenumber $q=(4\pi/\lambda)\sin\theta$ given by the wavelength λ of the probe (photon or neutron) and the Bragg angle θ. The oscillations of $S(q)$ express the *near order* of the liquid, i.e. the correlations between an atom and a few shells of nearest, second nearest, ... neighbors.

A more intuitive description of these correlations is given by the radial distribution function $g(r)$, which is basically the Fourier transform of $S(q)$. It represents a spatial average of a temporal snapshot of pair correlations in the liquid.

Structure of a classical monatomic liquid. Atoms have many nearest neighbors in contact, yet no long-range order is present.

8.6.2 Sound dispersion and structural relaxation

The above expression for the sound velocity $c = \sqrt{K/\rho}$ contains the bulk modulus K. If K is frequency independent then the liquid behaves as a linear medium, so that sound propagates without dissipation and without mode coupling. In reality, any liquid shows some dispersion: with increasing frequency, K crosses over from the low-frequency, liquid-like limit K_0 to the high-frequency, solid-like limit K_∞. In normal liquids, most of this cross over takes place at frequencies between GHz and THz, sometimes called hypersound.

At sub-GHz frequencies, a normal liquid cannot sustain shear waves: the zero-frequency limit of the shear modulus is $G_0 = 0$. This is sometimes seen as the defining property of a liquid.[14][15] However, just as the bulk modulus K, the shear modulus G is frequency dependent, and at hypersound frequencies it shows a similar cross over from the liquid-like limit G_0 to a solid-like, non-zero limit G_∞.

According to the Kramers-Kronig relation, the dispersion in the sound velocity (given by the real part of K or G) goes along with a maximum in the sound attenuation (dissipation, given by the imaginary part of K or G). According to linear response theory, the Fourier transform of K or G describes how the system returns to equilibrium after an external perturbation; for this reason, the dispersion step in the GHz..THz region is also called structural relaxation. According to the fluctuation-dissipation theorem, relaxation *towards* equilibrium is intimately connected to fluctuations *in* equilibrium. The density fluctuations associated with sound waves can be experimentally observed by Brillouin scattering.

On supercooling a liquid towards the glass transition, the crossover from liquid-like to solid-like response moves from GHz to MHz, kHz, Hz, ...; equivalently, the characteristic time of structural relaxation increases from ns to µs, ms, s, ... This is the microscopic explanation for the above-mentioned viscoelastic behaviour of glass-forming liquids.

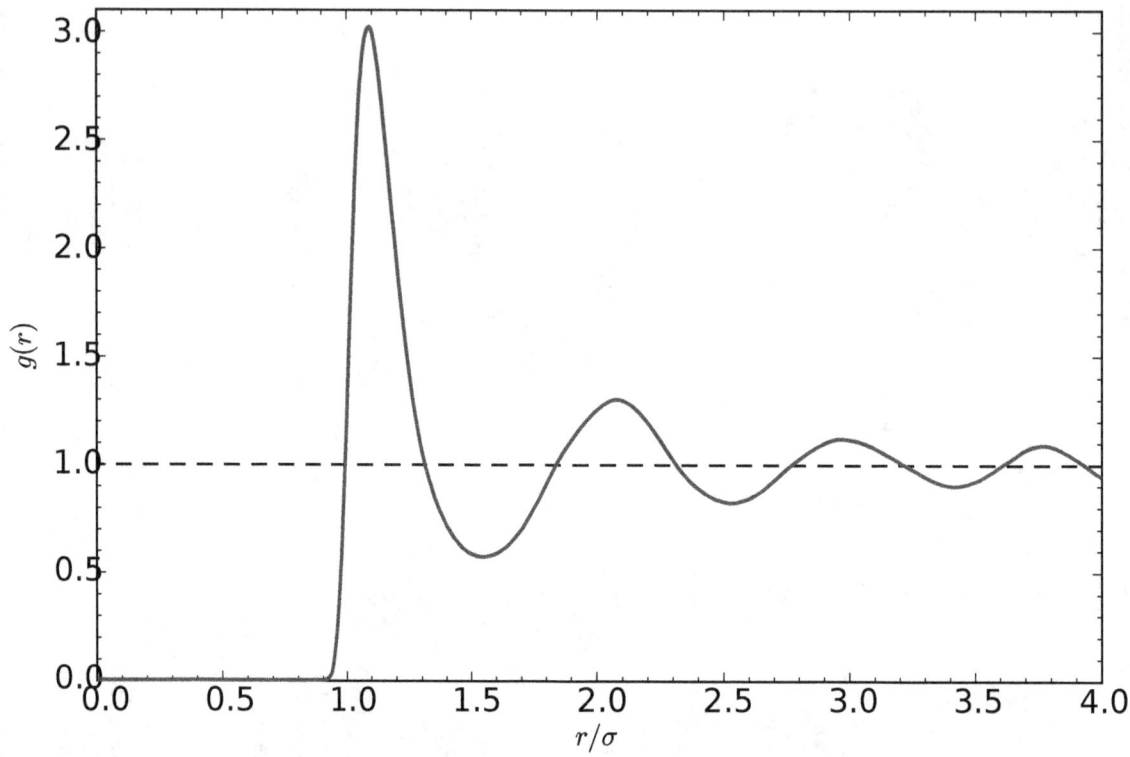

Radial distribution function of the Lennard-Jones model fluid.

8.6.3 Effects of association

The mechanisms of atomic/molecular diffusion (or particle displacement) in solids are closely related to the mechanisms of viscous flow and solidification in liquid materials. Descriptions of viscosity in terms of molecular "free space" within the liquid[16] were modified as needed in order to account for liquids whose molecules are known to be "associated" in the liquid state at ordinary temperatures. When various molecules combine together to form an associated molecule, they enclose within a semi-rigid system a certain amount of space which before was available as free space for mobile molecules. Thus, increase in viscosity upon cooling due to the tendency of most substances to become *associated* on cooling.[17]

Similar arguments could be used to describe the effects of pressure on viscosity, where it may be assumed that the viscosity is chiefly a function of the volume for liquids with a finite compressibility. An increasing viscosity with rise of pressure is therefore expected. In addition, if the volume is expanded by heat but reduced again by pressure, the viscosity remains the same.

The local tendency to orientation of molecules in small groups lends the liquid (as referred to previously) a certain degree of association. This association results in a considerable "internal pressure" within a liquid, which is due almost entirely to those molecules which, on account of their temporary low velocities (following the Maxwell distribution) have coalesced with other molecules. The internal pressure between several such molecules might correspond to that between a group of molecules in the solid form.

8.7 References

[1] Theodore Gray, The Elements: A Visual Exploration of Every Known Atom in the Universe New York: Workman Publishing, 2009 p. 127 ISBN 1-57912-814-9

[2] Theo Mang, Wilfried Dressel "Lubricants and lubrication", Wiley-VCH 2007 ISBN 3-527-31497-0

[3] George Wypych "Handbook of solvents" William Andrew Publishing 2001 pp. 847–881 ISBN 1-895198-24-0

[4] N. B. Vargaftik "Handbook of thermal conductivity of liquids and gases" CRC Press 1994 ISBN 0-8493-9345-0

[5] Jack Erjavec "Automotive technology: a systems approach" Delmar Learning 2000 p. 309 ISBN 1-4018-4831-1

[6] Gerald Wendt "The prospects of nuclear power and technology" D. Van Nostrand Company 1957 p. 266

[7] "Modern engineering for design of liquid-propellant rocket engines" by Dieter K. Huzel, David H. Huang – American Institute of Aeronautics and Astronautics 1992 p. 99 ISBN 1-56347-013-6

[8] Thomas E Mull "HVAC principles and applications manual" McGraw-Hill 1997 ISBN 0-07-044451-X

[9] R. Keith Mobley *Fluid power dynamics* Butterworth-Heinemann 2000 p. vii ISBN 0-7506-7174-2

[10] Bela G. Liptak "Instrument engineers' handbook: process control" CRC Press 1999 p. 807 ISBN 0-8493-1081-4

[11] http://hyperphysics.phy-astr.gsu.edu/hbase/tables/compress.html

[12] *Intelligent Energy Field Manufacturing: Interdisciplinary Process Innovations* By Wenwu Zhang -- CRC Press 2011 Page 144

[13] *Fluid Mechanics and Hydraulic Machines* by S. C. Gupta -- Dorling-Kindersley 2006 Page 85

[14] Born, Max (1940). "On the stability of crystal lattices". *Mathematical Proceedings* (Cambridge Philosophical Society) **36** (2): 160–172. doi:10.1017/S0305004100017138.

[15] Born, Max (1939). "Thermodynamics of Crystals and Melting". *Journal of Chemical Physics* **7**(8): 591–604. doi:10.1063/1.175

[16] D.B. Macleod (1923). "On a relation between the viscosity of a liquid and its coefficient of expansion". *Trans. Farad. Soc.* **19**: 6. doi:10.1039/tf9231900006.

[17] G.W Stewart (1930). "The Cybotactic (Molecular Group) Condition in Liquids; the Association of Molecules". *Phys. Rev.* **35** (7): 726. Bibcode:1930PhRv...35..726S. doi:10.1103/PhysRev.35.726.

Chapter 9

Liquid crystal

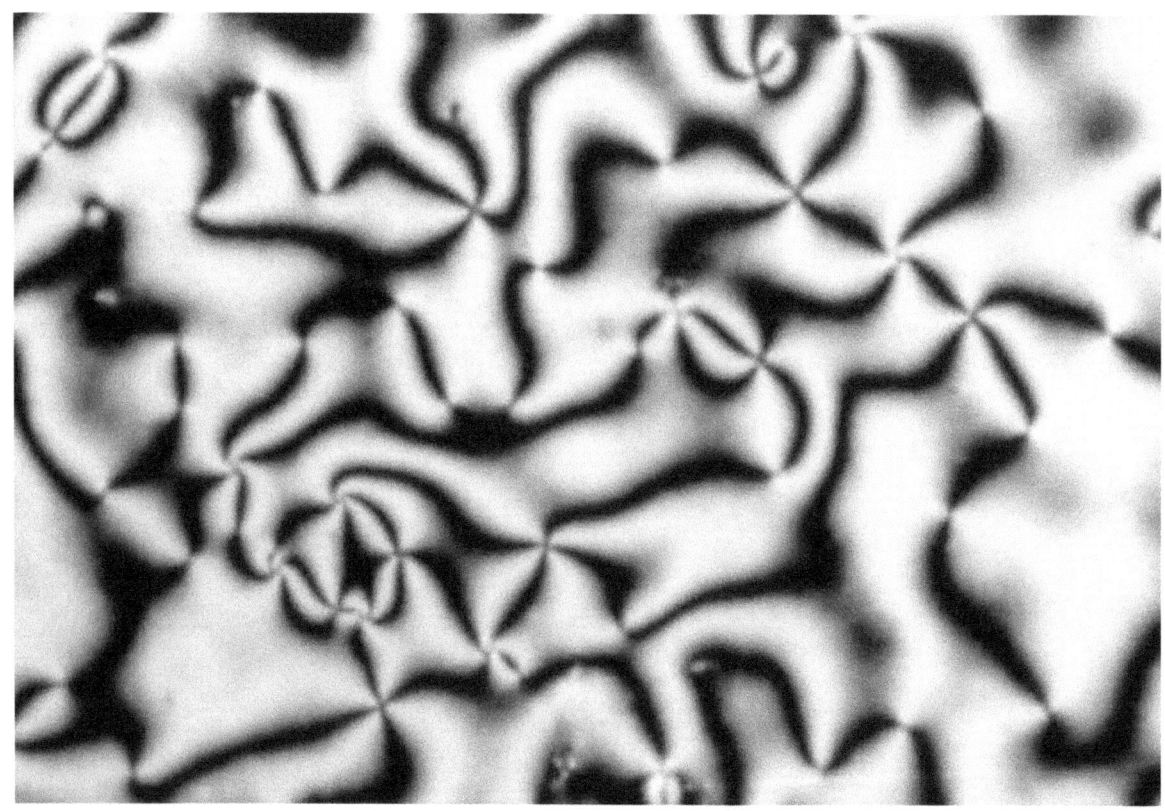

Schlieren texture of liquid crystal nematic phase

Liquid crystals (LCs) are matter in a state that has properties between those of conventional liquid and those of solid crystal.[1] For instance, a liquid crystal may flow like a liquid, but its molecules may be oriented in a crystal-like way. There are many different types of liquid-crystal phases, which can be distinguished by their different optical properties (such as birefringence). When viewed under a microscope using a polarized light source, different liquid crystal phases will appear to have distinct textures. The contrasting areas in the textures correspond to domains where the liquid-crystal molecules are oriented in different directions. Within a domain, however, the molecules are well ordered. LC materials may not always be in a liquid-crystal phase (just as water may turn into ice or steam).

Liquid crystals can be divided into thermotropic, lyotropic and metallotropic phases. Thermotropic and lyotropic liquid crystals consist of organic molecules. Thermotropic LCs exhibit a phase transition into the liquid-crystal phase as temperature is changed. Lyotropic LCs exhibit phase transitions as a function of both temperature and concentration of the liquid-crystal molecules in a solvent (typically water). Metallotropic LCs are composed of both organic and inorganic molecules; their liquid-crystal transition depends not only on temperature and concentration, but also on the inorganic-organic composition ratio.

Examples of liquid crystals can be found both in the natural world and in technological applications. Most contemporary electronic displays use liquid crystals. Lyotropic liquid-crystalline phases are abundant in living systems. For example, many proteins and cell membranes are liquid crystals. Other well-known examples of liquid crystals are solutions of soap and various related detergents, as well as the tobacco mosaic virus.

9.1   History

In 1888, Austrian botanical physiologist Friedrich Reinitzer, working at the Karl-Ferdinands-Universität, examined the physico-chemical properties of various derivatives of cholesterol which now belong to the class of materials known as cholesteric liquid crystals. Previously, other researchers had observed distinct color effects when cooling cholesterol derivatives just above the freezing point, but had not associated it with a new phenomenon. Reinitzer perceived that color changes in a derivative cholesteryl benzoate were not the most peculiar feature.

Chemical structure of cholesteryl benzoate molecule

He found that cholesteryl benzoate does not melt in the same manner as other compounds, but has two melting points. At 145.5 °C (293.9 °F) it melts into a cloudy liquid, and at 178.5 °C (353.3 °F) it melts again and the cloudy liquid becomes clear. The phenomenon is reversible. Seeking help from a physicist, on March 14, 1888, he wrote to Otto Lehmann, at that time a *Privatdozent* in Aachen. They exchanged letters and samples. Lehmann examined the intermediate cloudy fluid, and reported seeing crystallites. Reinitzer's Viennese colleague von Zepharovich also indicated that the intermediate "fluid" was crystalline. The exchange of letters with Lehmann ended on April 24, with many questions unanswered. Reinitzer presented his results, with credits to Lehmann and von Zepharovich, at a meeting of the Vienna Chemical Society on May 3, 1888.[2]

By that time, Reinitzer had discovered and described three important features of cholesteric liquid crystals (the name coined by Otto Lehmann in 1904): the existence of two melting points, the reflection of circularly polarized light, and the ability to rotate the polarization direction of light.

After his accidental discovery, Reinitzer did not pursue studying liquid crystals further. The research was continued by Lehmann, who realized that he had encountered a new phenomenon and was in a position to investigate it: In his postdoctoral years he had acquired expertise in crystallography and microscopy. Lehmann started a systematic study, first of cholesteryl benzoate, and then of related compounds which exhibited the double-melting phenomenon. He was able to make observations in polarized light, and his microscope was equipped with a hot stage (sample holder equipped with a heater) enabling high temperature observations. The intermediate cloudy phase clearly sustained flow, but other features, particularly the signature under a microscope, convinced Lehmann that he was dealing with a solid. By the end of August 1889 he had published his results in the Zeitschrift für Physikalische Chemie.[3]

Lehmann's work was continued and significantly expanded by the German chemist Daniel Vorländer, who from the beginning of 20th century until his retirement in 1935, had synthesized most of the liquid crystals known. However, liquid crystals were not popular among scientists and the material remained a pure scientific curiosity for about 80 years.[4]

After World War II work on the synthesis of liquid crystals was restarted at university research laboratories in Europe. George William Gray, a prominent researcher of liquid crystals, began investigating these materials in England in the late 1940s. His group synthesized many new materials that exhibited the liquid crystalline state and developed a better understanding of how to design molecules that exhibit the state. His book *Molecular Structure and the Properties of Liquid Crystals*[5] became a guidebook on the subject. One of the first U.S. chemists to study liquid crystals was Glenn H. Brown, starting in 1953 at the University of Cincinnati and later at Kent State University. In 1965, he organized the first international conference on liquid crystals, in Kent, Ohio, with about 100 of the world's top liquid crystal scientists in attendance. This conference marked the beginning of a worldwide effort to perform research in this field, which soon led to the development of practical applications for these unique materials.[6][7]

Liquid crystal materials became a topic of research into the development of flat panel electronic displays beginning in 1962 at RCA Laboratories.[8] When physical chemist Richard Williams applied an electric field to a thin layer of a nematic liquid crystal at 125 °C, he observed the formation of a regular pattern that he called domains (now known as Williams Domains). This led his colleague George H. Heilmeier to perform research on a liquid crystal-based flat panel display to replace the cathode ray vacuum tube used in televisions. But the para-Azoxyanisole that Williams and Heilmeier used exhibits the nematic liquid crystal state only above 116 °C, which made it impractical to use in a commercial display product. A material that could be operated at room temperature was clearly needed.

In 1966, Joel E. Goldmacher and Joseph A. Castellano, research chemists in Heilmeier group at RCA, discovered that mixtures made exclusively of nematic compounds that differed only in the number of carbon atoms in the terminal side chains could yield room-temperature nematic liquid crystals. A ternary mixture of Schiff base compounds resulted in a material that had a nematic range of 22–105 °C.[9] Operation at room temperature enabled the first practical display device to be made.[10] The team then proceeded to prepare numerous mixtures of nematic compounds many of which had much lower melting points. This technique of mixing nematic compounds to obtain wide operating temperature range eventually became the industry standard and is used to this very day to tailor materials to meet specific applications.

In 1969, Hans Kelker succeeded in synthesizing a substance that had a nematic phase at room temperature, MBBA, which is one of the most popular subjects of liquid crystal research.[11] The next step to commercialization of liquid crystal displays was the synthesis of further chemically stable substances (cyanobiphenyls) with low melting temperatures by George Gray.[12] That work with Ken Harrison and the UK MOD (RRE Malvern), in 1973, led to design of new materials resulting in rapid adoption of small area LCDs within electronic products.

These molecules are rod-shaped; some are created in the lab and some could appear spontaneously in nature. Since then, two new types of LC molecules were discovered, both are *man-make*: disc-shaped (created by S. Chandrasekhar's group in India, 1977) and bowl-shaped (invented by Lui Lam[13] in China, 1982, and synthesized in Europe three years later).

In 1991, when liquid crystal displays were already well established, Pierre-Gilles de Gennes working at the Université Paris-Sud received the Nobel Prize in physics "for discovering that methods developed for studying order phenomena in simple systems can be generalized to more complex forms of matter, in particular to liquid crystals and polymers".[14]

9.2   Design of liquid crystalline materials

A large number of chemical compounds are known to exhibit one or several liquid crystalline phases. Despite significant differences in chemical composition, these molecules have some common features in chemical and physical properties. There are three types of thermotropic liquid crystals: discotics, bowlics and rod-shaped molecules. Discotics are flat disc-like molecules consisting of a core of adjacent aromatic rings; the core in a bowlic is not flat but like a rice bowl (a three-dimensional object).[15] [16] This allows for two dimensional columnar ordering, for both discotics and bowlics.

Otto Lehmann

Rod-shaped molecules have an elongated, anisotropic geometry which allows for preferential alignment along one spatial direction.

Chemical structure of N-(4-Methoxybenzylidene)−4-butylaniline (MBBA) molecule

• The molecular shape should be relatively thin, flat or bowl-like, especially within rigid molecular frameworks.
• The molecular length should be at least 1.3 nm, consistent with the presence of long alkyl group on many room-temperature liquid crystals.
• The structure should not be branched or angular, except for the bowlics.
• A low melting point is preferable in order to avoid metastable, monotropic liquid crystalline phases. Low-temperature mesomorphic behavior in general is technologically more useful, and alkyl terminal groups promote this.

An extended, structurally rigid, highly anisotropic shape seems to be the main criterion for liquid crystalline behavior, and as a result many liquid crystalline materials are based on benzene rings.[17]

9.3   Liquid-crystal phases

The various liquid-crystal phases (called mesophases) can be characterized by the type of ordering. One can distinguish positional order (whether molecules are arranged in any sort of ordered lattice) and orientational order (whether molecules are mostly pointing in the same direction), and moreover order can be either short-range (only between molecules close to each other) or long-range (extending to larger, sometimes macroscopic, dimensions). Most thermotropic LCs will have an isotropic phase at high temperature. That is that heating will eventually drive them into a conventional liquid phase characterized by random and isotropic molecular ordering (little to no long-range order), and fluid-like flow behavior. Under other conditions (for instance, lower temperature), a LC might inhabit one or more phases with significant anisotropic orientational structure and short-range orientational order while still having an ability to flow.[1][18]

The ordering of liquid crystalline phases is extensive on the molecular scale. This order extends up to the entire domain size, which may be on the order of micrometers, but usually does not extend to the macroscopic scale as often occurs in classical crystalline solids. However some techniques, such as the use of boundaries or an applied electric field, can be used to enforce a single ordered domain in a macroscopic liquid crystal sample. The ordering in a liquid crystal might extend along only one dimension, with the material being essentially disordered in the other two directions.[19][20]

9.3.1 Thermotropic liquid crystals

See also: Thermotropic crystal

Thermotropic phases are those that occur in a certain temperature range. If the temperature rise is too high, thermal motion will destroy the delicate cooperative ordering of the LC phase, pushing the material into a conventional isotropic liquid phase. At too low temperature, most LC materials will form a conventional crystal.[1][18] Many thermotropic LCs exhibit a variety of phases as temperature is changed. For instance, on heating a particular type of LC molecule (called mesogen) may exhibit various smectic phases followed by the nematic phase and finally the isotropic phase as temperature is increased. An example of a compound displaying thermotropic LC behavior is para-azoxyanisole.[21]

Nematic phase

See also: Biaxial nematic and Twisted nematic field effect

One of the most common LC phases is the nematic. The word *nematic* comes from the Greek νῆμα (*Greek: nema*), which means "thread". This term originates from the thread-like topological defects observed in nematics, which are formally called 'disclinations'. Nematics also exhibit so-called "hedgehog" topological defects. In a nematic phase, the *calamitic* or rod-shaped organic molecules have no positional order, but they self-align to have long-range directional order with their long axes roughly parallel.[22] Thus, the molecules are free to flow and their center of mass positions are randomly distributed as in a liquid, but still maintain their long-range directional order. Most nematics are uniaxial: they have one axis that is longer and preferred, with the other two being equivalent (can be approximated as cylinders or rods). However, some liquid crystals are biaxial nematics, meaning that in addition to orienting their long axis, they also orient along a secondary axis.[23] Nematics have fluidity similar to that of ordinary (isotropic) liquids but they can be easily aligned by an external magnetic or electric field. Aligned nematics have the optical properties of uniaxial crystals and this makes them extremely useful in liquid crystal displays (LCD).[8]

Smectic phases

The smectic phases, which are found at lower temperatures than the nematic, form well-defined layers that can slide over one another in a manner similar to that of soap. The word "smectic" originates from the Latin word "smecticus", meaning cleaning, or having soap like properties.[24] The smectics are thus positionally ordered along one direction. In the Smectic A phase, the molecules are oriented along the layer normal, while in the Smectic C phase they are tilted away from the layer normal. These phases are liquid-like within the layers. There are many different smectic phases, all characterized by different types and degrees of positional and orientational order.[1][18]

Chiral phases

The chiral nematic phase exhibits chirality (handedness). This phase is often called the *cholesteric* phase because it was first observed for cholesterol derivatives. Only chiral molecules (i.e., those that have no internal planes of symmetry) can give rise to such a phase. This phase exhibits a twisting of the molecules perpendicular to the director, with the molecular axis parallel to the director. The finite twist angle between adjacent molecules is due to their asymmetric packing, which results in longer-range chiral order. In the smectic C* phase (an asterisk denotes a chiral phase), the molecules have positional ordering in a layered structure (as in the other smectic phases), with the molecules tilted by a finite angle with respect to the layer normal. The chirality induces a finite azimuthal twist from one layer to the next, producing a spiral twisting of the molecular axis along the layer normal.[18][19][20]

The *chiral pitch*, p, refers to the distance over which the LC molecules undergo a full 360° twist (but note that the structure of the chiral nematic phase repeats itself every half-pitch, since in this phase directors at 0° and ±180° are equivalent). The pitch, p, typically changes when the temperature is altered or when other molecules are added to the LC host (an achiral LC host material will form a chiral phase if doped with a chiral material), allowing the pitch of a given material to be tuned accordingly. In some liquid crystal systems, the pitch is of the same order as the wavelength of visible light. This causes these systems to exhibit unique optical properties, such as Bragg reflection and low-threshold laser emission,[25] and these properties are exploited in a number of optical applications.[4][19] For the case of Bragg reflection only the lowest-order reflection is allowed if the light is incident along the helical axis, whereas for oblique incidence higher-order reflections become permitted. Cholesteric liquid crystals also exhibit the unique property that they reflect circularly polarized light when it is incident along the helical axis and elliptically polarized if it comes in obliquely.[26]

Blue phases are liquid crystal phases that appear in the temperature range between a chiral nematic phase and an isotropic liquid phase. Blue phases have a regular three-dimensional cubic structure of defects with lattice periods of several hundred nanometers, and thus they exhibit selective Bragg reflections in the wavelength range of visible light corresponding to the cubic lattice. It was theoretically predicted in 1981 that these phases can possess icosahedral symmetry similar to quasicrystals.[27][28]

Although blue phases are of interest for fast light modulators or tunable photonic crystals, they exist in a very narrow temperature range, usually less than a few kelvin. Recently the stabilization of blue phases over a temperature range of more than 60 K including room temperature (260–326 K) has been demonstrated.[29][30] Blue phases stabilized at room temperature allow electro-optical switching with response times of the order of 10^{-4} s.[31]

In May 2008, the first Blue Phase Mode LCD panel had been developed.[32]

Discotic phases

Disk-shaped LC molecules can orient themselves in a layer-like fashion known as the discotic nematic phase. If the disks pack into stacks, the phase is called a discotic columnar. The columns themselves may be organized into rectangular or hexagonal arrays. Chiral discotic phases, similar to the chiral nematic phase, are also known.

Bowlic phases

Bowl-shaped LC molecules, like in discotics, can form columnar phases. Other phases, such as nonpolar nematic, polar nematic, stringbean, donut and onion phases, have been predicted. Bowlic phases, except nonpolar nematic, are polar phases.

9.3.2   Lyotropic liquid crystals

See also: Lyotropic liquid crystal and Columnar phase

A lyotropic liquid crystal consists of two or more components that exhibit liquid-crystalline properties in certain concentration ranges. In the lyotropic phases, solvent molecules fill the space around the compounds to provide fluidity to the system.[33] In contrast to thermotropic liquid crystals, these lyotropics have another degree of freedom of concentration that enables them to induce a variety of different phases.

A compound that has two immiscible hydrophilic and hydrophobic parts within the same molecule is called an amphiphilic molecule. Many amphiphilic molecules show lyotropic liquid-crystalline phase sequences depending on the volume balances between the hydrophilic part and hydrophobic part. These structures are formed through the micro-phase segregation of two incompatible components on a nanometer scale. Soap is an everyday example of a lyotropic liquid crystal.

The content of water or other solvent molecules changes the self-assembled structures. At very low amphiphile concentration, the molecules will be dispersed randomly without any ordering. At slightly higher (but still low) concentration, amphiphilic molecules will spontaneously assemble into micelles or vesicles. This is done so as to 'hide' the hydrophobic tail of the amphiphile inside the micelle core, exposing a hydrophilic (water-soluble) surface to aqueous solution. These spherical objects do not order themselves in solution, however. At higher concentration, the assemblies will become ordered. A typical phase is a hexagonal columnar phase, where the amphiphiles form long cylinders (again with a hydrophilic surface) that arrange themselves into a roughly hexagonal lattice. This is called the middle soap phase. At still higher concentration, a lamellar phase (neat soap phase) may form, wherein extended sheets of amphiphiles are separated by thin layers of water. For some systems, a cubic (also called viscous isotropic) phase may exist between the hexagonal and lamellar phases, wherein spheres are formed that create a dense cubic lattice. These spheres may also be connected to one another, forming a bicontinuous cubic phase.

The objects created by amphiphiles are usually spherical (as in the case of micelles), but may also be disc-like (bicelles), rod-like, or biaxial (all three micelle axes are distinct). These anisotropic self-assembled nano-structures can then order themselves in much the same way as thermotropic liquid crystals do, forming large-scale versions of all the thermotropic phases (such as a nematic phase of rod-shaped micelles).

For some systems, at high concentrations, inverse phases are observed. That is, one may generate an inverse hexagonal columnar phase (columns of water encapsulated by amphiphiles) or an inverse micellar phase (a bulk liquid crystal sample with spherical water cavities).

A generic progression of phases, going from low to high amphiphile concentration, is:

- Discontinuous cubic phase (micellar cubic phase)

- Hexagonal phase (hexagonal columnar phase) (middle phase)

- Lamellar phase

- Bicontinuous cubic phase

- Reverse hexagonal columnar phase

- Inverse cubic phase (Inverse micellar phase)

Even within the same phases, their self-assembled structures are tunable by the concentration: for example, in lamellar phases, the layer distances increase with the solvent volume. Since lyotropic liquid crystals rely on a subtle balance of intermolecular interactions, it is more difficult to analyze their structures and properties than those of thermotropic liquid crystals.

Similar phases and characteristics can be observed in immiscible diblock copolymers.

9.3.3 Metallotropic liquid crystals

Liquid crystal phases can also be based on low-melting *inorganic* phases like $ZnCl_2$ that have a structure formed of linked tetrahedra and easily form glasses. The addition of long chain soap-like molecules leads to a series of new phases that show a variety of liquid crystalline behavior both as a function of the inorganic-organic composition ratio and of temperature. This class of materials has been named metallotropic.[34]

9.3.4 Laboratory analysis of mesophases

Thermotropic mesophases are detected and characterized by two major methods, the original method was use of thermal optical microscopy, in which a small sample of the material was placed between two crossed polarizers; the sample was then heated and cooled. As the isotropic phase would not significantly affect the polarization of the light, it would appear very dark, whereas the crystal and liquid crystal phases will both polarize the light in a uniform way, leading to brightness and color gradients. This method allows for the characterization of the particular phase, as the different phases are defined by their particular order, which must be observed. The second method, Differential Scanning Calorimetry (DSC), allows for more precise determination of phase transitions and transition enthalpies. In DSC, a small sample is heated in a way that generates a very precise change in temperature with respect to time. During phase transitions, the heat flow required to maintain this heating or cooling rate will change. These changes can be observed and attributed to various phase transitions, such as key liquid crystal transitions.

Lyotropic mesophases are analyzed in a similar fashion, through these experiments are somewhat more complex, as the concentration of mesogen is a key factor. These experiments are run at various concentrations of mesogen in order to analyze that impact.

9.4   Biological liquid crystals

Lyotropic liquid-crystalline phases are abundant in living systems, the study of which is referred to as lipid polymorphism. Accordingly, lyotropic liquid crystals attract particular attention in the field of biomimetic chemistry. In particular, biological membranes and cell membranes are a form of liquid crystal. Their constituent molecules (e.g. phospholipids) are perpendicular to the membrane surface, yet the membrane is flexible. These lipids vary in shape (see page on lipid polymorphism). The constituent molecules can inter-mingle easily, but tend not to leave the membrane due to the high energy requirement of this process. Lipid molecules can flip from one side of the membrane to the other, this process being catalyzed by flippases and floppases (depending on the direction of movement). These liquid crystal membrane phases can also host important proteins such as receptors freely "floating" inside, or partly outside, the membrane, e.g. CCT.

Many other biological structures exhibit liquid-crystal behavior. For instance, the concentrated protein solution that is extruded by a spider to generate silk is, in fact, a liquid crystal phase. The precise ordering of molecules in silk is critical to its renowned strength. DNA and many polypeptides can also form LC phases and this too forms an important part of current academic research.

9.5   Pattern formation in liquid crystals

See also: Pattern formation

Anisotropy of liquid crystals is a property not observed in other fluids. This anisotropy makes flows of liquid crystals behave more differentially than those of ordinary fluids. For example, injection of a flux of a liquid crystal between two close parallel plates (viscous fingering) causes orientation of the molecules to couple with the flow, with the resulting emergence of dendritic patterns.[35] This anisotropy is also manifested in the interfacial energy (surface tension) between different liquid crystal phases. This anisotropy determines the equilibrium shape at the coexistence temperature, and is so strong that usually facets appear. When temperature is changed one of the phases grows, forming different morphologies depending on the temperature change.[36] Since growth is controlled by heat diffusion, anisotropy in thermal conductivity favors growth in specific directions, which has also an effect on the final shape.[37]

9.6   Theoretical treatment of liquid crystals

Microscopic theoretical treatment of fluid phases can become quite complicated, owing to the high material density, meaning that strong interactions, hard-core repulsions, and many-body correlations cannot be ignored. In the case of liquid crystals, anisotropy in all of these interactions further complicates analysis. There are a number of fairly simple theories, however, that can at least predict the general behavior of the phase transitions in liquid crystal systems.

9.6.1   Director

As we already saw above, the nematic liquid crystals are composed of rod-like molecules with the long axes of neighboring molecules aligned approximately to one another. To describe this anisotropic structure, a dimensionless unit vector n called the *director*, is introduced to represent the direction of preferred orientation of molecules in the neighborhood of any point. Because there is no physical polarity along the director axis, n and $-n$ are fully equivalent.[18]

9.6.2   Order parameter

The description of liquid crystals involves an analysis of order. A second rank symmetric traceless tensor order parameter is used to describe the orientational order of a nematic liquid crystal, although a scalar order parameter is usually sufficient to describe uniaxial nematic liquid crystals. To make this quantitative, an orientational order parameter is usually defined based on the average of the second Legendre polynomial:

$$S = \langle P_2(\cos\theta) \rangle = \left\langle \frac{3\cos^2\theta - 1}{2} \right\rangle$$

where θ is the angle between the liquid-crystal molecular axis and the *local director* (which is the 'preferred direction' in a volume element of a liquid crystal sample, also representing its *local optical axis*). The brackets denote both a temporal and spatial average. This definition is convenient, since for a completely random and isotropic sample, S=0, whereas for a perfectly aligned sample S=1. For a typical liquid crystal sample, S is on the order of 0.3 to 0.8, and generally decreases as the temperature is raised. In particular, a sharp drop of the order parameter to 0 is observed when the system undergoes a phase transition from an LC phase into the isotropic phase.[38] The order parameter can be measured experimentally in a number of ways; for instance, diamagnetism, birefringence, Raman scattering, NMR and EPR can be used to determine S.[20]

The order of a liquid crystal could also be characterized by using other even Legendre polynomials (all the odd polynomials average to zero since the director can point in either of two antiparallel directions). These higher-order averages are more difficult to measure, but can yield additional information about molecular ordering.[1]

A positional order parameter is also used to describe the ordering of a liquid crystal. It is characterized by the variation of the density of the center of mass of the liquid crystal molecules along a given vector. In the case of positional variation along the z-axis the density $\rho(z)$ is often given by:

$$\rho(\mathbf{r}) = \rho(z) = \rho_0 + \rho_1 \cos\left(q_s z - \phi\right) + \cdots$$

The complex positional order parameter is defined as $\psi(\mathbf{r}) = \rho_1(\mathbf{r})e^{i\phi(\mathbf{r})}$ and ρ_0 the average density. Typically only the first two terms are kept and higher order terms are ignored since most phases can be described adequately using sinusoidal functions. For a perfect nematic $\psi = 0$ and for a smectic phase ψ will take on complex values. The complex nature of this order parameter allows for many parallels between nematic to smectic phase transitions and conductor to superconductor transitions.[18]

9.6.3 Onsager hard-rod model

A simple model which predicts lyotropic phase transitions is the hard-rod model proposed by Lars Onsager. This theory considers the volume excluded from the center-of-mass of one idealized cylinder as it approaches another. Specifically, if the cylinders are oriented parallel to one another, there is very little volume that is excluded from the center-of-mass of the approaching cylinder (it can come quite close to the other cylinder). If, however, the cylinders are at some angle to one another, then there is a large volume surrounding the cylinder which the approaching cylinder's center-of-mass cannot enter (due to the hard-rod repulsion between the two idealized objects). Thus, this angular arrangement sees a *decrease* in the net positional entropy of the approaching cylinder (there are fewer states available to it).[39][40]

The fundamental insight here is that, whilst parallel arrangements of anisotropic objects lead to a decrease in orientational entropy, there is an increase in positional entropy. Thus in some case greater positional order will be entropically favorable. This theory thus predicts that a solution of rod-shaped objects will undergo a phase transition, at sufficient concentration, into a nematic phase. Although this model is conceptually helpful, its mathematical formulation makes several assumptions that limit its applicability to real systems.[40]

9.6.4 Maier–Saupe mean field theory

This statistical theory, proposed by Alfred Saupe and Wilhelm Maier, includes contributions from an attractive intermolecular potential from an induced dipole moment between adjacent liquid crystal molecules. The anisotropic attraction stabilizes parallel alignment of neighboring molecules, and the theory then considers a mean-field average of the interaction. Solved self-consistently, this theory predicts thermotropic nematic-isotropic phase transitions, consistent with experiment.[41][42][43]

9.6.5   McMillan's model

McMillan's model, proposed by William McMillan,[44] is an extension of the Maier–Saupe mean field theory used to describe the phase transition of a liquid crystal from a nematic to a smectic A phase. It predicts that the phase transition can be either continuous or discontinuous depending on the strength of the short-range interaction between the molecules. As a result, it allows for a triple critical point where the nematic, isotropic, and smectic A phase meet. Although it predicts the existence of a triple critical point, it does not successfully predict its value. The model utilizes two order parameters that describe the orientational and positional order of the liquid crystal. The first is simply the average of the second Legendre polynomial and the second order parameter is given by:

$$\sigma = \left\langle \cos\left(\frac{2\pi z_i}{d}\right)\left(\frac{3}{2}\cos^2\theta_i - \frac{1}{2}\right)\right\rangle$$

The values z_i, θ_i, and d are the position of the molecule, the angle between the molecular axis and director, and the layer spacing. The postulated potential energy of a single molecule is given by:

$$U_i(\theta_i, z_i) = -U_0\left(S + \alpha\sigma\cos\left(\frac{2\pi z_i}{d}\right)\right)\left(\frac{3}{2}\cos^2\theta_i - \frac{1}{2}\right)$$

Here constant α quantifies the strength of the interaction between adjacent molecules. The potential is then used to derive the thermodynamic properties of the system assuming thermal equilibrium. It results in two self-consistency equations that must be solved numerically, the solutions of which are the three stable phases of the liquid crystal.[20]

9.6.6   Elastic continuum theory

In this formalism, a liquid crystal material is treated as a continuum; molecular details are entirely ignored. Rather, this theory considers perturbations to a presumed oriented sample. The distortions of the liquid crystal are commonly described by the Frank free energy density. One can identify three types of distortions that could occur in an oriented sample: (1) **twists** of the material, where neighboring molecules are forced to be angled with respect to one another, rather than aligned; (2) **splay** of the material, where bending occurs perpendicular to the director; and (3) **bend** of the material, where the distortion is parallel to the director and molecular axis. All three of these types of distortions incur an energy penalty. They are distortions that are induced by the boundary conditions at domain walls or the enclosing container. The response of the material can then be decomposed into terms based on the elastic constants corresponding to the three types of distortions. Elastic continuum theory is a particularly powerful tool for modeling liquid crystal devices [45] and lipid bilayers.[46]

9.7   External influences on liquid crystals

Scientists and engineers are able to use liquid crystals in a variety of applications because external perturbation can cause significant changes in the macroscopic properties of the liquid crystal system. Both electric and magnetic fields can be used to induce these changes. The magnitude of the fields, as well as the speed at which the molecules align are important characteristics industry deals with. Special surface treatments can be used in liquid crystal devices to force specific orientations of the director.

9.7.1   Electric and magnetic field effects

The ability of the director to align along an external field is caused by the electric nature of the molecules. Permanent electric dipoles result when one end of a molecule has a net positive charge while the other end has a net negative charge. When an external electric field is applied to the liquid crystal, the dipole molecules tend to orient themselves along the direction of the field.

Even if a molecule does not form a permanent dipole, it can still be influenced by an electric field. In some cases, the field produces slight re-arrangement of electrons and protons in molecules such that an induced electric dipole results. While not as strong as permanent dipoles, orientation with the external field still occurs. The effects of magnetic fields on liquid crystal molecules are analogous to electric fields. Because magnetic fields are generated by moving electric charges, permanent magnetic dipoles are produced by electrons moving about atoms. When a magnetic field is applied, the molecules will tend to align with or against the field.

9.7.2 Surface preparations

In the absence of an external field, the director of a liquid crystal is free to point in any direction. It is possible, however, to force the director to point in a specific direction by introducing an outside agent to the system. For example, when a thin polymer coating (usually a polyimide) is spread on a glass substrate and rubbed in a single direction with a cloth, it is observed that liquid crystal molecules in contact with that surface align with the rubbing direction. The currently accepted mechanism for this is believed to be an epitaxial growth of the liquid crystal layers on the partially aligned polymer chains in the near surface layers of the polyimide.

9.7.3 Fredericks transition

The competition between orientation produced by surface anchoring and by electric field effects is often exploited in liquid crystal devices. Consider the case in which liquid crystal molecules are aligned parallel to the surface and an electric field is applied perpendicular to the cell. At first, as the electric field increases in magnitude, no change in alignment occurs. However at a threshold magnitude of electric field, deformation occurs. Deformation occurs where the director changes its orientation from one molecule to the next. The occurrence of such a change from an aligned to a deformed state is called a Fredericks transition and can also be produced by the application of a magnetic field of sufficient strength.

The Fredericks transition is fundamental to the operation of many liquid crystal displays because the director orientation (and thus the properties) can be controlled easily by the application of a field.

9.8 Effect of chirality

As already described, chiral liquid-crystal molecules usually give rise to chiral mesophases. This means that the molecule must possess some form of asymmetry, usually a stereogenic center. An additional requirement is that the system not be racemic: a mixture of right- and left-handed molecules will cancel the chiral effect. Due to the cooperative nature of liquid crystal ordering, however, a small amount of chiral dopant in an otherwise achiral mesophase is often enough to select out one domain handedness, making the system overall chiral.

Chiral phases usually have a helical twisting of the molecules. If the pitch of this twist is on the order of the wavelength of visible light, then interesting optical interference effects can be observed. The chiral twisting that occurs in chiral LC phases also makes the system respond differently from right- and left-handed circularly polarized light. These materials can thus be used as polarization filters.[47]

It is possible for chiral LC molecules to produce essentially achiral mesophases. For instance, in certain ranges of concentration and molecular weight, DNA will form an achiral line hexatic phase. An interesting recent observation is of the formation of chiral mesophases from achiral LC molecules. Specifically, bent-core molecules (sometimes called banana liquid crystals) have been shown to form liquid crystal phases that are chiral.[48] In any particular sample, various domains will have opposite handedness, but within any given domain, strong chiral ordering will be present. The appearance mechanism of this macroscopic chirality is not yet entirely clear. It appears that the molecules stack in layers and orient themselves in a tilted fashion inside the layers. These liquid crystals phases may be ferroelectric or anti-ferroelectric, both of which are of interest for applications.[49][50]

Chirality can also be incorporated into a phase by adding a chiral dopant, which may not form LCs itself. Twisted-nematic or super-twisted nematic mixtures often contain a small amount of such dopants.

9.9   Applications of liquid crystals

See also: Liquid crystal display

Liquid crystals find wide use in liquid crystal displays, which rely on the optical properties of certain liquid crystalline substances in the presence or absence of an electric field. In a typical device, a liquid crystal layer (typically 4 μm thick) sits between two polarizers that are crossed (oriented at 90° to one another). The liquid crystal alignment is chosen so that its relaxed phase is a twisted one (see Twisted nematic field effect).[8] This twisted phase reorients light that has passed through the first polarizer, allowing its transmission through the second polarizer (and reflected back to the observer if a reflector is provided). The device thus appears transparent. When an electric field is applied to the LC layer, the long molecular axes tend to align parallel to the electric field thus gradually untwisting in the center of the liquid crystal layer. In this state, the LC molecules do not reorient light, so the light polarized at the first polarizer is absorbed at the second polarizer, and the device loses transparency with increasing voltage. In this way, the electric field can be used to make a pixel switch between transparent or opaque on command. Color LCD systems use the same technique, with color filters used to generate red, green, and blue pixels.[8] Chiral smectic liquid crystals are used in ferroelectric LCDs which are fast-switching binary light modulators. Similar principles can be used to make other liquid crystal based optical devices.[51]

Liquid crystal tunable filters are used as electrooptical devices, e.g., in hyperspectral imaging.

Thermotropic chiral LCs whose pitch varies strongly with temperature can be used as crude liquid crystal thermometers, since the color of the material will change as the pitch is changed. Liquid crystal color transitions are used on many aquarium and pool thermometers as well as on thermometers for infants or baths.[52] Other liquid crystal materials change color when stretched or stressed. Thus, liquid crystal sheets are often used in industry to look for hot spots, map heat flow, measure stress distribution patterns, and so on. Liquid crystal in fluid form is used to detect electrically generated hot spots for failure analysis in the semiconductor industry.[53]

Liquid crystal lasers use a liquid crystal in the lasing medium as a distributed feedback mechanism instead of external mirrors. Emission at a photonic bandgap created by the periodic dielectric structure of the liquid crystal gives a low-threshold high-output device with stable monochromatic emission.[54][55]

Polymer Dispersed Liquid Crystal (PDLC) sheets and rolls are available as adhesive backed Smart film which can be applied to windows and electrically switched between transparent and opaque to provide privacy.

Many common fluids, such as soapy water, are in fact liquid crystals. Soap forms a variety of LC phases depending on its concentration in water.[56]

Bowlic columns could be used for fast switches.[57]

9.10   See also

- Biaxial nematic

- Columnar phase

- Chromonic

- LCD classification

- Liquid crystal display

- Liquid Crystal on Silicon

- Liquid crystal polymer

- Liquid crystal tunable filter

- Lyotropic liquid crystal

- Pattern formation

- Plastic crystallinity

- Smart glass

- Thermochromics

- Thermotropic crystal

- Twisted nematic field effect

- Nematicon

- Liquid crystal thermometer

- Mood ring

9.11 References

[1] Chandrasekhar, S. (1992). *Liquid Crystals* (2nd ed.). Cambridge: Cambridge University Press. ISBN 0-521-41747-3.

[2] Reinitzer, Friedrich (1888). "Beiträge zur Kenntniss des Cholesterins". *Monatshefte für Chemie (Wien)* **9** (1): 421–441. doi:10.1007/BF01516710.

[3] Lehmann, O. (1889). "Über fliessende Krystalle". *Zeitschrift für Physikalische Chemie* **4**: 462–72.

[4] Sluckin, T. J.; Dunmur, D. A. and Stegemeyer, H. (2004). *Crystals That Flow – classic papers from the history of liquid crystals*. London: Taylor & Francis. ISBN 0-415-25789-1.

[5] Gray, G. W. (1962) *Molecular Structure and the Properties of Liquid Crystals*, Academic Press

[6] Stegemeyer, H (1994). "Professor Horst Sackmann, 1921 – 1993". *Liquid Crystals Today* **4**: 1. doi:10.1080/13583149408628630.

[7] Liquid Crystals. kfupm.edu.sa

[8] Castellano, Joseph A. (2005). *Liquid Gold: The Story of Liquid Crystal Displays and the Creation of an Industry*. World Scientific Publishing. ISBN 978-981-238-956-5.

[9] Goldmacher, Joel E. and Castellano, Joseph A. "Electro-optical Compositions and Devices," U.S. Patent 3,540,796, Issue date: November 17, 1970.

[10] Heilmeier, G. H.; Zanoni, L. A.; Barton, L. A. (1968). "Dynamic Scattering in Nematic Liquid Crystals". *Applied Physics Letters* **13**: 46. Bibcode:1968ApPhL..13...46H. doi:10.1063/1.1652453.

[11] Kelker, H.; Scheurle, B. (1969). "A Liquid-crystalline (Nematic) Phase with a Particularly Low Solidification Point". *Angew. Chem. Int. Ed.* **8** (11): 884. doi:10.1002/anie.196908841.

[12] Gray, G.W.; Harrison, K.J.; Nash, J.A. (1973). "New family of nematic liquid crystals for displays". *Electronics Lett.* **9** (6): 130. doi:10.1049/el:19730096.

[13] Lin, Lei (Lam, Lui) (1982). "Liquid crystal phases and the 'dimensionality' of molecules". *Wuli (Physics)* **11**, 171-178.

[14] "History and Properties of Liquid Crystals". Nobelprize.org. Retrieved June 6, 2009.

[15] Lam, Lui (1994). "Bowlics". *Liquid Crystalline and Mesomorphic Polymers*, eds. Valery P. Shibaev and Lui Lam. New York: Springer.

[16] Lin, Lei (Lam, Lui) (1987). "Bowlic liquid crytals". *Mol. Cryst. Liq. Cryst* **146**: 41-54.

[17]

[18] de Gennes, P.G. and Prost, J (1993). *The Physics of Liquid Crystals*. Oxford: Clarendon Press. ISBN 0-19-852024-7.

[19] Dierking, I. (2003). *Textures of Liquid Crystals*. Weinheim: Wiley-VCH. ISBN 3-527-30725-7.

[20] Collings, P.J. and Hird, M (1997). *Introduction to Liquid Crystals*. Bristol, PA: Taylor & Francis. ISBN 0-7484-0643-3.

[21] Shao, Y.; Zerda, T. W. (1998). "Phase Transitions of Liquid Crystal PAA in Confined Geometries". *Journal of Physical Chemistry B* **102** (18): 3387–3394. doi:10.1021/jp9734437.

[22] Rego, J.A.; Harvey, Jamie A.A.; MacKinnon, Andrew L.; Gatdula, Elysse (January 2010). "Asymmetric synthesis of a highly soluble 'trimeric' analogue of the chiral nematic liquid crystal twist agent Merck S1011" (PDF). *Liquid Crystals* **37** (1): 37–43. doi:10.1080/02678290903359291.

[23] Madsen, L. A.; Dingemans, T. J.; Nakata, M.; Samulski, E. T. (2004). "Thermotropic Biaxial Nematic Liquid Crystals". *Phys. Rev. Lett.* **92** (14): 145505. Bibcode:2004PhRvL..92n5505M. doi:10.1103/PhysRevLett.92.145505. PMID 15089552.

[24] "smectic". Merriam-Webster Dictionary.

[25] Kopp, V. I.; Fan, B.; Vithana, H. K. M.; Genack, A. Z.; Fan; Vithana; Genack (1998). "Low threshold lasing at the edge of a photonic stop band in cholesteric liquid crystals". *Opt. Lett* **23** (21): 1707–1709. Bibcode:1998OptL...23.1707K. doi:10.1364/OL.23.001707. PMID 18091891.

[26] Priestley, E. B.; Wojtowicz, P. J. and Sheng, P. (1974). *Introduction to Liquid Crystals*. Plenum Press. ISBN 0-306-30858-4.

[27] Kleinert H. and Maki K. (1981). "Lattice Textures in Cholesteric Liquid Crystals" (PDF). *Fortschritte der Physik* **29** (5): 219–259. Bibcode:1981ForPh..29...219K. doi:10.1002/prop.19810290503.

[28] Seideman, T (1990). "The liquid-crystalline blue phases" (PDF). *Rep. Prog. Phys.* **53** (6): 659–705. Bibcode:1990RPPh...53..65 doi:10.1088/0034-4885/53/6/001.

[29] Coles, Harry J.; Pivnenko, Mikhail N. (2005). "Liquid crystal 'blue phases' with a wide temperature range". *Nature* **436** (7053): 997–1000. Bibcode:2005Natur.436..997C. doi:10.1038/nature03932. PMID 16107843.

[30] Yamamoto, Jun; Nishiyama, Isa; Inoue, Miyoshi; Yokoyama, Hiroshi (2005). "Optical isotropy and iridescence in a smectic blue phase". *Nature* **437** (7058): 525. Bibcode:2005Natur.437..525Y. doi:10.1038/nature04034.

[31] Kikuchi H, Yokota M, Hisakado Y, Yang H, Kajiyama T.; Yokota; Hisakado; Yang; Kajiyama (2002). "Polymer-stabilized liquid crystal blue phases". *Nature Materials* **1** (1): 64–8. Bibcode:2002NatMa...1...64K. doi:10.1038/nmat712. PMID 12618852.

[32] "Samsung Develops World's First 'Blue Phase' Technology to Achieve 240 Hz Driving Speed for High-Speed Video". Retrieved April 23, 2009.

[33] Qizhen Liang, Pengtao Liu, Cheng Liu, Xigao Jian, Dingyi Hong, Yang Li. (2005). "Synthesis and Properties of Lyotropic Liquid Crystalline Copolyamides Containing Phthalazinone Moieties and Ether Linkages". *Polymer* **46** (16): 6258–6265. doi:10.1016/j.polymer.2005.05.059.

[34] Martin, James D.; Keary, Cristin L.; Thornton, Todd A.; Novotnak, Mark P.; Knutson, Jeremey W.; Folmer, Jacob C. W. (2006). "Metallotropic liquid crystals formed by surfactant templating of molten metal halides". *Nature Materials* **5** (4): 271–5. Bibcode:2006NatMa...5..271M. doi:10.1038/nmat1610. PMID 16547520.

[35] Buka, A.; Palffy-Muhoray, P.; Rácz, Z. (1987). "Viscous fingering in liquid crystals". *Phys. Rev.* A**36**(8): 3984.Bibcode:1987P doi:10.1103/PhysRevA.36.3984.

[36] González-Cinca, R.; Ramírez-Piscina, L.; Casademunt, J.; Hernández-Machado, A.; Kramer, L.; Tóth Katona, T.; Börzsönyi, T.; Buka, Á. (1996). "Phase-field simulations and experiments of faceted growth in liquid crystal". *Physica D* **99** (2–3): 359. Bibcode:1996PhyD...99..359G. doi:10.1016/S0167-2789(96)00162-5.

[37] González-Cinca, R; RamíRez-Piscina, L; Casademunt, J; Hernández-Machado, A; Tóth-Katona, T; Börzsönyi, T; Buka, Á (1998). "Heat diffusion anisotropy in dendritic growth: phase field simulations and experiments in liquid crystals". *Journal of Crystal Growth* **193** (4): 712. Bibcode:1998JCrGr.193..712G. doi:10.1016/S0022-0248(98)00505-3.

[38] Ghosh, S. K. (1984). "A model for the orientational order in liquid crystals". *Il Nuovo Cimento D***4**(3): 229.Bibcode:1984NCim doi:10.1007/BF02453342.

[39] Onsager, Lars (1949). "The effects of shape on the interaction of colloidal particles". *Annals of the New York Academy of Sciences* **51** (4): 627. Bibcode:1949NYASA..51..627O. doi:10.1111/j.1749-6632.1949.tb27296.x.

[40] Vroege, G J; Lekkerkerker, H N W (1992). "Phase transitions in lyotropic colloidal and polymer liquid crystals". *Rep. Progr. Phys.* **55** (8): 1241. Bibcode:1992RPPh...55.1241V. doi:10.1088/0034-4885/55/8/003.

[41] Maier W. and Saupe A.; Saupe (1958). "Eine einfache molekulare theorie des nematischen kristallinflussigen zustandes". *Z. Naturforsch. A* (in German) **13**: 564. Bibcode:1958ZNatA..13..564M. doi:10.1515/zna-1958-0716.

[42] Maier W. and Saupe A.; Saupe (1959). "Eine einfache molekular-statistische theorie der nematischen kristallinflussigen phase .1". *Z. Naturforsch. A* (in German) **14**: 882. Bibcode:1959ZNatA..14..882M. doi:10.1515/zna-1959-1005.

[43] Maier W. and Saupe A.; Saupe (1960). "Eine einfache molekular-statistische theorie der nematischen kristallinflussigen phase .2". *Z. Naturforsch. A* (in German) **15**: 287. Bibcode:1960ZNatA..15..287M. doi:10.1515/zna-1960-0401.

[44] McMillan, W. (1971). "Simple Molecular Model for the Smectic A Phase of Liquid Crystals". *Phys. Rev. A* **4** (3): 1238. Bibcode:1971PhRvA...4.1238M. doi:10.1103/PhysRevA.4.1238.

[45] Leslie, F. M. (1992). "Continuum theory for nematic liquid crystals". *Continuum Mechanics and Thermodynamics* **4** (3): 167. Bibcode:1992CMT.....4..167L. doi:10.1007/BF01130288.

[46] Watson, M. C.; Brandt, E. G.; Welch, P. M.; Brown, F. L. H. (2012). "Determining Biomembrane Bending Rigidities from Simulations of Modest Size".*Physical Review Letters***109**(2): 028102.Bibcode:2012PhRvL.109b8102W.doi:10.1103/PhysRevLe

[47] Fujikake, H.; Takizawa, K.; Aida, T.; Negishi, T.; Kobayashi, M. (1998). "Video camera system using liquid-crystal polarizing filter toreduce reflected light". *IEEE Transactions on Broadcasting* **44** (4): 419. doi:10.1109/11.735903.

[48] Achard, M.F.; Bedel, J.Ph.; Marcerou, J.P.; Nguyen, H.T.; Rouillon, J.C. (2003). "Switching of banana liquid crystal mesophases under field". *European Physical Journal E* **10** (2): 129–34. Bibcode:2003EPJE...10..129A. doi:10.1140/epje/e2003-00016-y. PMID 15011066.

[49] Baus, Marc; Colot, Jean-Louis (1989). "Ferroelectric nematic liquid-crystal phases of dipolar hard ellipsoids". *Phys. Rev. A* **40** (9): 5444. Bibcode:1989PhRvA..40.5444B. doi:10.1103/PhysRevA.40.5444.

[50] Uehara, Hiroyuki; Hatano, Jun (2002). "Pressure-Temperature Phase Diagrams of Ferroelectric Liquid Crystals". *J. Phys. Soc. Jpn.* **71** (2): 509. Bibcode:2002JPSJ...71..509U. doi:10.1143/JPSJ.71.509.

[51] Alkeskjold, Thomas Tanggaard; Scolari, Lara; Noordegraaf, Danny; Lægsgaard, Jesper; Weirich, Johannes; Wei, Lei; Tartarini, Giovanni; Bassi, Paolo; Gauza, Sebastian; Wu, Shin-Tson; Bjarklev, Anders (2007). "Integrating liquid crystal based optical devices in photonic crystal". *Optical and Quantum Electronics* **39** (12–13): 1009. doi:10.1007/s11082-007-9139-8.

[52] Plimpton, R. Gregory "Pool thermometer" U.S. Patent 4,738,549 Issued on April 19, 1988

[53] "Hot-spot detection techniques for ICs". *acceleratedanalysis.com*. Retrieved May 5, 2009.

[54] Kopp, V. I.; Fan, B.; Vithana, H. K. M.; Genack, A. Z. (1998). "Low-threshold lasing at the edge of a photonic stop band in cholesteric liquid crystals". *Optics Express* **23** (21): 1707–1709. Bibcode:1998OptL...23.1707K. doi:10.1364/OL.23.001707. PMID 18091891.

[55] Dolgaleva, Ksenia; Simon K.H. Wei; Svetlana G. Lukishova; Shaw H. Chen; Katie Schwertz; Robert W. Boyd (2008). "Enhanced laser performance of cholesteric liquid crystals doped with oligofluorene dye". *Journal of the Optical Society of America* **25** (9): 1496–1504. Bibcode:2008JOSAB..25.1496D. doi:10.1364/JOSAB.25.001496.

[56] Luzzati, V.; Mustacchi, H.; Skoulios, A. (1957). "Structure of the Liquid-Crystal Phases of the Soap–water System: Middle Soap and Neat Soap". *Nature* **180** (4586): 600. Bibcode:1957Natur.180..600L. doi:10.1038/180600a0.

[57] Bock, H.; Helfrich, W.; Heppke, G. (1992). "Switchable columnar liquid crystalline systems". European Patent EP0529439B1 (filing date: 08/14/1992; publication date: 02/14/1996).

9.12 External links

- "History and Properties of Liquid Crystals". Nobelprize.org. Retrieved June 6, 2009.

- Definitions of basic terms relating to low-molar-mass and polymer liquid crystals (IUPAC Recommendations 2001)

- An intelligible introduction to liquid crystals from Case Western Reserve University

- Liquid Crystal Physics tutorial from the Liquid Crystals Group, University of Colorado

- Liquid Crystals & Photonics Group – Ghent University (Belgium), good tutorial

- Simulation of light propagation in liquid crystals, free program

- Liquid Crystals Interactive Online

- Liquid Crystal Institute Kent State University

- Liquid Crystals a journal by Taylor&Francis

- Molecular Crystals and Liquid Crystals a journal by Taylor & Francis

- Hot-spot detection techniques for ICs

- What are liquid crystals? from Chalmers University of Technology, Sweden

- H. Kleinert and K. Maki (1981). "Lattice Textures in Cholesteric Liquid Crystals" (PDF). *Fortschritte der Physik* **29** (5): 219. Bibcode:1981ForPh..29..219K. doi:10.1002/prop.19810290503.

- Progress in liquid crystal chemistry Thematic series in the Open Access Beilstein Journal of Organic Chemistry

- DoITPoMS Teaching and Learning Package- "Liquid Crystals"

- Bowlic liquid crystal from San Jose State University

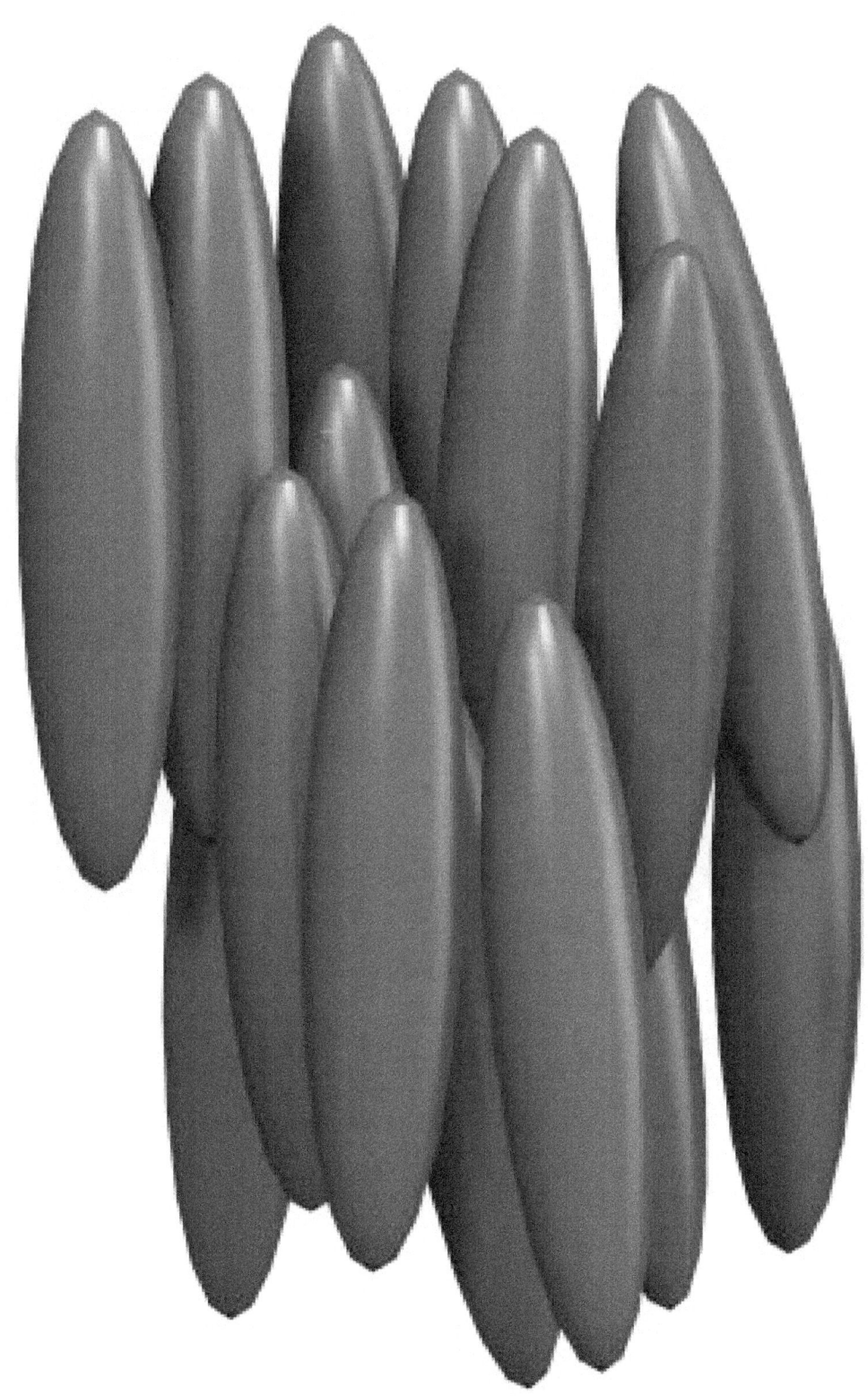

Alignment in a nematic phase.

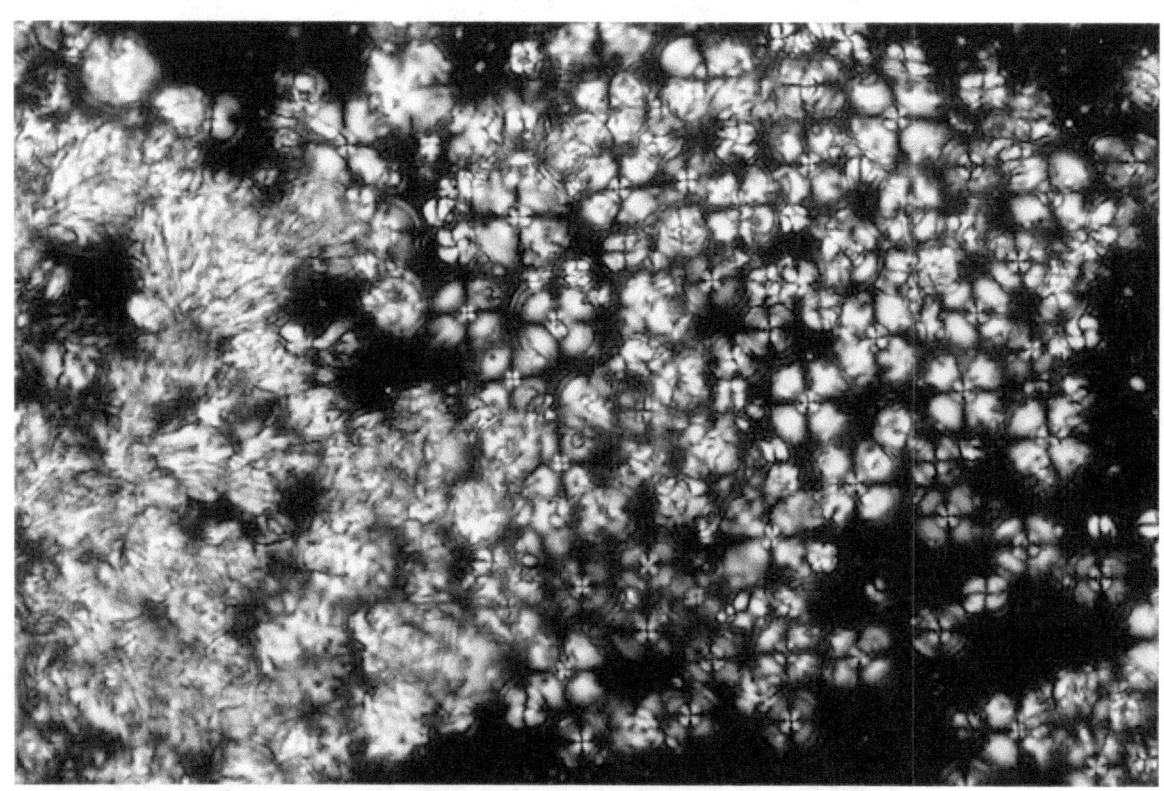

Phase transition between a nematic (left) and smectic A (right) phases observed between crossed polarizers. The black color corresponds to isotropic medium.

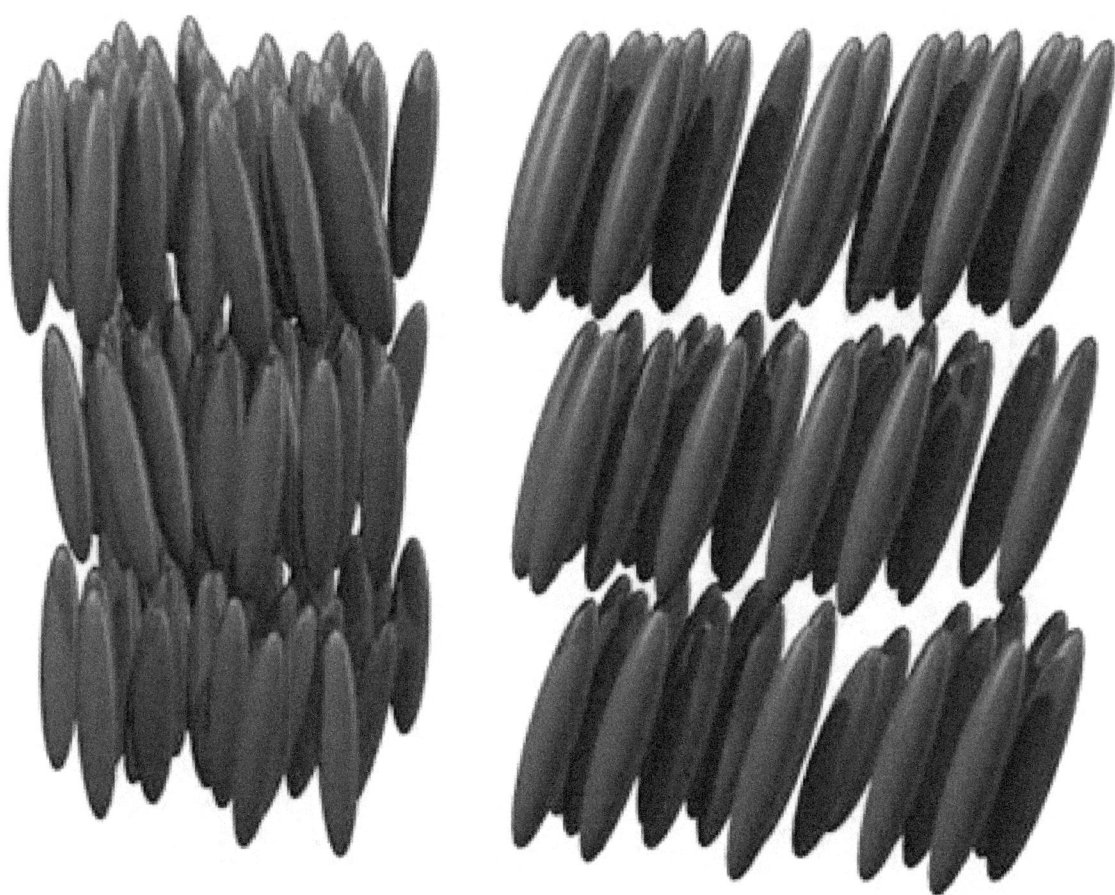

Schematic of alignment in the smectic phases. The smectic A phase (left) has molecules organized into layers. In the smectic C phase (right), the molecules are tilted inside the layers.

Schematic of ordering in chiral liquid crystal phases. The chiral nematic phase (left), also called the cholesteric phase, and the smectic C* phase (right).

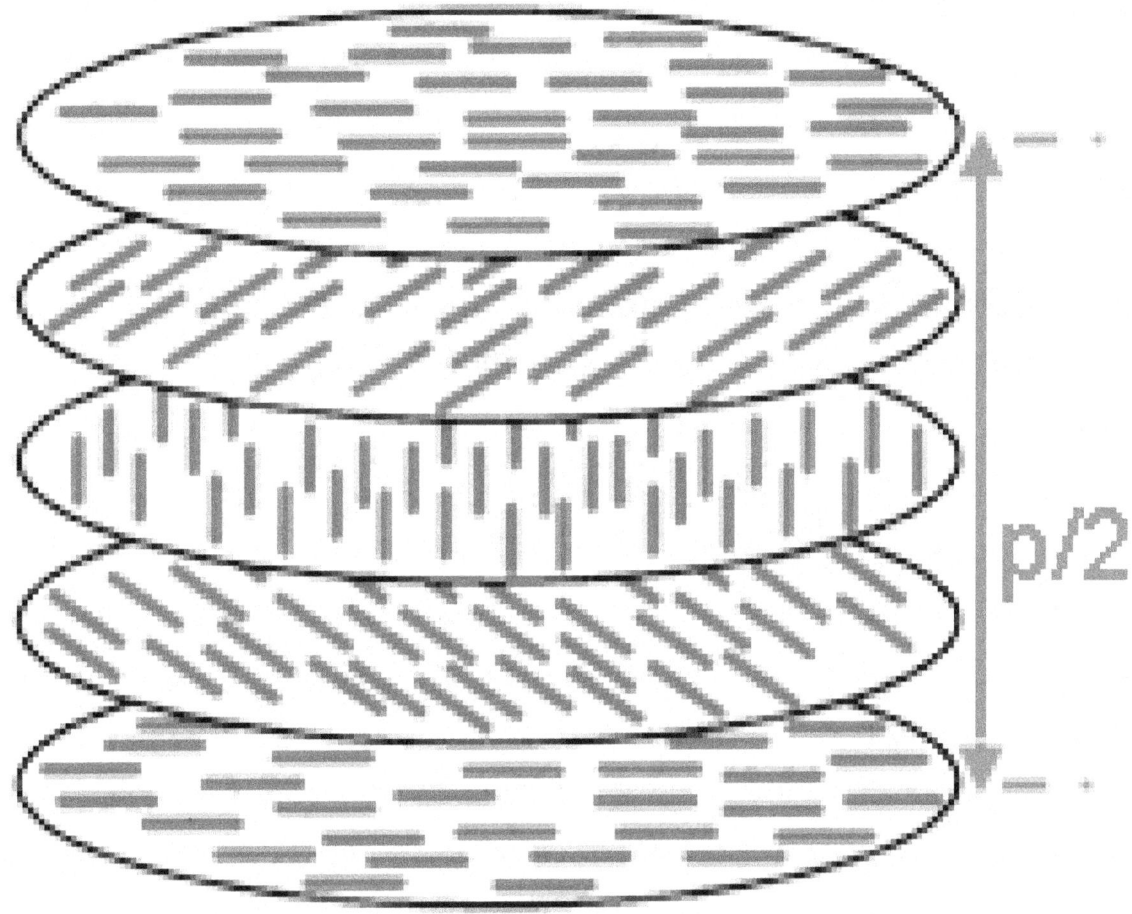

Chiral nematic phase; p refers to the chiral pitch (see text)

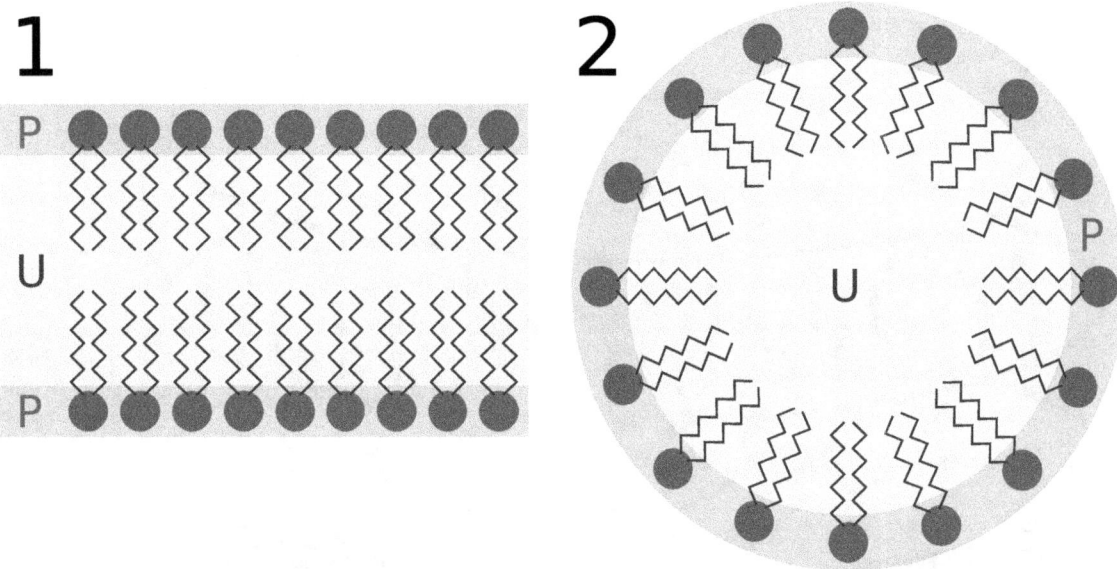

Structure of lyotropic liquid crystal. The red heads of surfactant molecules are in contact with water, whereas the tails are immersed in oil (blue): bilayer (left) and micelle (right).

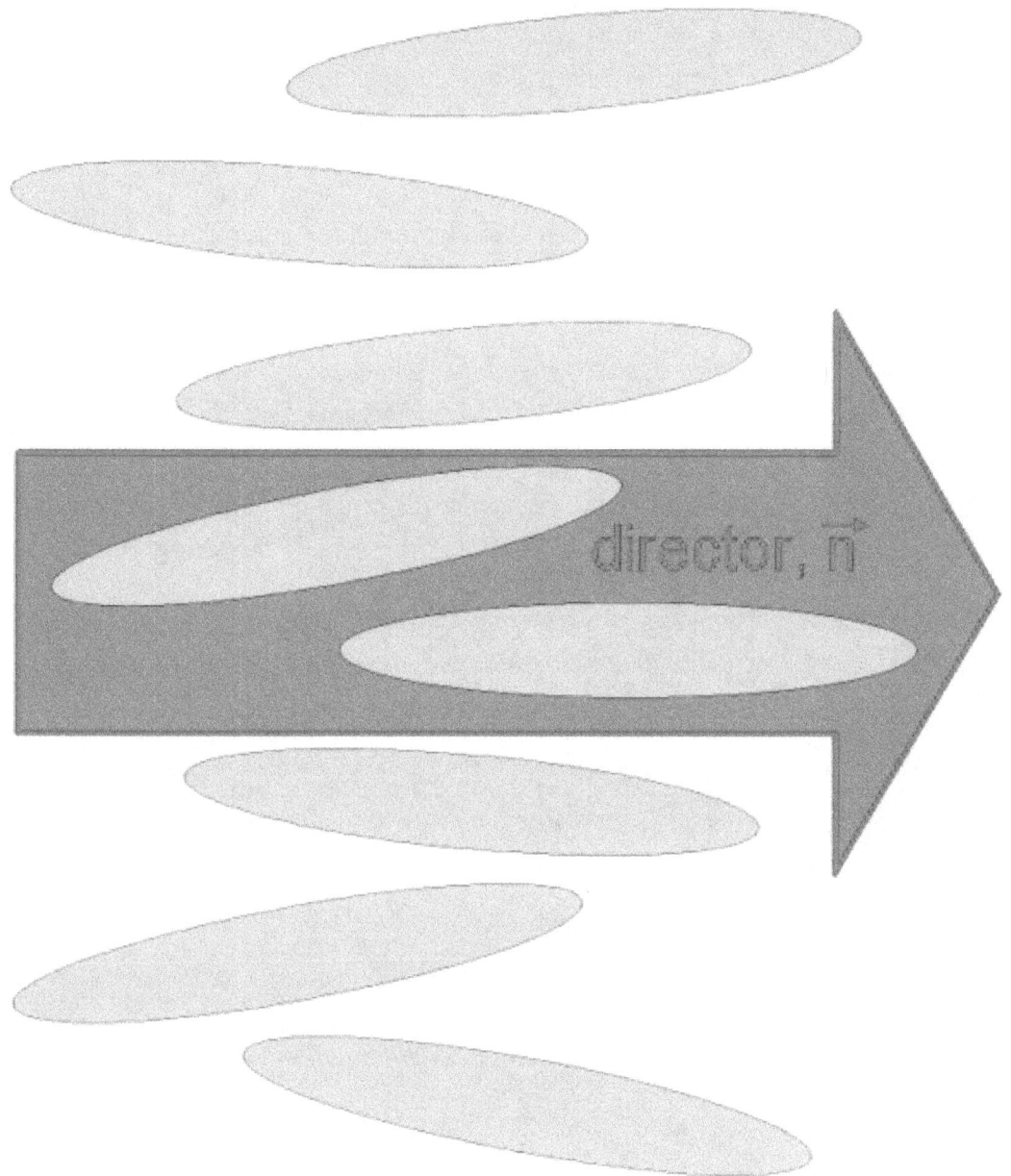

The local nematic director, *which is also the* local optical axis, *is given by the spatial and temporal average of the long molecular axes*

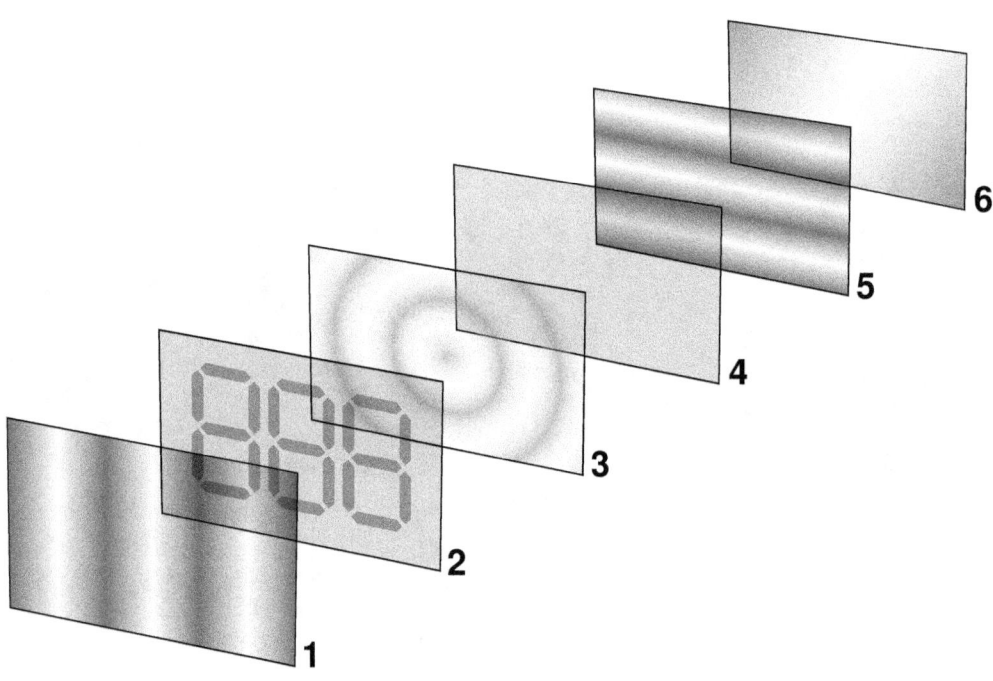

Structure of liquid crystal display: 1 – vertical polarization filter, 2,4 – glass with electrodes, 3 – liquid crystals, 5 – horizontal polarization filter, 6 – reflector

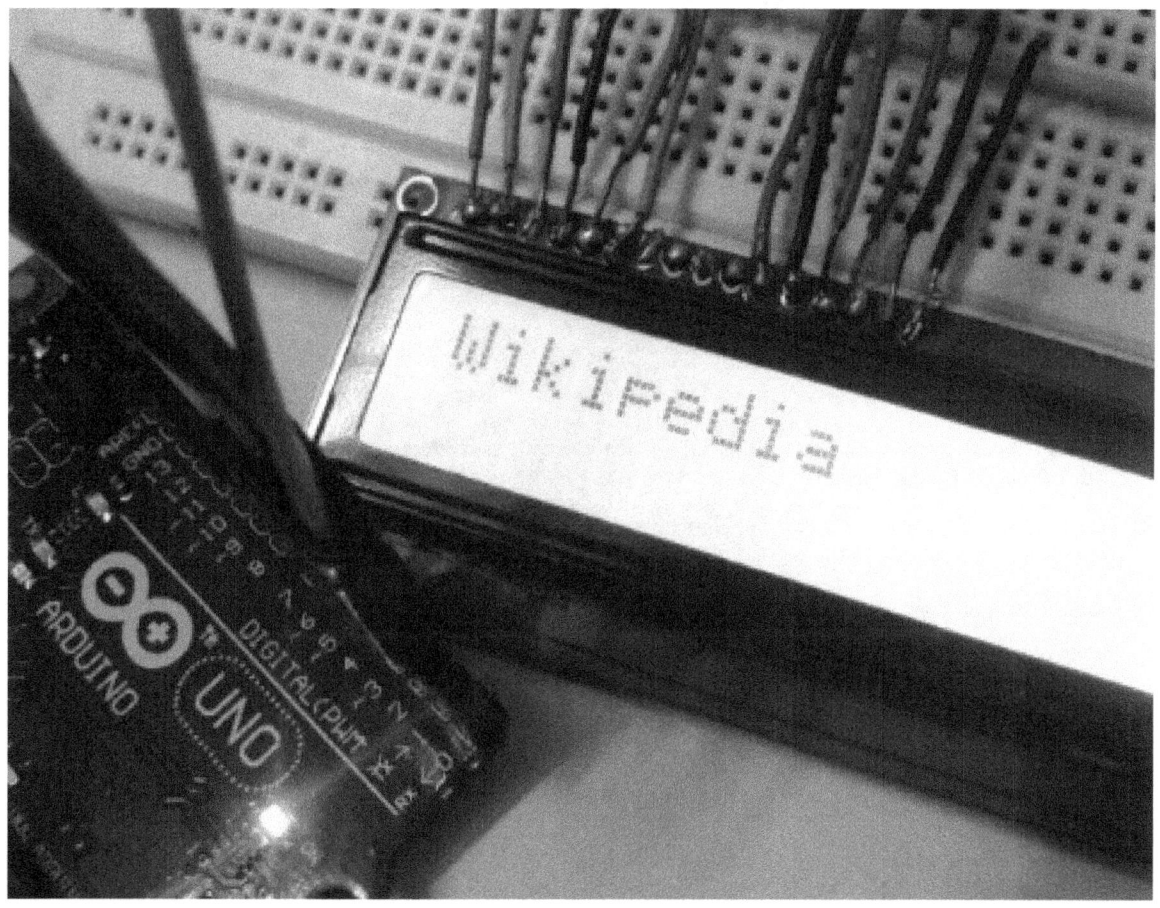

"Wikipedia" displayed on an LCD

Chapter 10

Disordered Hyperuniformity

Disordered Hyperuniformity is a type of Liquid which has crystal properties. It greatly suppresses variations in the density of particles like a crystal and the particles have the same physical properties in all directions at shorter distances, like a liquid. It was discovered in the eyes of chickens. This is thought to be the case because chicken eyes can not support the ordered, complex system best for eyesight.[1][2]This may eventually be used for self-organizing colloids or optics with the ability to transmit light with crystal efficiency while still retaining liquid flexibility.[3]

10.1 References

[1] Melissa - TodayIFoundOut.com (March 21, 2014). "Disordered Hyperuniformity: A Weird New State of Matter in Chicken Eyes". *Gizmodo*. Gawker Media.

[2] David Freeman (26 February 2014). "Scientists Look In Chicken's Eye And Discover Weird New State Of Matter". *The Huffington Post*. Retrieved 20 December 2015.

[3] https://www.princeton.edu/main/news/archive/S39/32/02E70/index.xml?section=topstories

Chapter 11

Gas

This article is about the physical properties of gas as a state of matter. For the automotive fuel, see gasoline. For the uses of gases, and other meanings, see Gas (disambiguation).

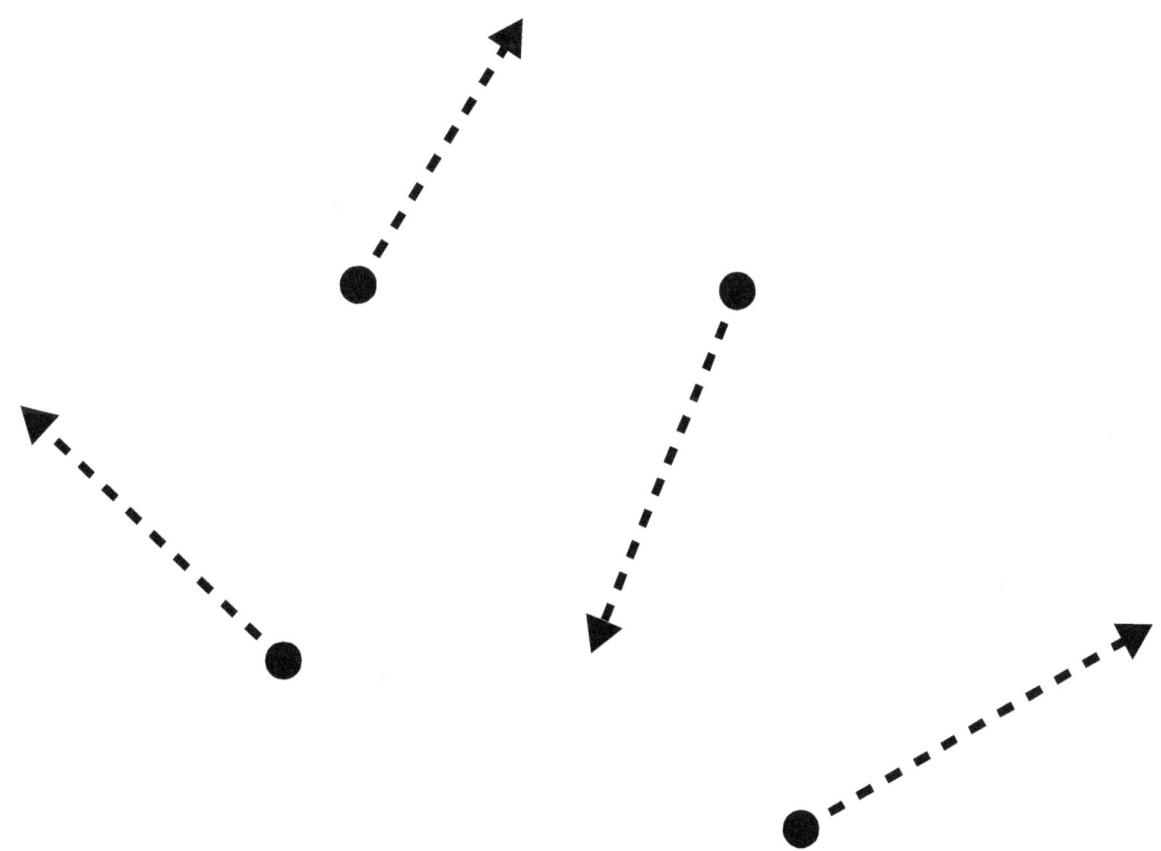

Gas phase particles (atoms, molecules, or ions) move around freely in the absence of an applied electric field.

Gas is one of the four fundamental states of matter (the others being solid, liquid, and plasma). A pure gas may be made up of individual atoms (e.g. a noble gas like neon), elemental molecules made from one type of atom (e.g. oxygen), or compound molecules made from a variety of atoms (e.g. carbon dioxide). A gas mixture would contain a variety of pure gases much like the air. What distinguishes a gas from liquids and solids is the vast separation of the individual gas particles. This separation usually makes a colorless gas invisible to the human observer. The interaction of gas particles in the presence of electric and gravitational fields are considered negligible as indicated by the constant velocity vectors in the image. One type of commonly known gas is steam.

The gaseous state of matter is found between the liquid and plasma states,[1] the latter of which provides the upper temperature boundary for gases. Bounding the lower end of the temperature scale lie degenerative quantum gases[2] which are gaining increasing attention.[3] High-density atomic gases super cooled to incredibly low temperatures are classified by their statistical behavior as either a Bose gas or a Fermi gas. For a comprehensive listing of these exotic states of matter see list of states of matter.

11.1   Elemental gases

The only chemical elements which are stable multi atom homonuclear molecules at standard temperature and pressure (STP), are hydrogen (H_2), nitrogen (N_2) and oxygen (O_2); plus two halogens, fluorine (F_2) and chlorine (Cl_2). These gases, when grouped together with the monatomic noble gases; which are helium (He), neon (Ne), argon (Ar), krypton (Kr), xenon (Xe) and radon (Rn) ; are called "elemental gases". Alternatively they are sometimes known as "molecular gases" to distinguish them from molecules that are also chemical compounds.

11.2   Etymology

The word *gas* is a neologism first used by the early 17th-century Flemish chemist J.B. van Helmont.[4] Van Helmont's word appears to have been simply a phonetic transcription of the Greek word χάος *Chaos* – the g in Dutch being pronounced like *ch* in "loch" – in which case Van Helmont was simply following the established alchemical usage first attested in the works of Paracelsus. According to Paracelsus's terminology, *chaos* meant something like "ultra-rarefied water".[5]

An alternative story[6] is that Van Helmont's word is corrupted from *gahst* (or *geist*), signifying a ghost or spirit. This was because certain gases suggested a supernatural origin, such as from their ability to cause death, extinguish flames, and to occur in "mines, bottom of wells, churchyards and other lonely places".

11.3   Physical characteristics

Because most gases are difficult to observe directly, they are described through the use of four physical properties or macroscopic characteristics: pressure, volume, number of particles (chemists group them by moles) and temperature. These four characteristics were repeatedly observed by scientists such as Robert Boyle, Jacques Charles, John Dalton, Joseph Gay-Lussac and Amedeo Avogadro for a variety of gases in various settings. Their detailed studies ultimately led to a mathematical relationship among these properties expressed by the ideal gas law (see simplified models section below).

Gas particles are widely separated from one another, and consequently have weaker intermolecular bonds than liquids or solids. These intermolecular forces result from electrostatic interactions between gas particles. Like-charged areas of different gas particles repel, while oppositely charged regions of different gas particles attract one another; gases that contain permanently charged ions are known as plasmas. Gaseous compounds with polar covalent bonds contain permanent charge imbalances and so experience relatively strong intermolecular forces, although the molecule while the compound's net charge remains neutral. Transient, randomly induced charges exist across non-polar covalent bonds of molecules and electrostatic interactions caused by them are referred to as Van der Waals forces. The interaction of these intermolecular forces varies within a substance which determines many of the physical properties unique to each gas.[7][8] A comparison of *boiling points* for compounds formed by ionic and covalent bonds leads us to this conclusion.[9] The drifting smoke particles in the image provides some insight into low pressure gas behavior.

Compared to the other states of matter, gases have low density and viscosity. Pressure and temperature influence the particles within a certain volume. This variation in particle separation and speed is referred to as *compressibility*. This particle separation and size influences optical properties of gases as can be found in the following list of refractive indices. Finally, gas particles spread apart or diffuse in order to homogeneously distribute themselves throughout any container.

Drifting smoke particles provide clues to the movement of the surrounding gas.

11.4 Macroscopic

When observing a gas, it is typical to specify a frame of reference or length scale. A *larger* length scale corresponds to a macroscopic or global point of view of the gas. This region (referred to as a volume) must be sufficient in size to contain a large sampling of gas particles. The resulting statistical analysis of this sample size produces the **"average"** behavior (i.e. velocity, temperature or pressure) of all the gas particles within the region. In contrast, a *smaller* length scale corresponds to a microscopic or particle point of view.

Macroscopically, the gas characteristics measured are either in terms of the gas particles themselves (velocity, pressure, or temperature) or their surroundings (volume). For example, Robert Boyle studied pneumatic chemistry for a small portion of his career. One of his experiments related the macroscopic properties of pressure and volume of a gas. His experiment used a J-tube manometer which looks like a test tube in the shape of the letter J. Boyle trapped an inert gas in the closed end of the test tube with a column of mercury, thereby making the number of particles and the temperature constant. He observed that when the pressure was increased in the gas, by adding more mercury to the column, the trapped gas' volume decreased (this is known as an inverse relationship). Furthermore, when Boyle multiplied the pressure and volume of each observation, the product was constant. This relationship held for every gas that Boyle observed leading to the law, (PV=k), named to honor his work in this field.

There are many mathematical tools available for analyzing gas properties. As gases are subjected to extreme conditions, these tools become a bit more complex, from the Euler equations for inviscid flow to the Navier–Stokes equations[10] that fully account for viscous effects. These equations are adapted to the conditions of the gas system in question. Boyle's lab equipment allowed the use of algebra to obtain his analytical results. His results were possible because he was studying gases in relatively low pressure situations where they behaved in an "ideal" manner. These ideal relationships apply to

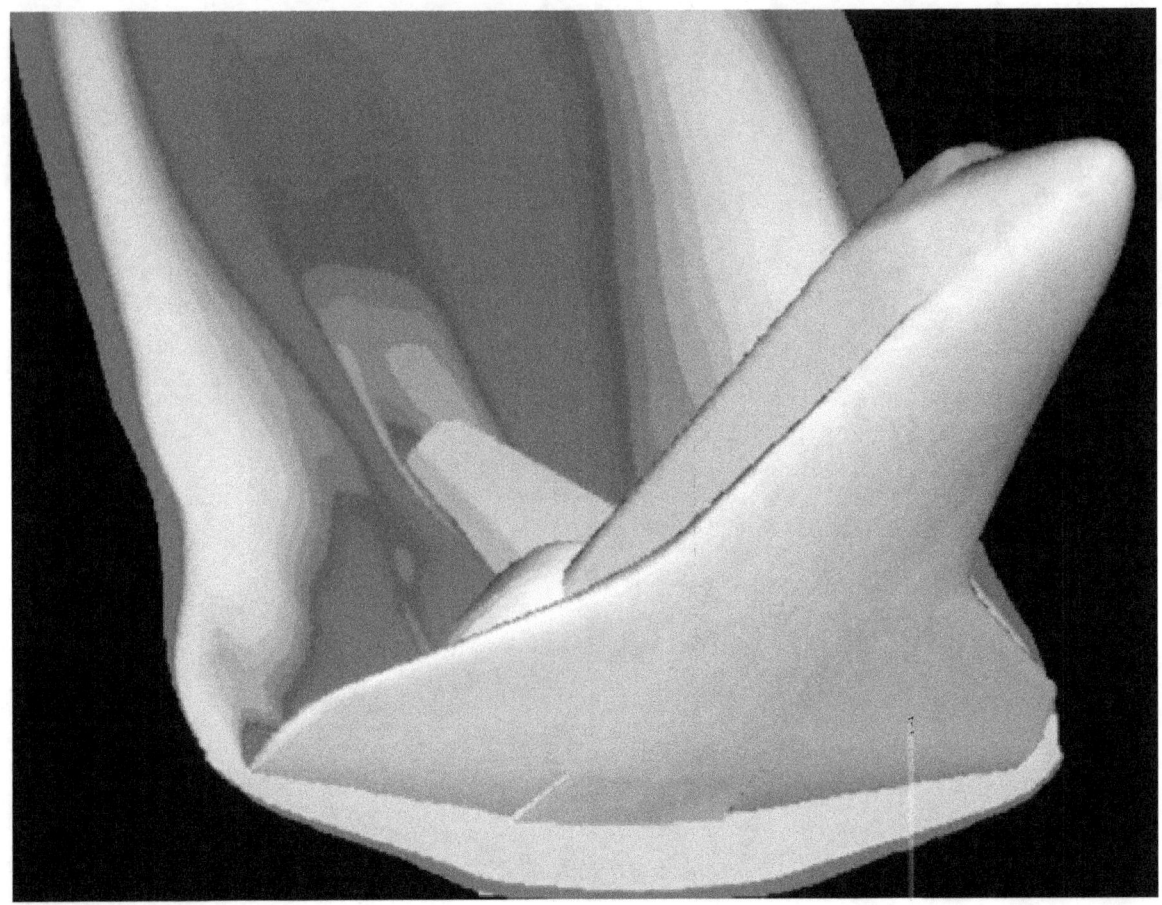

Shuttle imagery of re-entry phase.

safety calculations for a variety of flight conditions on the materials in use. The high technology equipment in use today was designed to help us safely explore the more exotic operating environments where the gases no longer behave in an "ideal" manner. This advanced math, including statistics and multivariable calculus, makes possible the solution to such complex dynamic situations as space vehicle reentry. An example is the analysis of the space shuttle reentry pictured to ensure the material properties under this loading condition are appropriate. In this flight regime, the gas is no longer behaving ideally.

11.4.1 Pressure

Main article: Pressure

The symbol used to represent *pressure* in equations is **"p"** or **"P"** with SI units of pascals.

When describing a container of gas, the term pressure (or absolute pressure) refers to the average force per unit area that the gas exerts on the surface of the container. Within this volume, it is sometimes easier to visualize the gas particles moving in straight lines until they collide with the container (see diagram at top of the article). The force imparted by a gas particle into the container during this collision is the change in momentum of the particle.[11] During a collision only the normal component of velocity changes. A particle traveling parallel to the wall does not change its momentum. Therefore, the average force on a surface must be the average change in linear momentum from all of these gas particle collisions.

Pressure is the sum of all the normal components of force exerted by the particles impacting the walls of the container divided by the surface area of the wall.

11.4.2 Temperature

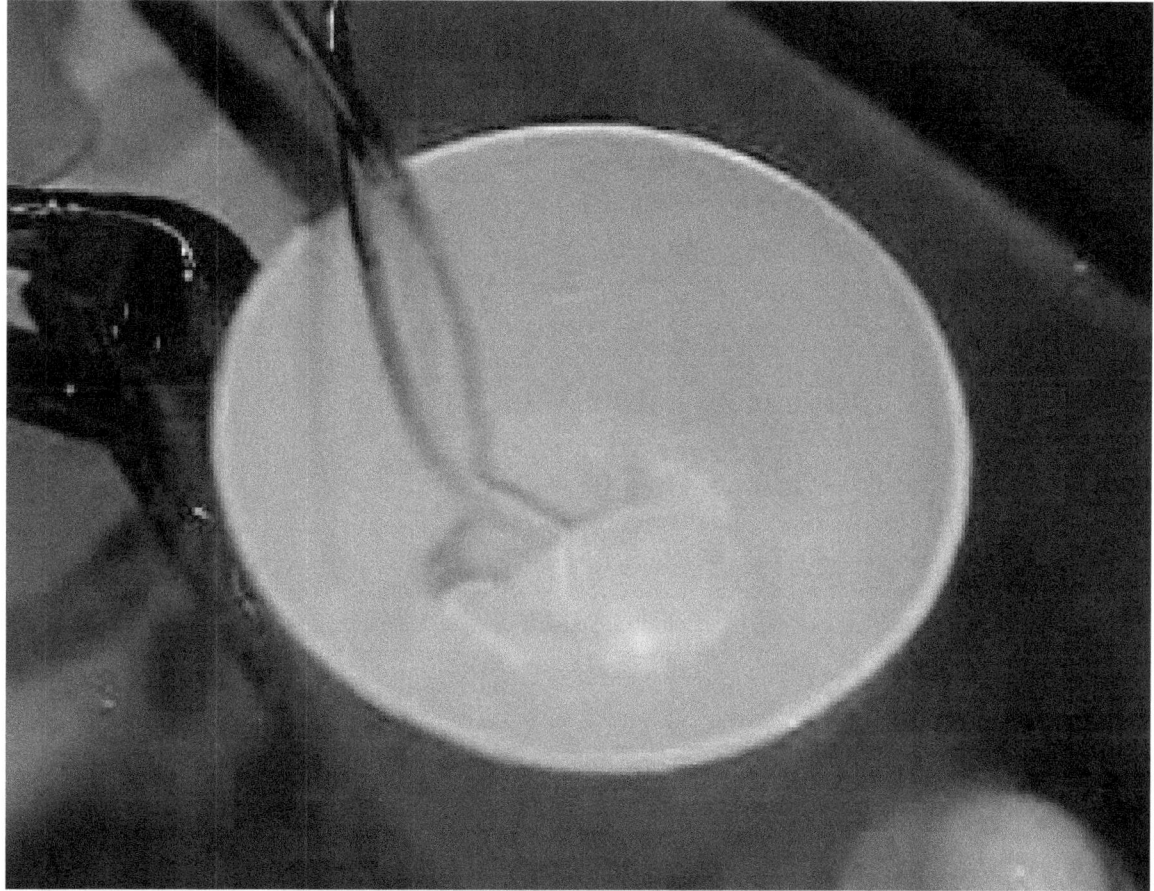

Air balloon shrinks after submersion in liquid nitrogen

Main article: Thermodynamic temperature

The symbol used to represent *temperature* in equations is T with SI units of kelvins.

The speed of a gas particle is proportional to its absolute temperature. The volume of the balloon in the video shrinks when the trapped gas particles slow down with the addition of extremely cold nitrogen. The temperature of any physical system is related to the motions of the particles (molecules and atoms) which make up the [gas] system.[12] In statistical mechanics, temperature is the measure of the average kinetic energy stored in a particle. The methods of storing this energy are dictated by the degrees of freedom of the particle itself (energy modes). Kinetic energy added (endothermic process) to gas particles by way of collisions produces linear, rotational, and vibrational motion. In contrast, a molecule in a solid can only increase its vibrational modes with the addition of heat as the lattice crystal structure prevents both linear and rotational motions. These heated gas molecules have a greater speed range which constantly varies due to constant collisions with other particles. The speed range can be described by the Maxwell–Boltzmann distribution. Use of this distribution implies ideal gases near thermodynamic equilibrium for the system of particles being considered.

11.4.3   Specific volume

Main article: Specific volume

The symbol used to represent *specific volume* in equations is **"v"** with SI units of cubic meters per kilogram.

See also: Gas volume

The symbol used to represent **volume** in equations is **"V"** with SI units of cubic meters.

When performing a thermodynamic analysis, it is typical to speak of intensive and extensive properties. Properties which depend on the amount of gas (either by mass or volume) are called *extensive* properties, while properties that do not depend on the amount of gas are called *intensive* properties. **Specific volume** is an example of an *intensive* property because it is the ratio of volume occupied by a *unit of mass* of a gas that is identical throughout a system at equilibrium.[13] 1000 atoms a gas occupy the same space as any other 1000 atoms for any given temperature and pressure. This concept is easier to visualize for solids such as iron which are incompressible compared to gases. Since a gas fills any container in which it is placed, **volume** is an *extensive property*.

11.4.4   Density

Main article: Density

The symbol used to represent **density** in equations is ρ (rho) with SI units of kilograms per cubic meter. This term is the reciprocal of specific volume.

Since gas molecules can move freely within a container, their mass is normally characterized by **density**. Density is the amount of mass per unit volume of a substance, or the inverse of specific volume. For gases, the density can vary over a wide range because the particles are free to move closer together when constrained by pressure or volume. This variation of density is referred to as compressibility. Like pressure and temperature, density is a state variable of a gas and the change in density during any process is governed by the laws of thermodynamics. For a static gas, the density is the same throughout the entire container. Density is therefore a scalar quantity. It can be shown by **kinetic theory** that the density is *inversely* proportional to the size of the container in which a fixed mass of gas is confined. In this case of a fixed mass, the density decreases as the volume increases.

11.5   Microscopic

If one could observe a gas under a powerful microscope, one would see a collection of particles (molecules, atoms, ions, electrons, etc.) without any definite shape or volume that are in more or less random motion. These neutral gas particles only change direction when they collide with another particle or with the sides of the container. In an ideal gas, these collisions are perfectly elastic. This particle or microscopic view of a gas is described by the Kinetic-molecular theory. The assumptions behind this theory can be found in the postulates section of Kinetic Theory.

11.5.1   Kinetic theory

Main article: Kinetic theory of gases

Kinetic theory provides insight into the macroscopic properties of gases by considering their molecular composition and motion. Starting with the definitions of momentum and kinetic energy,[14] one can use the conservation of momentum and geometric relationships of a cube to relate macroscopic system properties of temperature and pressure to the microscopic property of kinetic energy per molecule. The theory provides averaged values for these two properties.

The theory also explains how the gas system responds to change. For example, as a gas is heated from absolute zero, when it is (in theory) perfectly still, its internal energy (temperature) is increased. As a gas is heated, the particles speed up and its temperature rises. This results in greater numbers of collisions with the container per unit time due to the higher particle speeds associated with elevated temperatures. The pressure increases in proportion to the number of collisions per unit time.

11.5.2 Brownian motion

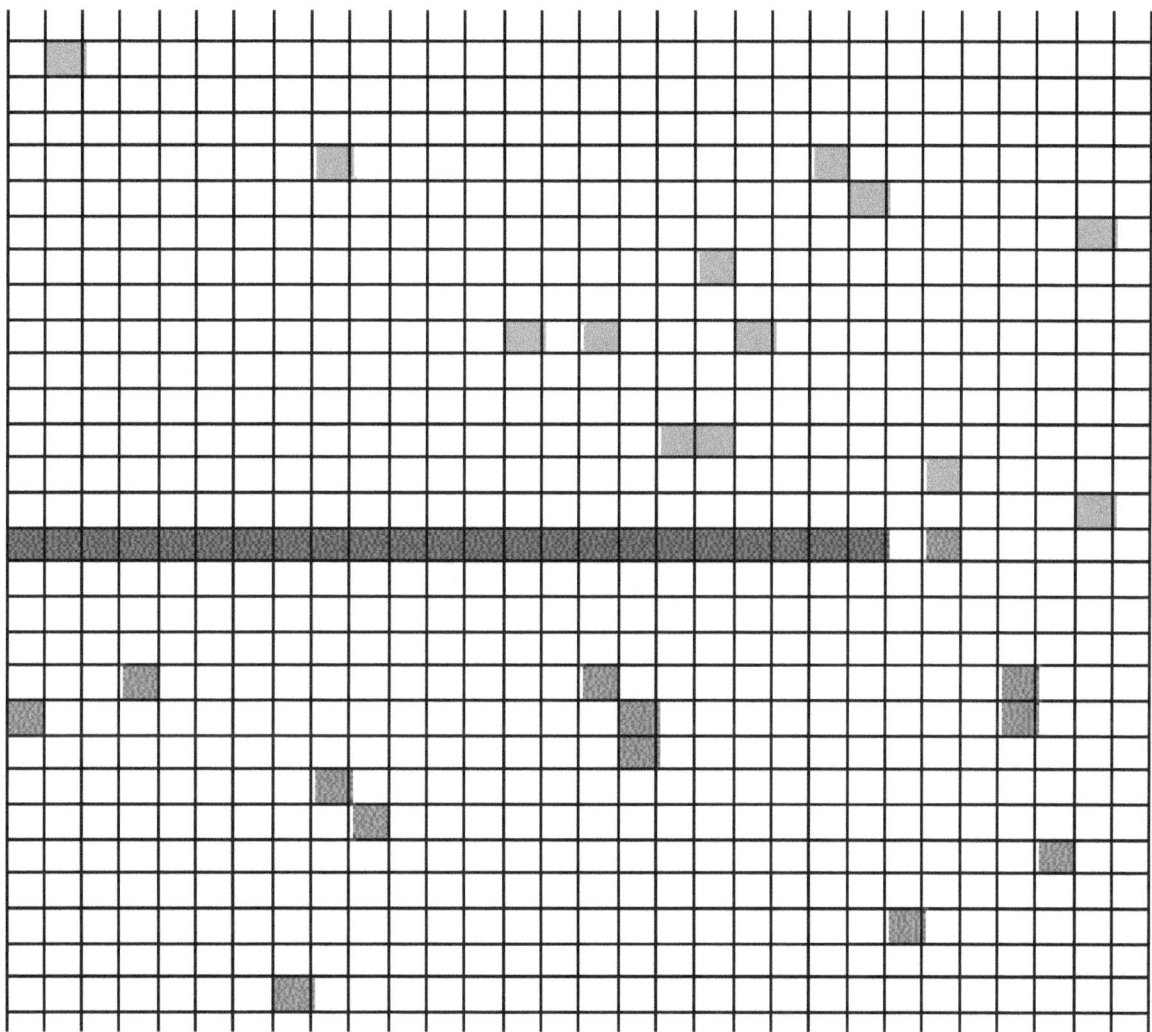

Random motion of gas particles results in diffusion.

Main article: Brownian motion

Brownian motion is the mathematical model used to describe the random movement of particles suspended in a fluid. The gas particle animation, using pink and green particles, illustrates how this behavior results in the spreading out of gases (entropy). These events are also described by particle theory.

Since it is at the limit of (or beyond) current technology to observe individual gas particles (atoms or molecules), only theoretical calculations give suggestions about how they move, but their motion is different from Brownian motion because Brownian motion involves a smooth drag due to the frictional force of many gas molecules, punctuated by violent collisions of an individual (or several) gas molecule(s) with the particle. The particle (generally consisting of millions or billions of atoms) thus moves in a jagged course, yet not so jagged as would be expected if an individual gas molecule were examined.

11.5.3   Intermolecular forces

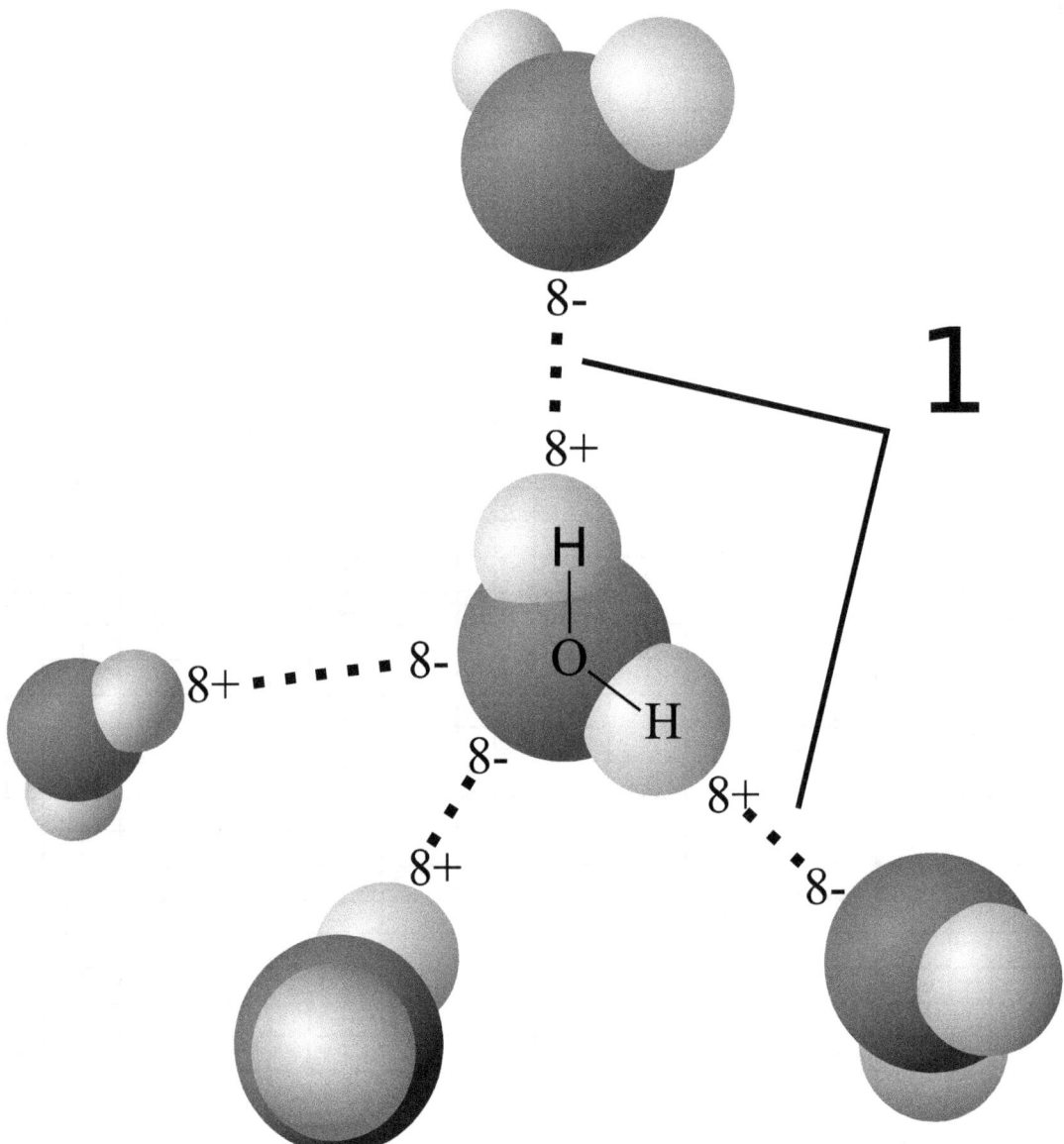

When gases are compressed, intermolecular forces like those shown here start to play a more active role.

Main articles: van der Waals force and Intermolecular force

As discussed earlier, momentary attractions (or repulsions) between particles have an effect on gas dynamics. In physical chemistry, the name given to these intermolecular forces is *van der Waals force*. These forces play a key role in determining physical properties of a gas such as viscosity and flow rate (see physical characteristics section). Ignoring these forces in certain conditions (see Kinetic-molecular theory) allows a real gas to be treated like an ideal gas. This assumption allows the use of ideal gas laws which greatly simplifies calculations.

Proper use of these gas relationships requires the Kinetic-molecular theory (KMT). When gas particles possess a magnetic charge or Intermolecular force they gradually influence one another as the spacing between them is reduced (the hydrogen bond model illustrates one example). In the absence of any charge, at some point when the spacing between gas particles is greatly reduced they can no longer avoid collisions between themselves at normal gas temperatures. Another case for increased collisions among gas particles would include a fixed volume of gas, which upon heating would contain very fast particles. *This means that these ideal equations provide reasonable results* except *for extremely high pressure (compressible) or high temperature (ionized) conditions.* Notice that all of these excepted conditions allow energy transfer to take place within the gas system. The absence of these internal transfers is what is referred to as ideal conditions in which the energy exchange occurs only at the boundaries of the system. Real gases experience some of these collisions and intermolecular forces. When these collisions are statistically negligible (incompressible), results from these ideal equations are still meaningful. If the gas particles are compressed into close proximity they behave more like a liquid (see fluid dynamics).

11.6   Simplified models

Main article: Equation of state

An *equation of state* (for gases) is a mathematical model used to roughly describe or predict the state properties of a gas. At present, there is no single equation of state that accurately predicts the properties of all gases under all conditions. Therefore, a number of much more accurate equations of state have been developed for gases in specific temperature and pressure ranges. The "gas models" that are most widely discussed are "perfect gas", "ideal gas" and "real gas". Each of these models has its own set of assumptions to facilitate the analysis of a given thermodynamic system.[15] Each successive model expands the temperature range of coverage to which it applies.

11.6.1   Ideal and perfect gas models

Main article: Perfect gas

The equation of state for an ideal or perfect gas is the ideal gas law and reads

$$PV = nRT,$$

where P is the pressure, V is the volume, n is amount of gas (in mol units), R is the universal gas constant, 8.314 J/(mol K), and T is the temperature. Written this way, it is sometimes called the "chemist's version", since it emphasizes the number of molecules n. It can also be written as

$$P = \rho R_s T,$$

where R_s is the specific gas constant for a particular gas, in units J/(kg K), and ρ = m/V is density. This notation is the "gas dynamicist's" version, which is more practical in modeling of gas flows involving acceleration without chemical reactions.

The ideal gas law does not make an assumption about the specific heat of a gas. In the most general case, the specific heat is a function of both temperature and pressure. If the pressure-dependence is neglected (and possibly the temperature-dependence as well) in a particular application, sometimes the gas is said to be a perfect gas, although the exact assumptions may vary depending on the author and/or field of science.

For an ideal gas, the ideal gas law applies without restrictions on the specific heat. An ideal gas is a simplified "real gas" with the assumption that the compressibility factor Z is set to 1 meaning that this pneumatic ratio remains constant. A compressibility factor of one also requires the four state variables to follow the ideal gas law.

This approximation is more suitable for applications in engineering although simpler models can be used to produce a "ball-park" range as to where the real solution should lie. An example where the "ideal gas approximation" would be suitable would be inside a combustion chamber of a jet engine.[16] It may also be useful to keep the elementary reactions and chemical dissociations for calculating emissions.

11.6.2   Real gas

21 April 1990 eruption of Mount Redoubt, Alaska, illustrating real gases not in thermodynamic equilibrium.

Main article: Real gas

Each one of the assumptions listed below adds to the complexity of the problem's solution. As the density of a gas increases with rising pressure, the intermolecular forces play a more substantial role in gas behavior which results in the ideal gas law no longer providing "reasonable" results. At the upper end of the engine temperature ranges (e.g. combustor sections – 1300 K), the complex fuel particles absorb internal energy by means of rotations and vibrations that cause their specific heats to vary from those of diatomic molecules and noble gases. At more than double that temperature, electronic excitation and dissociation of the gas particles begins to occur causing the pressure to adjust to a greater number of particles (transition from gas to plasma).[17] Finally, all of the thermodynamic processes were presumed to describe uniform gases whose velocities varied according to a fixed distribution. Using a non-equilibrium situation implies the flow field must be characterized in some manner to enable a solution. One of the first attempts to expand the boundaries of the ideal gas law was to include coverage for different thermodynamic processes by adjusting the equation to read $pV^n = constant$ and then varying the n through different values such as the specific heat ratio, γ.

Real gas effects include those adjustments made to account for a greater range of gas behavior:

- Compressibility effects (Z allowed to vary from 1.0)

- Variable heat capacity (specific heats vary with temperature)

- Van der Waals forces (related to compressibility, can substitute other equations of state)

- Non-equilibrium thermodynamic effects

- Issues with molecular dissociation and elementary reactions with variable composition.

For most applications, such a detailed analysis is excessive. Examples where "Real Gas effects" would have a significant impact would be on the Space Shuttle re-entry where extremely high temperatures and pressures are present or the gases produced during geological events as in the image of the 1990 eruption of Mount Redoubt.

11.7 Historical synthesis

See also: Gas laws

11.7.1 Boyle's law

Main article: Boyle's law

Boyle's Law was perhaps the first expression of an equation of state. In 1662 Robert Boyle performed a series of experiments employing a J-shaped glass tube, which was sealed on one end. Mercury was added to the tube, trapping a fixed quantity of air in the short, sealed end of the tube. Then the volume of gas was carefully measured as additional mercury was added to the tube. The pressure of the gas could be determined by the difference between the mercury level in the short end of the tube and that in the long, open end. The image of Boyle's Equipment shows some of the exotic tools used by Boyle during his study of gases.

Through these experiments, Boyle noted that the pressure exerted by a gas held at a constant temperature varies inversely with the volume of the gas.[18] For example, if the volume is halved, the pressure is doubled; and if the volume is doubled, the pressure is halved. Given the inverse relationship between pressure and volume, the product of pressure (P) and volume (V) is a constant (k) for a given mass of confined gas as long as the temperature is constant. Stated as a formula, thus is:

$$PV = k$$

Because the before and after volumes and pressures of the fixed amount of gas, where the before and after temperatures are the same both equal the constant k, they can be related by the equation:

$$P_1 V_1 = P_2 V_2.$$

11.7.2 Charles's Law

Main article: Charles's law

In 1787, the French physicist and balloon pioneer, Jacques Charles, found that oxygen, nitrogen, hydrogen, carbon dioxide, and air expand to the same extent over the same 80 kelvin interval. He noted that, for an ideal gas at constant pressure, the volume is directly proportional to its temperature:

$$\frac{V_1}{T_1} = \frac{V_2}{T_2}$$

11.7.3 Gay-Lussac's Law

Main article: Gay-Lussac's Law

In 1802, Joseph Louis Gay-Lussac published results of similar, though more extensive experiments.[19] Gay-Lussac credited Charle's earlier work by naming the law in his honor. Gay-Lussac himself is credited with the law describing pressure, which he found in 1809. It states that the pressure exerted on a container's sides by an ideal gas is proportional to its temperature.

$$\frac{P_1}{T_1} = \frac{P_2}{T_2}$$

11.7.4 Avogadro's law

Main article: Avogadro's law

In 1811, Amedeo Avogadro verified that equal volumes of pure gases contain the same number of particles. His theory was not generally accepted until 1858 when another Italian chemist Stanislao Cannizzaro was able to explain non-ideal exceptions. For his work with gases a century prior, the number that bears his name Avogadro's constant represents the number of atoms found in 12 grams of elemental carbon-12 (6.022×10^{23} mol^{-1}). This specific number of gas particles, at standard temperature and pressure (ideal gas law) occupies 22.40 liters, which is referred to as the molar volume.

Avogadro's law states that the volume occupied by an ideal gas is proportional to the number of moles (or molecules) present in the container. This gives rise to the molar volume of a gas, which at STP is 22.4 dm^3 (or litres). The relation is given by

$$\frac{V_1}{n_1} = \frac{V_2}{n_2}$$

where n is equal to the number of moles of gas (the number of molecules divided by Avogadro's Number).

11.7.5 Dalton's law

Main article: Dalton's law

In 1801, John Dalton published the **Law of Partial Pressures** from his work with ideal gas law relationship: The pressure of a mixture of non reactive gases is equal to the sum of the pressures of all of the constituent gases alone. Mathematically, this can be represented for *n* species as:

Pressuretotal = Pressure$_1$ + Pressure$_2$ + ... + Pressuren

The image of Dalton's journal depicts symbology he used as shorthand to record the path he followed. Among his key journal observations upon mixing unreactive "elastic fluids" (gases) were the following:[20]

- Unlike liquids, heavier gases did not drift to the bottom upon mixing.

- Gas particle identity played no role in determining final pressure (they behaved as if their size was negligible).

11.8 Special topics

11.8.1 Compressibility

Main article: Compressibility factor

Thermodynamicists use this factor (Z) to alter the ideal gas equation to account for compressibility effects of real gases. This factor represents the ratio of actual to ideal specific volumes. It is sometimes referred to as a "fudge-factor" or correction to expand the useful range of the ideal gas law for design purposes. *Usually* this Z value is very close to unity. The compressibility factor image illustrates how Z varies over a range of very cold temperatures.

11.8.2 Reynolds number

Main article: Reynolds number

In fluid mechanics, the Reynolds number is the ratio of inertial forces ($v s\varrho$) to viscous forces (μ/L). It is one of the most important dimensionless numbers in fluid dynamics and is used, usually along with other dimensionless numbers, to provide a criterion for determining dynamic similitude. As such, the Reynolds number provides the link between modeling results (design) and the full-scale actual conditions. It can also be used to characterize the flow.

11.8.3 Viscosity

Main article: Viscosity

Viscosity, a physical property, is a measure of how well adjacent molecules stick to one another. A solid can withstand a shearing force due to the strength of these sticky intermolecular forces. A fluid will continuously deform when subjected to a similar load. While a gas has a lower value of viscosity than a liquid, it is still an observable property. If gases had no viscosity, then they would not stick to the surface of a wing and form a boundary layer. A study of the delta wing in the Schlieren image reveals that the gas particles stick to one another (see Boundary layer section).

11.8.4 Turbulence

Main article: Turbulence

In fluid dynamics, **turbulence** or turbulent flow is a flow regime characterized by chaotic, stochastic property changes. This includes low momentum diffusion, high momentum convection, and rapid variation of pressure and velocity in space and time. The Satellite view of weather around Robinson Crusoe Islands illustrates just one example.

11.8.5 Boundary layer

Main article: Boundary layer

Particles will, in effect, "stick" to the surface of an object moving through it. This layer of particles is called the **boundary layer**. At the surface of the object, it is essentially static due to the friction of the surface. The object, with its boundary layer is effectively the new shape of the object that the rest of the molecules "see" as the object approaches. This boundary layer *can* separate from the surface, essentially creating a new surface and completely changing the flow path. The classical example of this is a stalling airfoil. The delta wing image clearly shows the boundary layer thickening as the gas flows from right to left along the leading edge.

11.8.6 Maximum entropy principle

Main article: Principle of maximum entropy

As the total number of degrees of freedom approaches infinity, the system will be found in the macrostate that corresponds to the highest multiplicity. In order to illustrate this principle, observe the skin temperature of a frozen metal bar. Using a thermal image of the skin temperature, note the temperature distribution on the surface. This initial observation of temperature represents a "microstate." At some future time, a second observation of the skin temperature produces a second microstate. By continuing this observation process, it is possible to produce a series of microstates that illustrate the thermal history of the bar's surface. Characterization of this historical series of microstates is possible by choosing the macrostate that successfully classifies them all into a single grouping.

11.8.7   Thermodynamic equilibrium

Main article: Thermodynamic equilibrium

When energy transfer ceases from a system, this condition is referred to as thermodynamic equilibrium. Usually this condition implies the system and surroundings are at the same temperature so that heat no longer transfers between them. It also implies that external forces are balanced (volume does not change), and all chemical reactions within the system are complete. The timeline varies for these events depending on the system in question. A container of ice allowed to melt at room temperature takes hours, while in semiconductors the heat transfer that occurs in the device transition from an on to off state could be on the order of a few nanoseconds.

11.9   See also

- Quasi-solid
- Greenhouse gas
- Natural gas
- Volcanic gas
- Breathing
- Wind

11.10   Notes

[1] This early 20th century discussion infers what is regarded as the plasma state. See page 137 of American Chemical Society, Faraday Society, Chemical Society (Great Britain) *The Journal of physical chemistry, Volume 11* Cornell (1907).

[2] The work by T. Zelevinski provides another link to latest research about Strontium in this new field of study. See Tanya Zelevinsky (2009). "84Sr—just right for forming a Bose-Einstein condensate". *Physics* 2: 94. Bibcode:2009PhyOJ...2...94Z. doi:10.1103/physics.2.94.

[3] for links material on the Bose–Einstein condensate see Quantum Gas Microscope Offers Glimpse Of Quirky Ultracold Atoms. ScienceDaily. 4 November 2009.

[4] J. B. van Helmont, *Ortus medicinae.* ... (Amsterdam, (Netherlands): Louis Elzevir, 1652 (first edition: 1648)). The word "gas" first appears on page 58, where he mentions: "... Gas (meum scil. inventum) ..." (... gas (namely, my discovery) ...). On page 59, he states: "... in nominis egestate, halitum illum, Gas vocavi, non longe a Chao ..." (... in need of a name, I called this vapor "gas", not far from "chaos" ...)

[5] Harper, Douglas. "gas". *Online Etymology Dictionary.*

[6] Draper, John William (1861). *A textbook on chemistry.* New York: Harper and Sons. p. 178.

[7] The authors make the connection between molecular forces of metals and their corresponding physical properties. By extension, this concept would apply to gases as well, though not universally. Cornell (1907) pp. 164–5.

[8] One noticeable exception to this physical property connection is conductivity which varies depending on the state of matter (ionic compounds in water) as described by Michael Faraday in the 1833 when he noted that ice does not conduct a current. See page 45 of John Tyndall's *Faraday as a Discoverer* (1868).

[9] John S. Hutchinson (2008). *Concept Development Studies in Chemistry.* p. 67.

[10] Anderson, p.501

[11] J. Clerk Maxwell (1904). *Theory of Heat.* Mineola: Dover Publications. pp. 319–20. ISBN 0-486-41735-2.

[12] See pages 137–8 of Society, Cornell (1907).

[13] Kenneth Wark (1977). *Thermodynamics* (3 ed.). McGraw-Hill. p. 12. ISBN 0-07-068280-1.

[14] For assumptions of Kinetic Theory see McPherson, pp.60–61

[15] Anderson, pp. 289–291

[16] John, p.205

[17] John, pp. 247–56

[18] McPherson, pp.52–55

[19] McPherson, pp.55–60

[20] John P. Millington (1906). *John Dalton.* pp. 72, 77–78.

11.11 References

- Anderson, John D. (1984). *Fundamentals of Aerodynamics.* McGraw-Hill Higher Education. ISBN 0-07-001656-9.

- John, James (1984). *Gas Dynamics.* Allyn and Bacon. ISBN 0-205-08014-6.

- McPherson, William and Henderson, William (1917). *An Elementary study of chemistry.*

11.12 Further reading

- Philip Hill and Carl Peterson. *Mechanics and Thermodynamics of Propulsion: Second Edition* Addison-Wesley, 1992. ISBN 0-201-14659-2

- National Aeronautics and Space Administration (NASA). Animated Gas Lab. Accessed February 2008.

- Georgia State University. HyperPhysics. Accessed February 2008.

- Antony Lewis WordWeb. Accessed February 2008.

- Northwestern Michigan College The Gaseous State. Accessed February 2008.

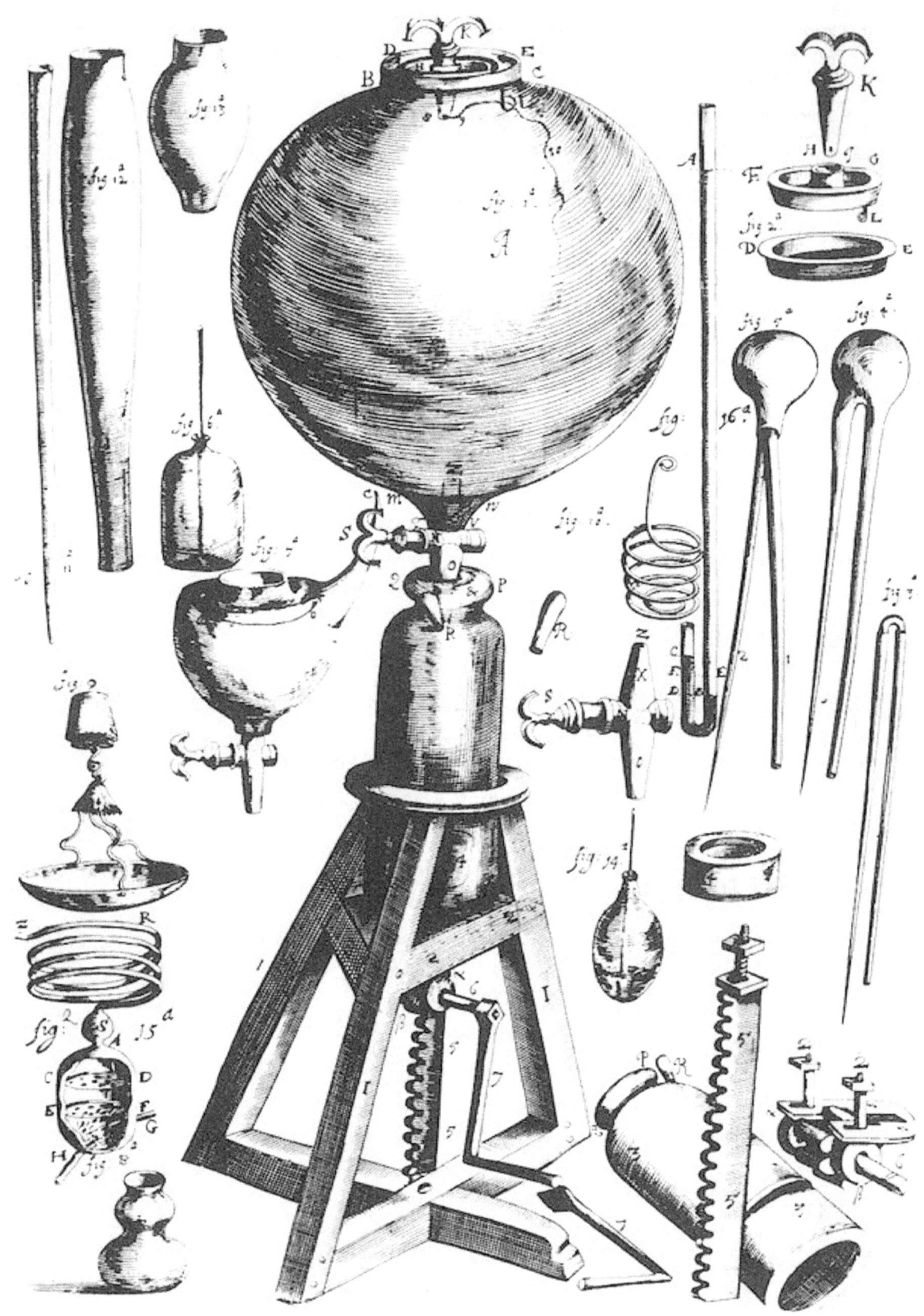

Boyle's equipment.

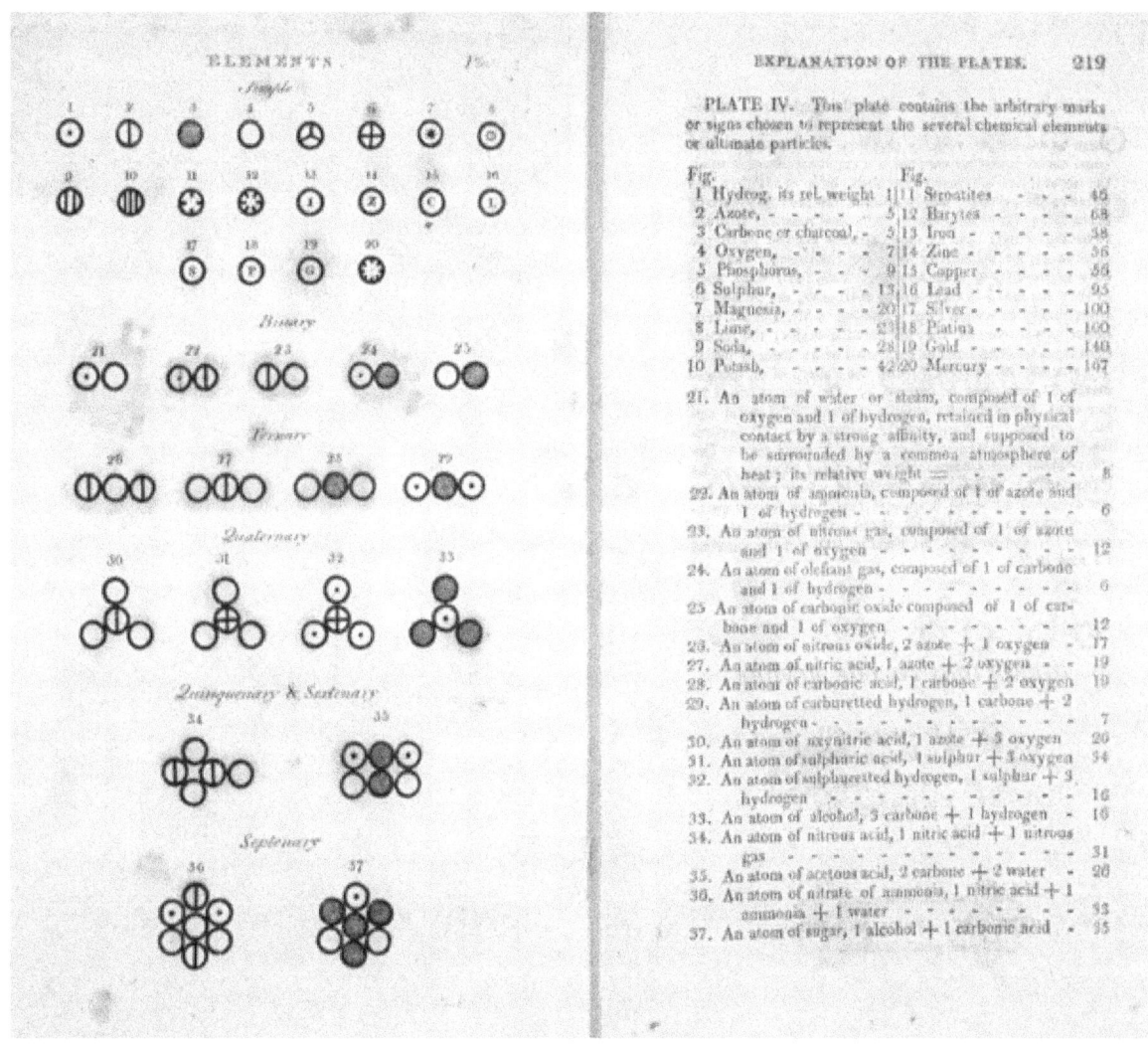

Dalton's notation.

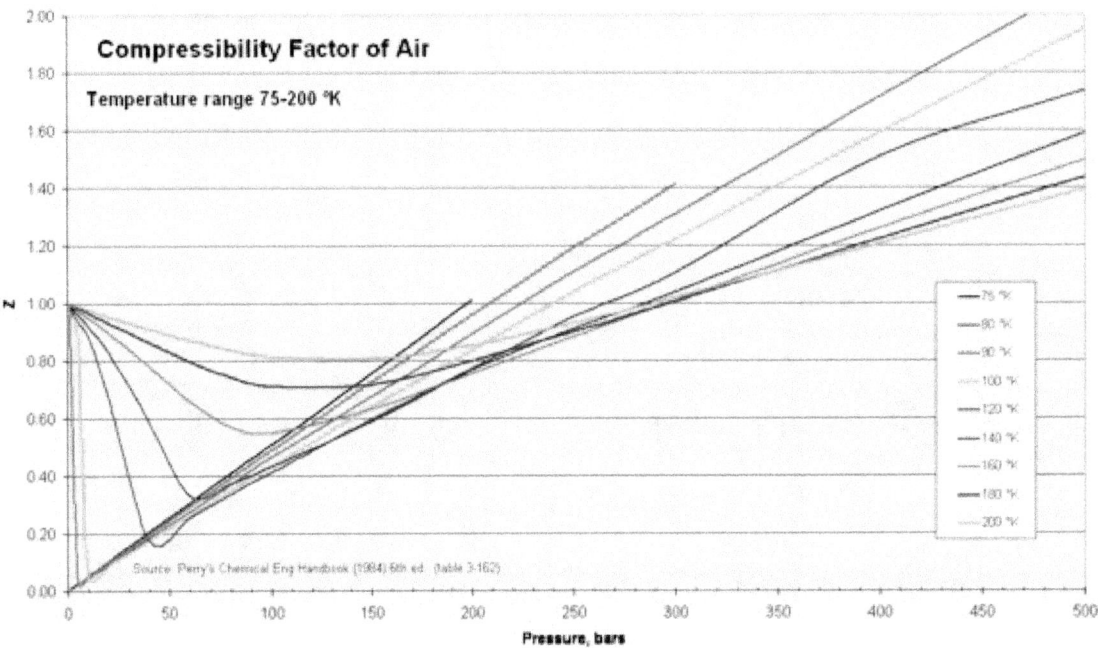

Compressibility factors for air.

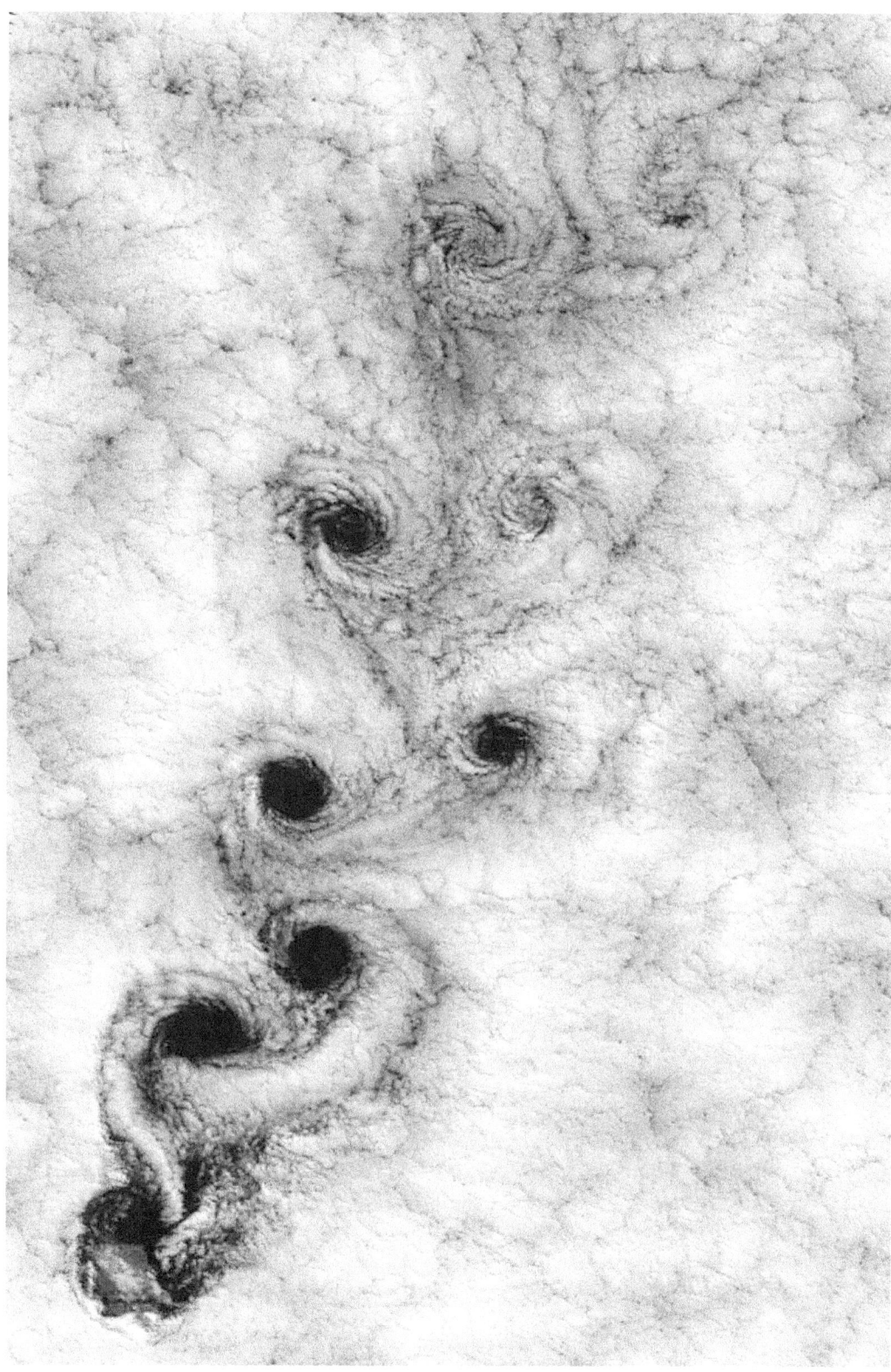

Satellite view of weather pattern in vicinity of Robinson Crusoe Islands on 15 September 1999, shows a unique turbulent cloud pattern called a Kármán vortex street

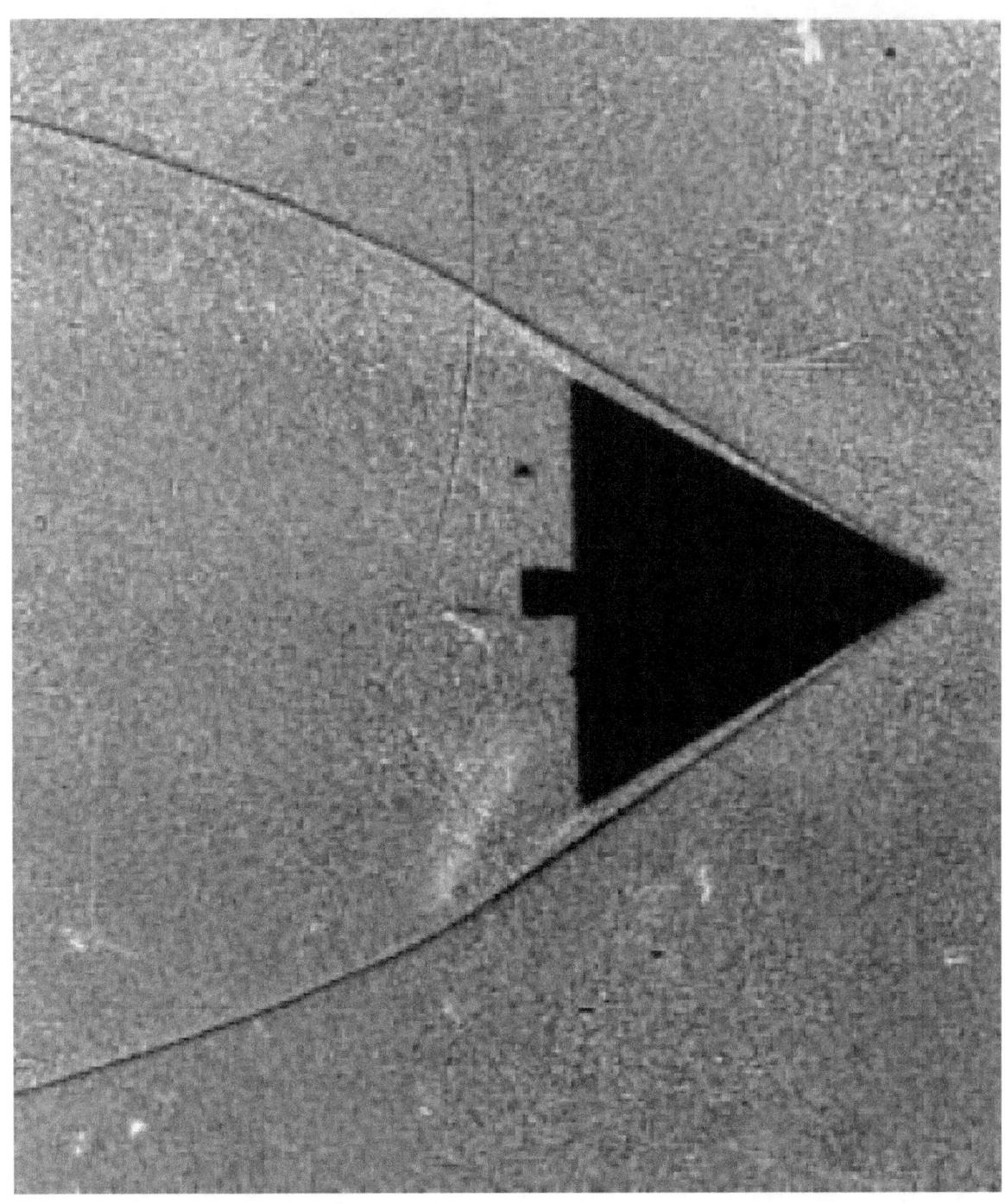

Delta wing in wind tunnel. The shadows form as the indices of refraction change within the gas as it compresses on the leading edge of this wing.

Chapter 12

Plasma (physics)

For other uses, see Plasma.

Plasma (from Greek πλάσμα, "anything formed"[1]) is one of the four fundamental states of matter, the others being solid, liquid, and gas. A plasma has properties unlike those of the other states.

A plasma can be created by heating a gas or subjecting it to a strong electromagnetic field applied with a laser or microwave generator. This decreases or increases the number of electrons, creating positive or negative charged particles called ions,[2] and is accompanied by the dissociation of molecular bonds, if present.[3]

The presence of a significant number of charge carriers makes plasma electrically conductive so that it responds strongly to electromagnetic fields. Like gas, plasma does not have a definite shape or a definite volume unless enclosed in a container. Unlike gas, under the influence of a magnetic field, it may form structures such as filaments, beams and double layers.

Plasma is the most abundant form of ordinary matter in the Universe (the only matter known to exist for sure, the more abundant dark matter is hypothetical and may or may not be explained by ordinary matter), most of which is in the rarefied intergalactic regions, particularly the intracluster medium, and in stars, including the Sun.[4][5] A common form of plasmas on Earth is seen in neon signs.

Much of the understanding of plasmas has come from the pursuit of controlled nuclear fusion and fusion power, for which plasma physics provides the scientific basis.

12.1 Properties and parameters

12.1.1 Definition

Plasma is loosely described as an electrically neutral medium of unbound positive and negative particles (i.e. the overall charge of a plasma is roughly zero). It is important to note that although they are unbound, these particles are not 'free' in the sense of not experiencing forces. When the charges move, they generate electric currents with magnetic fields, and as a result, they are affected by each other's fields. This governs their collective behavior with many degrees of freedom.[3][7] A definition can have three criteria:[8][9]

1. **The plasma approximation**: Charged particles must be close enough together that each particle influences many nearby charged particles, rather than just interacting with the closest particle (these collective effects are a distinguishing feature of a plasma). The plasma approximation is valid when the number of charge carriers within the sphere of influence (called the *Debye sphere* whose radius is the Debye screening length) of a particular particle is higher than unity to provide collective behavior of the charged particles. The average number of particles in the Debye sphere is given by the plasma parameter, "Λ" (the Greek uppercase letter Lambda).

2. **Bulk interactions**: The Debye screening length (defined above) is short compared to the physical size of the

Artist's rendition of the Earth's plasma fountain, showing oxygen, helium, and hydrogen ions that gush into space from regions near the Earth's poles. The faint yellow area shown above the north pole represents gas lost from Earth into space; the green area is the aurora borealis, where plasma energy pours back into the atmosphere.[6]

plasma. This criterion means that interactions in the bulk of the plasma are more important than those at its edges, where boundary effects may take place. When this criterion is satisfied, the plasma is quasineutral.

3. **Plasma frequency**: The electron plasma frequency (measuring plasma oscillations of the electrons) is large compared to the electron-neutral collision frequency (measuring frequency of collisions between electrons and neutral particles). When this condition is valid, electrostatic interactions dominate over the processes of ordinary gas kinetics.

12.1.2 Ranges of parameters

Plasma parameters can take on values varying by many orders of magnitude, but the properties of plasmas with apparently disparate parameters may be very similar (see plasma scaling). The following chart considers only conventional atomic plasmas and not exotic phenomena like quark gluon plasmas:

12.1.3 Degree of ionization

For plasma to exist, ionization is necessary. The term "plasma density" by itself usually refers to the "electron density", that is, the number of free electrons per unit volume. The degree of ionization of a plasma is the proportion of atoms that have lost or gained electrons, and is controlled mostly by the temperature. Even a partially ionized gas in which as little as 1% of the particles are ionized can have the characteristics of a plasma (i.e., response to magnetic fields and high electrical conductivity). The degree of ionization, α , is defined as $\alpha = \frac{n_i}{n_i+n_n}$, where n_i is the number density of ions and n_n is the number density of neutral atoms. The *electron density* is related to this by the average charge state $\langle Z \rangle$ of the ions through $n_e = \langle Z \rangle n_i$, where n_e is the number density of electrons.

12.1.4 Temperatures

See also: Nonthermal plasma

Plasma temperature is commonly measured in kelvins or electronvolts and is, informally, a measure of the thermal kinetic energy per particle. Very high temperatures are usually needed to sustain ionization, which is a defining feature of a plasma. The degree of plasma ionization is determined by the electron temperature relative to the ionization energy (and more weakly by the density), in a relationship called the Saha equation. At low temperatures, ions and electrons tend to recombine into bound states—atoms[12]—and the plasma will eventually become a gas.

In most cases the electrons are close enough to thermal equilibrium that their temperature is relatively well-defined, even when there is a significant deviation from a Maxwellian energy distribution function, for example, due to UV radiation, energetic particles, or strong electric fields. Because of the large difference in mass, the electrons come to thermodynamic equilibrium amongst themselves much faster than they come into equilibrium with the ions or neutral atoms. For this reason, the ion temperature may be very different from (usually lower than) the electron temperature. This is especially common in weakly ionized technological plasmas, where the ions are often near the ambient temperature.

Thermal vs. nonthermal plasmas

Based on the relative temperatures of the electrons, ions and neutrals, plasmas are classified as "thermal" or "non-thermal". Thermal plasmas have electrons and the heavy particles at the same temperature, i.e. they are in thermal equilibrium with each other. Nonthermal plasmas on the other hand have the ions and neutrals at a much lower temperature (sometimes room temperature), whereas electrons are much "hotter" ($T_e \gg T_n$).

A plasma is sometimes referred to as being "hot" if it is nearly fully ionized, or "cold" if only a small fraction (for example 1%) of the gas molecules are ionized, but other definitions of the terms "hot plasma" and "cold plasma" are common. Even in a "cold" plasma, the electron temperature is still typically several thousand degrees Celsius. Plasmas utilized in

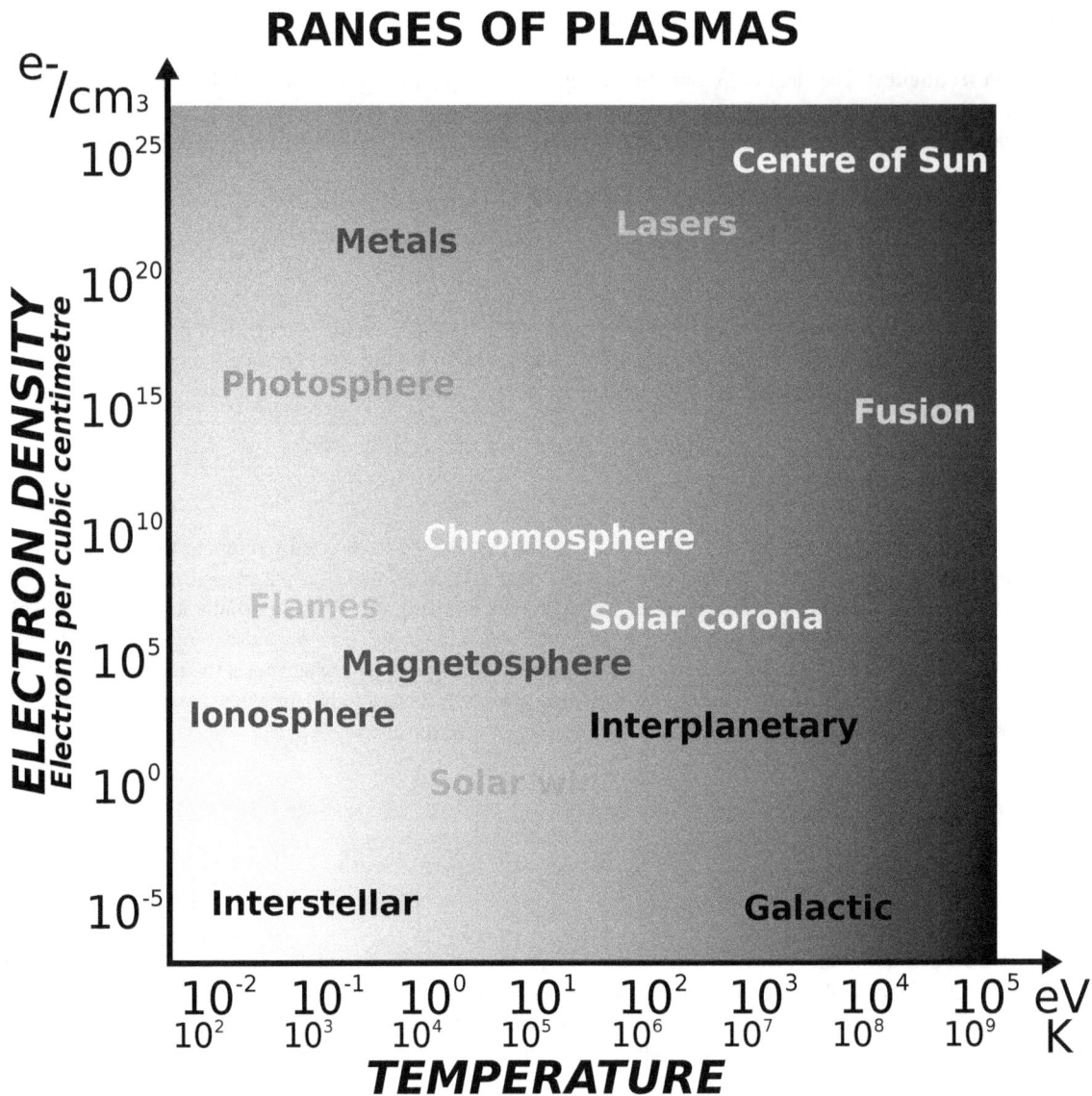

Range of plasmas. *Density increases upwards, temperature increases towards the right. The free electrons in a metal may be considered an electron plasma.*[10]

"plasma technology" ("technological plasmas") are usually cold plasmas in the sense that only a small fraction of the gas molecules are ionized.

12.1.5   Plasma potential

Since plasmas are very good electrical conductors, electric potentials play an important role. The potential as it exists on average in the space between charged particles, independent of the question of how it can be measured, is called the "plasma potential", or the "space potential". If an electrode is inserted into a plasma, its potential will generally lie considerably below the plasma potential due to what is termed a Debye sheath. The good electrical conductivity of plasmas makes their electric fields very small. This results in the important concept of "quasineutrality", which says the density of negative charges is approximately equal to the density of positive charges over large volumes of the plasma ($n_e = \langle Z \rangle n_i$

), but on the scale of the Debye length there can be charge imbalance. In the special case that *double layers* are formed, the charge separation can extend some tens of Debye lengths.

The magnitude of the potentials and electric fields must be determined by means other than simply finding the net charge density. A common example is to assume that the electrons satisfy the Boltzmann relation:

$$n_e \propto e^{e\Phi/k_B T_e}.$$

Differentiating this relation provides a means to calculate the electric field from the density:

$$\vec{E} = (k_B T_e/e)(\nabla n_e/n_e).$$

It is possible to produce a plasma that is not quasineutral. An electron beam, for example, has only negative charges. The density of a non-neutral plasma must generally be very low, or it must be very small, otherwise it will be dissipated by the repulsive electrostatic force.

In astrophysical plasmas, Debye screening prevents electric fields from directly affecting the plasma over large distances, i.e., greater than the Debye length. However, the existence of charged particles causes the plasma to generate, and be affected by, magnetic fields. This can and does cause extremely complex behavior, such as the generation of plasma double layers, an object that separates charge over a few tens of Debye lengths. The dynamics of plasmas interacting with external and self-generated magnetic fields are studied in the academic discipline of magnetohydrodynamics.

12.1.6   Magnetization

Plasma with a magnetic field strong enough to influence the motion of the charged particles is said to be magnetized. A common quantitative criterion is that a particle on average completes at least one gyration around the magnetic field before making a collision, i.e., $\omega_{ce}/v_{coll} > 1$, where ω_{ce} is the "electron gyrofrequency" and v_{coll} is the "electron collision rate". It is often the case that the electrons are magnetized while the ions are not. Magnetized plasmas are *anisotropic*, meaning that their properties in the direction parallel to the magnetic field are different from those perpendicular to it. While electric fields in plasmas are usually small due to the high conductivity, the electric field associated with a plasma moving in a magnetic field is given by $\mathbf{E} = -v \times \mathbf{B}$ (where $\mathbf{E}$ is the electric field, $\mathbf{v}$ is the velocity, and $\mathbf{B}$ is the magnetic field), and is not affected by Debye shielding.[14]

12.1.7   Comparison of plasma and gas phases

Plasma is often called the *fourth state of matter* after solid, liquids and gases.[15][16] It is distinct from these and other lower-energy states of matter. Although it is closely related to the gas phase in that it also has no definite form or volume, it differs in a number of ways, including the following:

12.2   Common plasmas

Further information: Astrophysical plasma, Interstellar medium and Intergalactic space

Plasmas are by far the most common phase of ordinary matter in the universe, both by mass and by volume.[18] Essentially, all of the visible light from space comes from stars, which are plasmas with a temperature such that they radiate strongly at visible wavelengths. Most of the ordinary (or baryonic) matter in the universe, however, is found in the intergalactic medium, which is also a plasma, but much hotter, so that it radiates primarily as X-rays.

In 1937, Hannes Alfvén argued that if plasma pervaded the universe, it could then carry electric currents capable of generating a galactic magnetic field.[19] After winning the Nobel Prize, he emphasized that:

In order to understand the phenomena in a certain plasma region, it is necessary to map not only the magnetic but also the electric field and the electric currents. Space is filled with a network of currents which transfer energy and momentum over large or very large distances. The currents often pinch to filamentary or surface currents. The latter are likely to give space, as also interstellar and intergalactic space, a cellular structure.[20]

By contrast the current scientific consensus is that about 96% of the total energy density in the universe is not plasma or any other form of ordinary matter, but a combination of cold dark matter and dark energy. Our Sun, and all stars, are made of plasma, much of interstellar space is filled with a plasma, albeit a very sparse one, and intergalactic space too. Even black holes, which are not directly visible, are thought to be fuelled by accreting ionising matter (i.e. plasma),[21] and they are associated with astrophysical jets of luminous ejected plasma,[22] such as M87's jet that extends 5,000 light-years.[23]

In our solar system, interplanetary space is filled with the plasma of the Solar Wind that extends from the Sun out to the heliopause. However, the density of ordinary matter is much higher than average and much higher than that of either dark matter or dark energy. The planet Jupiter accounts for most of the *non*-plasma, only about 0.1% of the mass and 10^{-15}% of the volume within the orbit of Pluto.

Dust and small grains within a plasma will also pick up a net negative charge, so that they in turn may act like a very heavy negative ion component of the plasma (see dusty plasmas).

12.3   Complex plasma phenomena

Although the underlying equations governing plasmas are relatively simple, plasma behavior is extraordinarily varied and subtle: the emergence of unexpected behavior from a simple model is a typical feature of a complex system. Such systems lie in some sense on the boundary between ordered and disordered behavior and cannot typically be described either by simple, smooth, mathematical functions, or by pure randomness. The spontaneous formation of interesting spatial features on a wide range of length scales is one manifestation of plasma complexity. The features are interesting, for example, because they are very sharp, spatially intermittent (the distance between features is much larger than the features themselves), or have a fractal form. Many of these features were first studied in the laboratory, and have subsequently been recognized throughout the universe. Examples of complexity and complex structures in plasmas include:

12.3.1   Filamentation

Striations or string-like structures,[27] also known as birkeland currents, are seen in many plasmas, like the plasma ball, the aurora,[28] lightning,[29] electric arcs, solar flares,[30] and supernova remnants.[31] They are sometimes associated with larger current densities, and the interaction with the magnetic field can form a magnetic rope structure.[32] High power microwave breakdown at atmospheric pressure also leads to the formation of filamentary structures.[33] (See also Plasma pinch)

Filamentation also refers to the self-focusing of a high power laser pulse. At high powers, the nonlinear part of the index of refraction becomes important and causes a higher index of refraction in the center of the laser beam, where the laser is brighter than at the edges, causing a feedback that focuses the laser even more. The tighter focused laser has a higher peak brightness (irradiance) that forms a plasma. The plasma has an index of refraction lower than one, and causes a defocusing of the laser beam. The interplay of the focusing index of refraction, and the defocusing plasma makes the formation of a long filament of plasma that can be micrometers to kilometers in length.[34] One interesting aspect of the filamentation generated plasma is the relatively low ion density due to defocusing effects of the ionized electrons.[35] (See also Filament propagation)

12.3.2   Shocks or double layers

Plasma properties change rapidly (within a few Debye lengths) across a two-dimensional sheet in the presence of a (moving) shock or (stationary) double layer. Double layers involve localized charge separation, which causes a large potential difference across the layer, but does not generate an electric field outside the layer. Double layers separate adjacent plasma regions with different physical characteristics, and are often found in current carrying plasmas. They accelerate both ions and electrons.

12.3.3 Electric fields and circuits

Quasineutrality of a plasma requires that plasma currents close on themselves in electric circuits. Such circuits follow Kirchhoff's circuit laws and possess a resistance and inductance. These circuits must generally be treated as a strongly coupled system, with the behavior in each plasma region dependent on the entire circuit. It is this strong coupling between system elements, together with nonlinearity, which may lead to complex behavior. Electrical circuits in plasmas store inductive (magnetic) energy, and should the circuit be disrupted, for example, by a plasma instability, the inductive energy will be released as plasma heating and acceleration. This is a common explanation for the heating that takes place in the solar corona. Electric currents, and in particular, magnetic-field-aligned electric currents (which are sometimes generically referred to as "Birkeland currents"), are also observed in the Earth's aurora, and in plasma filaments.

12.3.4 Cellular structure

Narrow sheets with sharp gradients may separate regions with different properties such as magnetization, density and temperature, resulting in cell-like regions. Examples include the magnetosphere, heliosphere, and heliospheric current sheet. Hannes Alfvén wrote: "From the cosmological point of view, the most important new space research discovery is probably the cellular structure of space. As has been seen in every region of space accessible to in situ measurements, there are a number of 'cell walls', sheets of electric currents, which divide space into compartments with different magnetization, temperature, density, etc."[36]

12.3.5 Critical ionization velocity

The critical ionization velocity is the relative velocity between an ionized plasma and a neutral gas, above which a runaway ionization process takes place. The critical ionization process is a quite general mechanism for the conversion of the kinetic energy of a rapidly streaming gas into ionization and plasma thermal energy. Critical phenomena in general are typical of complex systems, and may lead to sharp spatial or temporal features.

12.3.6 Ultracold plasma

Ultracold plasmas are created in a magneto-optical trap (MOT) by trapping and cooling neutral atoms, to temperatures of 1 mK or lower, and then using another laser to ionize the atoms by giving each of the outermost electrons just enough energy to escape the electrical attraction of its parent ion.

One advantage of ultracold plasmas are their well characterized and tunable initial conditions, including their size and electron temperature. By adjusting the wavelength of the ionizing laser, the kinetic energy of the liberated electrons can be tuned as low as 0.1 K, a limit set by the frequency bandwidth of the laser pulse. The ions inherit the millikelvin temperatures of the neutral atoms, but are quickly heated through a process known as disorder induced heating (DIH). This type of non-equilibrium ultracold plasma evolves rapidly, and displays many other interesting phenomena.[37]

One of the metastable states of a strongly nonideal plasma is Rydberg matter, which forms upon condensation of excited atoms.

12.3.7 Non-neutral plasma

The strength and range of the electric force and the good conductivity of plasmas usually ensure that the densities of positive and negative charges in any sizeable region are equal ("quasineutrality"). A plasma with a significant excess of charge density, or, in the extreme case, is composed of a single species, is called a non-neutral plasma. In such a plasma, electric fields play a dominant role. Examples are charged particle beams, an electron cloud in a Penning trap and positron plasmas.[38]

12.3.8 Dusty plasma/grain plasma

A dusty plasma contains tiny charged particles of dust (typically found in space). The dust particles acquire high charges and interact with each other. A plasma that contains larger particles is called grain plasma. Under laboratory conditions, dusty plasmas are also called *complex plasmas*.[39]

12.3.9 Impermeable plasma

Impermeable plasma is a type of thermal plasma which acts like an impermeable solid with respect to gas or cold plasma and can be physically pushed. Interaction of cold gas and thermal plasma was briefly studied by a group led by Hannes Alfvén in 1960s and 1970s for its possible applications in insulation of fusion plasma from the reactor walls.[40] However, later it was found that the external magnetic fields in this configuration could induce kink instabilities in the plasma and subsequently lead to an unexpectedly high heat loss to the walls.[41] In 2013, a group of materials scientists reported that they have successfully generated stable impermeable plasma with no magnetic confinement using only an ultrahigh-pressure blanket of cold gas. While spectroscopic data on the characteristics of plasma were claimed to be difficult to obtain due to the high pressure, the passive effect of plasma on synthesis of different nanostructures clearly suggested the effective confinement. They also showed that upon maintaining the impermeability for a few tens of seconds, screening of ions at the plasma-gas interface could give rise to a strong secondary mode of heating (known as viscous heating) leading to different kinetics of reactions and formation of complex nanomaterials.[42]

12.4 Mathematical descriptions

Main article: Plasma modeling

To completely describe the state of a plasma, we would need to write down all the particle locations and velocities and describe the electromagnetic field in the plasma region. However, it is generally not practical or necessary to keep track of all the particles in a plasma. Therefore, plasma physicists commonly use less detailed descriptions, of which there are two main types:

12.4.1 Fluid model

Fluid models describe plasmas in terms of smoothed quantities, like density and averaged velocity around each position (see Plasma parameters). One simple fluid model, magnetohydrodynamics, treats the plasma as a single fluid governed by a combination of Maxwell's equations and the Navier–Stokes equations. A more general description is the two-fluid plasma picture, where the ions and electrons are described separately. Fluid models are often accurate when collisionality is sufficiently high to keep the plasma velocity distribution close to a Maxwell–Boltzmann distribution. Because fluid models usually describe the plasma in terms of a single flow at a certain temperature at each spatial location, they can neither capture velocity space structures like beams or double layers, nor resolve wave-particle effects.

12.4.2 Kinetic model

Kinetic models describe the particle velocity distribution function at each point in the plasma and therefore do not need to assume a Maxwell–Boltzmann distribution. A kinetic description is often necessary for collisionless plasmas. There are two common approaches to kinetic description of a plasma. One is based on representing the smoothed distribution function on a grid in velocity and position. The other, known as the particle-in-cell (PIC) technique, includes kinetic information by following the trajectories of a large number of individual particles. Kinetic models are generally more computationally intensive than fluid models. The Vlasov equation may be used to describe the dynamics of a system of charged particles interacting with an electromagnetic field. In magnetized plasmas, a gyrokinetic approach can substantially reduce the computational expense of a fully kinetic simulation.

12.5 Artificial plasmas

Most artificial plasmas are generated by the application of electric and/or magnetic fields through a gas. Plasma generated in a laboratory setting and for industrial use can be generally categorized by:

- The type of power source used to generate the plasma—DC, RF and microwave

- The pressure they operate at—vacuum pressure (< 10 mTorr or 1 Pa), moderate pressure (~ 1 Torr or 100 Pa), atmospheric pressure (760 Torr or 100 kPa)

- The degree of ionization within the plasma—fully, partially, or weakly ionized

- The temperature relationships within the plasma—thermal plasma ($T_e = T_i = T_{gas}$), non-thermal or "cold" plasma ($T_e \gg T_i = T_{gas}$)

- The electrode configuration used to generate the plasma

- The magnetization of the particles within the plasma—magnetized (both ion and electrons are trapped in Larmor orbits by the magnetic field), partially magnetized (the electrons but not the ions are trapped by the magnetic field), non-magnetized (the magnetic field is too weak to trap the particles in orbits but may generate Lorentz forces)

- The application.

12.5.1 Generation of artificial plasma

Just like the many uses of plasma, there are several means for its generation, however, one principle is common to all of them: there must be energy input to produce and sustain it.[44] For this case, plasma is generated when an electric current is applied across a dielectric gas or fluid (an electrically non-conducting material) as can be seen in the image to the right, which shows a discharge tube as a simple example (DC used for simplicity).

The potential difference and subsequent electric field pull the bound electrons (negative) toward the anode (positive electrode) while the cathode (negative electrode) pulls the nucleus.[45] As the voltage increases, the current stresses the material (by electric polarization) beyond its dielectric limit (termed strength) into a stage of electrical breakdown, marked by an electric spark, where the material transforms from being an insulator into a conductor (as it becomes increasingly ionized). The underlying process is the Townsend avalanche, where collisions between electrons and neutral gas atoms create more ions and electrons (as can be seen in the figure on the right). The first impact of an electron on an atom results in one ion and two electrons. Therefore, the number of charged particles increases rapidly (in the millions) only "after about 20 successive sets of collisions",[46] mainly due to a small mean free path (average distance travelled between collisions).

Electric arc

With ample current density and ionization, this forms a luminous electric arc (a continuous electric discharge similar to lightning) between the electrodes.[Note 1] Electrical resistance along the continuous electric arc creates heat, which dissociates more gas molecules and ionizes the resulting atoms (where degree of ionization is determined by temperature), and as per the sequence: solid-liquid-gas-plasma, the gas is gradually turned into a thermal plasma.[Note 2] A thermal plasma is in thermal equilibrium, which is to say that the temperature is relatively homogeneous throughout the heavy particles (i.e. atoms, molecules and ions) and electrons. This is so because when thermal plasmas are generated, electrical energy is given to electrons, which, due to their great mobility and large numbers, are able to disperse it rapidly and by elastic collision (without energy loss) to the heavy particles.[47][Note 3]

12.5.2 Examples of industrial/commercial plasma

Because of their sizable temperature and density ranges, plasmas find applications in many fields of research, technology and industry. For example, in: industrial and extractive metallurgy,[47] surface treatments such as plasma spraying (coating), etching in microelectronics,[48] metal cutting[49] and welding; as well as in everyday vehicle exhaust cleanup and fluorescent/luminescent lamps,[44] while even playing a part in supersonic combustion engines for aerospace engineering.[50]

Low-pressure discharges

- *Glow discharge plasmas*: non-thermal plasmas generated by the application of DC or low frequency RF (<100 kHz) electric field to the gap between two metal electrodes. Probably the most common plasma; this is the type of plasma generated within fluorescent light tubes.[51]

- *Capacitively coupled plasma (CCP)*: similar to glow discharge plasmas, but generated with high frequency RF electric fields, typically 13.56 MHz. These differ from glow discharges in that the sheaths are much less intense. These are widely used in the microfabrication and integrated circuit manufacturing industries for plasma etching and plasma enhanced chemical vapor deposition.[52]

- *Cascaded Arc Plasma Source*: a device to produce low temperature (~1eV) high density plasmas (HDP).

- *Inductively coupled plasma (ICP)*: similar to a CCP and with similar applications but the electrode consists of a coil wrapped around the chamber where plasma is formed.[53]

- *Wave heated plasma*: similar to CCP and ICP in that it is typically RF (or microwave). Examples include helicon discharge and electron cyclotron resonance (ECR).[54]

Atmospheric pressure

- *Arc discharge:* this is a high power thermal discharge of very high temperature (~10,000 K). It can be generated using various power supplies. It is commonly used in metallurgical processes. For example, it is used to smelt minerals containing Al_2O_3 to produce aluminium.

- *Corona discharge:* this is a non-thermal discharge generated by the application of high voltage to sharp electrode tips. It is commonly used in ozone generators and particle precipitators.

- *Dielectric barrier discharge (DBD):* this is a non-thermal discharge generated by the application of high voltages across small gaps wherein a non-conducting coating prevents the transition of the plasma discharge into an arc. It is often mislabeled 'Corona' discharge in industry and has similar application to corona discharges. It is also widely used in the web treatment of fabrics.[55] The application of the discharge to synthetic fabrics and plastics functionalizes the surface and allows for paints, glues and similar materials to adhere.[56]

- *Capacitive discharge:* this is a nonthermal plasma generated by the application of RF power (e.g., 13.56 MHz) to one powered electrode, with a grounded electrode held at a small separation distance on the order of 1 cm. Such discharges are commonly stabilized using a noble gas such as helium or argon.[57]

- "Piezoelectric direct discharge plasma:" is a nonthermal plasma generated at the high-side of a piezoelectric transformer (PT). This generation variant is particularly suited for high efficient and compact devices where a separate high voltage power supply is not desired.

12.6   History

Plasma was first identified in a Crookes tube, and so described by Sir William Crookes in 1879 (he called it "radiant matter").[58] The nature of the Crookes tube "cathode ray" matter was subsequently identified by British physicist Sir J.J. Thomson in 1897.[59] The term "plasma" was coined by Irving Langmuir in 1928,[60] perhaps because the glowing discharge molds itself to the shape of the Crookes tube (Gr. πλάσμα – a thing moulded or formed).[61] Langmuir described his observations as:

Except near the electrodes, where there are *sheaths* containing very few electrons, the ionized gas contains ions and electrons in about equal numbers so that the resultant space charge is very small. We shall use the name *plasma* to describe this region containing balanced charges of ions and electrons.[60]

12.7 Fields of active research

This is just a partial list of topics. See list of plasma (physics) articles. A more complete and organized list can be found on web sites on plasma science and technology.[62]

12.8 Science fiction

String physicist Michio Kaku thinks that plasma saber is the closest practical possibility of awesome weapon cause we cant have star wars light saber. However lots of research and work is required before we can realise that. We would need the power source to be extremely mobile something you can expect to not see for at least 20 years and the materials that can withstand extreme temperatures .These materials(ceramic fibres) exist but major modifications and improvements are needed. The plasma would be controlled by electromagnetic waves. A simple solution is to wound coils around it and pass current so that it is contained by magnetic field.

12.9 See also

- Plasma torch

- Ambipolar diffusion

- Hannes Alfvén Prize

- Plasma channel

- Plasma parameters

- Plasma nitriding

- Magnetohydrodynamics (MHD)

- Electric field screening

- List of plasma physicists

- List of plasma (physics) articles

- Important publications in plasma physics

- IEEE Nuclear and Plasma Sciences Society

- Quark-gluon plasma

- Nikola Tesla

- Space physics

- Total electron content

12.10   Notes

[1] The material undergoes various 'regimes' or stages (e.g. saturation, breakdown, glow, transition and thermal arc) as the voltage is increased under the voltage-current relationship. The voltage rises to its maximum value in the saturation stage, and thereafter it undergoes fluctuations of the various stages; while the current progressively increases throughout.[46]

[2] Across literature, there appears to be no strict definition on where the boundary is between a gas and plasma. Nevertheless, it is enough to say that at 2,000°C the gas molecules become atomized, and ionized at 3,000 °C and "in this state, [the] gas has a liquid like viscosity at atmospheric pressure and the free electric charges confer relatively high electrical conductivities that can approach those of metals."[47]

[3] Note that non-thermal, or non-equilibrium plasmas are not as ionized and have lower energy densities, and thus the temperature is not dispersed evenly among the particles, where some heavy ones remain 'cold'.

12.11   References

[1] πλάσμα, Henry George Liddell, Robert Scott, *A Greek–English Lexicon*, on Perseus

[2] Luo, Q-Z; D'Angelo, N; Merlino, R. L. (1998). "Shock formation in a negative ion plasma" (PDF) **5** (8). Department of Physics and Astronomy. Retrieved 2011-11-20.

[3] Sturrock, Peter A. (1994). *Plasma Physics: An Introduction to the Theory of Astrophysical, Geophysical & Laboratory Plasmas*. Cambridge University Press. ISBN 978-0-521-44810-9.

[4] "Ionization and Plasmas". The University of Tennessee, Knoxville Department of Physics and Astronomy.

[5] "How Lightning Works". HowStuffWorks.

[6] Plasma fountain Source, press release: Solar Wind Squeezes Some of Earth's Atmosphere into Space

[7] Hazeltine, R.D.; Waelbroeck, F.L. (2004). *The Framework of Plasma Physics*. Westview Press. ISBN 978-0-7382-0047-7.

[8] Dendy, R. O. (1990). *Plasma Dynamics*. Oxford University Press. ISBN 978-0-19-852041-2.

[9] Hastings, Daniel & Garrett, Henry (2000). *Spacecraft-Environment Interactions*. Cambridge University Press. ISBN 978-0-521-47128-2.

[10] Peratt, A. L. (1996). "Advances in Numerical Modeling of Astrophysical and Space Plasmas". *Astrophysics and Space Science* **242** (1–2): 93–163. Bibcode:1996Ap&SS.242...93P. doi:10.1007/BF00645112.

[11] See The Nonneutral Plasma Group at the University of California, San Diego

[12] Nicholson, Dwight R. (1983). *Introduction to Plasma Theory*. John Wiley & Sons. ISBN 978-0-471-09045-8.

[13] See Flashes in the Sky: Earth's Gamma-Ray Bursts Triggered by Lightning

[14] Richard Fitzpatrick, *Introduction to Plasma Physics*, Magnetized plasmas

[15] Yaffa Eliezer, Shalom Eliezer, *The Fourth State of Matter: An Introduction to the Physics of Plasma*, Publisher: Adam Hilger, 1989, ISBN 978-0-85274-164-1, 226 pages, page 5

[16] Bittencourt, J.A. (2004). *Fundamentals of Plasma Physics*. Springer. p. 1. ISBN 9780387209753.

[17] Hong, Alice (2000). "Dielectric Strength of Air". *The Physics Factbook*.

[18] It is often stated that more than 99% of the material in the visible universe is plasma. See, for example, Gurnett, D. A. & Bhattacharjee, A. (2005). *Introduction to Plasma Physics: With Space and Laboratory Applications*. Cambridge, UK: Cambridge University Press. p. 2. ISBN 978-0-521-36483-6. and Scherer, K; Fichtner, H & Heber, B (2005). *Space Weather: The Physics Behind a Slogan*. Berlin: Springer. p. 138. ISBN 978-3-540-22907-0..

[19] Alfvén, Hannes (1937). "Cosmic Radiation as an Intra-galactic Phenomenon". *Ark. f. mat., astr. o. fys.* **25B**: 29.

[20] Hannes, A (1990). "Cosmology in the Plasma Universe: An Introductory Exposition". *IEEE Transactions on Plasma Science* **18**: 5–10. Bibcode:1990ITPS...18....5P. doi:10.1109/27.45495. ISSN 0093-3813.

[21] Mészáros, Péter (2010) *The High Energy Universe: Ultra-High Energy Events in Astrophysics and Cosmology*, Publisher Cambridge University Press, ISBN 978-0-521-51700-3, p. 99.

[22] Raine, Derek J. and Thomas, Edwin George (2010) *Black Holes: An Introduction*, Publisher: Imperial College Press, ISBN 978-1-84816-382-9, p. 160

[23] Nemiroff, Robert and Bonnell, Jerry (11 December 2004) Astronomy Picture of the Day, nasa.gov

[24] IPPEX Glossary of Fusion Terms. Ippex.pppl.gov. Retrieved on 2011-11-19.

[25] "Plasma and Flames – The Burning Question", from the Coalition for Plasma Science, retrieved 8 November 2012

[26] von Engel, A. and Cozens, J.R. (1976) "Flame Plasma" in *Advances in electronics and electron physics*, L. L. Marton (ed.), Academic Press, ISBN 978-0-12-014520-1, p. 99

[27] Dickel, J. R. (1990). "The Filaments in Supernova Remnants: Sheets, Strings, Ribbons, or?". *Bulletin of the American Astronomical Society* **22**: 832. Bibcode:1990BAAS...22..832D.

[28] Grydeland, T. (2003). "Interferometric observations of filamentary structures associated with plasma instability in the auroral ionosphere". *Geophysical Research Letters* **30** (6). doi:10.1029/2002GL016362.

[29] Moss, G. D.; Pasko, V. P.; Liu, N.; Veronis, G. (2006). "Monte Carlo model for analysis of thermal runaway electrons in streamer tips in transient luminous events and streamer zones of lightning leaders". *Journal of Geophysical Research* **111**. doi:10.1029/2005JA011350.

[30] Doherty, Lowell R.; Menzel, Donald H. (1965). "Filamentary Structure in Solar Prominences". *The Astrophysical Journal* **141**: 251. Bibcode:1965ApJ...141..251D. doi:10.1086/148107.

[31] Hubble views the Crab Nebula M1: The Crab Nebula Filaments at the Wayback Machine (archived 5 October 2009). The University of Arizona

[32] Zhang, Y. A.; Song, M. T.; Ji, H. S. (2002). "A rope-shaped solar filament and a IIIb flare". *Chinese Astronomy and Astrophysics* **26** (4): 442. doi:10.1016/S0275-1062(02)00095-4.

[33] Boeuf, J. P.; Chaudhury, B.; Zhu, G. Q. (2010). "Theory and Modeling of Self-Organization and Propagation of Filamentary Plasma Arrays in Microwave Breakdown at Atmospheric Pressure".*Physical Review Letters***104**.doi:10.1103/PhysRevLett.104.0

[34] Chin, S. L. (2006). "Some Fundamental Concepts of Femtosecond Laser Filamentation" (PDF). *Journal of the Korean Physical Society* **49**: 281.

[35] Talebpour, A.; Abdel-Fattah, M.; Chin, S. L. (2000). "Focusing limits of intense ultrafast laser pulses in a high pressure gas: Road to new spectroscopic source". *Optics Communications* **183** (5–6): 479. doi:10.1016/S0030-4018(00)00903-2.

[36] Alfvén, Hannes (1981). "section VI.13.1. Cellular Structure of Space". *Cosmic Plasma*. Dordrecht. ISBN 978-90-277-1151-9.

[37] National Research Council (U.S.). Plasma 2010 Committee (2007). *Plasma science: advancing knowledge in the national interest*. National Academies Press. pp. 190–193. ISBN 978-0-309-10943-7.

[38] Greaves, R. G.; Tinkle, M. D.; Surko, C. M. (1994). "Creation and uses of positron plasmas". *Physics of Plasmas* **1** (5): 1439. doi:10.1063/1.870693.

[39] Morfill, G. E.; Ivlev, Alexei V. (2009). "Complex plasmas: An interdisciplinary research field". *Review of Modern Physics* **81** (4): 1353–1404. Bibcode:2009RvMP...81.1353M. doi:10.1103/RevModPhys.81.1353.

[40] Alfvén, H.; Smårs, E. (1960). "Gas-Insulation of a Hot Plasma". *Nature* **188** (4753): 801–802. Bibcode:1960Natur.188..801A. doi:10.1038/188801a0.

[41] Braams, C.M. (1966). "Stability of Plasma Confined by a Cold-Gas Blanket". *Physical Review Letters* **17** (9): 470–471. Bibcode:1966PhRvL..17..470B. doi:10.1103/PhysRevLett.17.470.

[42] Yaghoubi, A.; Mélinon, P. (2013). "Tunable synthesis and in situ growth of silicon-carbon mesostructures using impermeable plasma". *Scientific Reports* **3**. Bibcode:2013NatSR...3E1083Y. doi:10.1038/srep01083. PMC 3547321. PMID 23330064.

12.12 External links

- Free plasma physics books and notes

- Plasmas: the Fourth State of Matter

- Plasma Science and Technology

- Plasma on the Internet – a list of plasma related links.

- Introduction to Plasma Physics: Graduate course given by Richard Fitzpatrick|M.I.T. Introduction by I.H.Hutchinson

- Plasma Material Interaction

- How to make a glowing ball of plasma in your microwave with a grape|More (Video)

- How to make plasma in your microwave with only one match (video)

- OpenPIC3D – 3D Hybrid Particle-In-Cell simulation of plasma dynamics

- Plasma Formulary Interactive

Lightning is an example of plasma present at Earth's surface. Typically, lightning discharges 30,000 amperes at up to 100 million volts, and emits light, radio waves, X-rays and even gamma rays. Plasma temperatures in lightning can approach 28,000 K (28,000 °C;

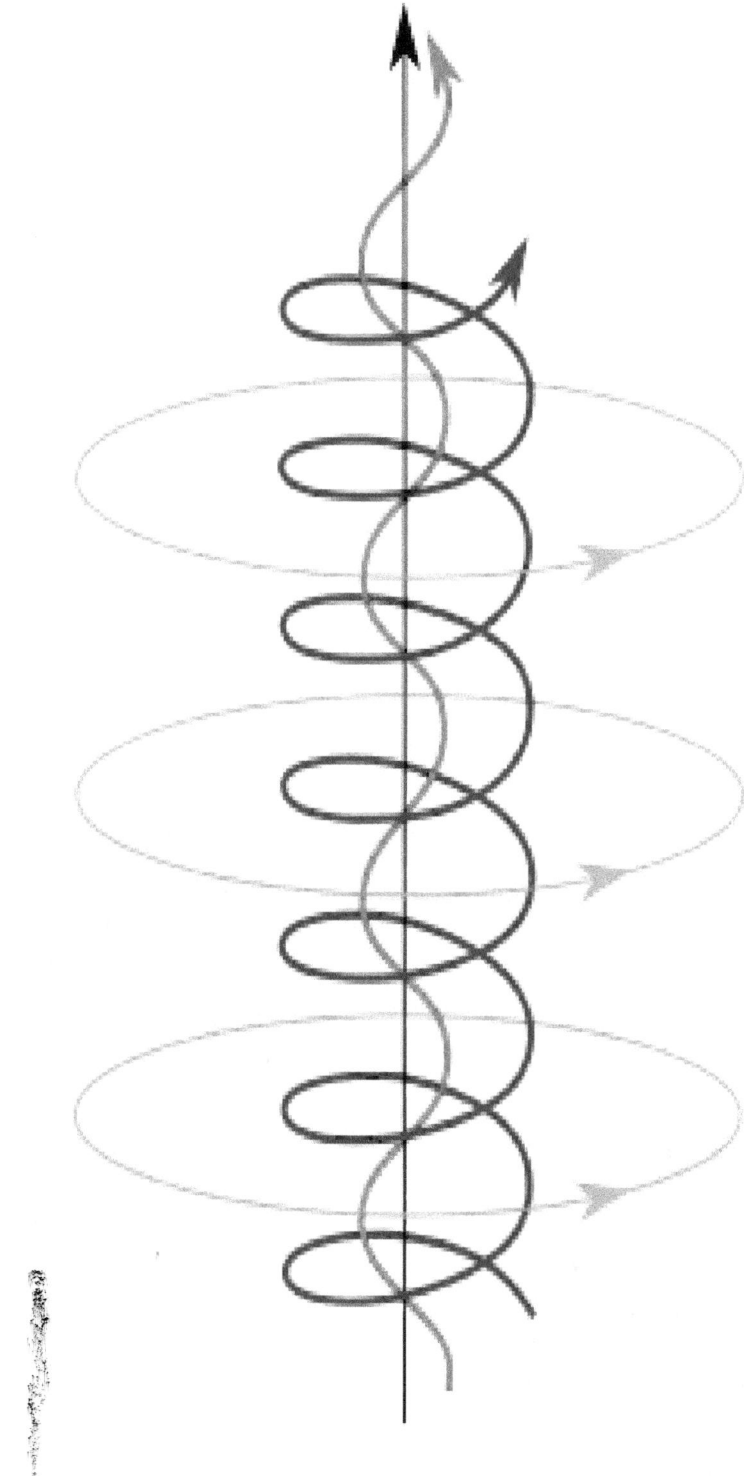

The complex self-constricting magnetic field lines and current paths in a field-aligned Birkeland current that can develop in a plasma.[43]

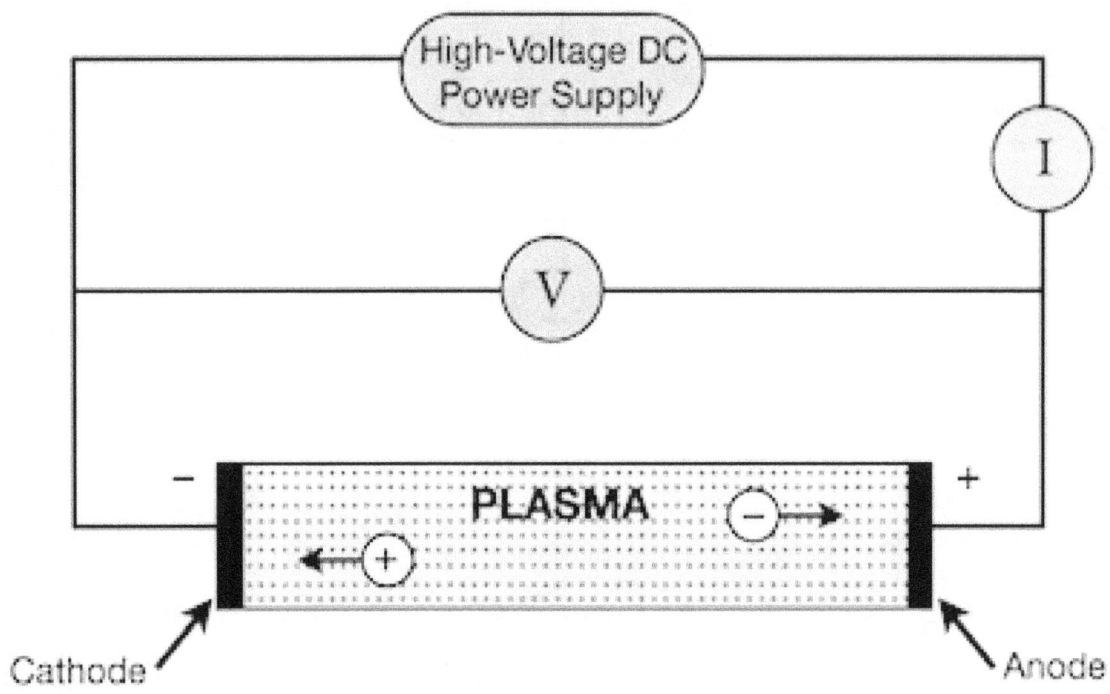

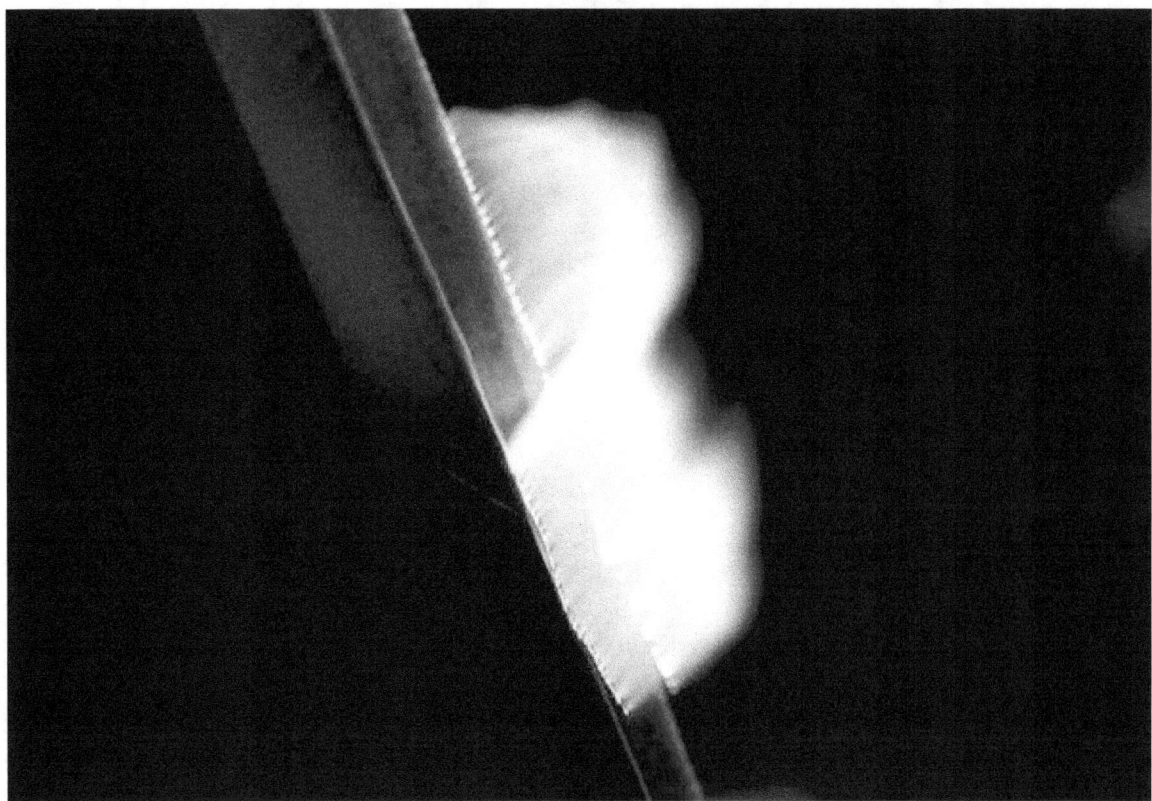

Artificial plasma produced in air by a Jacob's Ladder

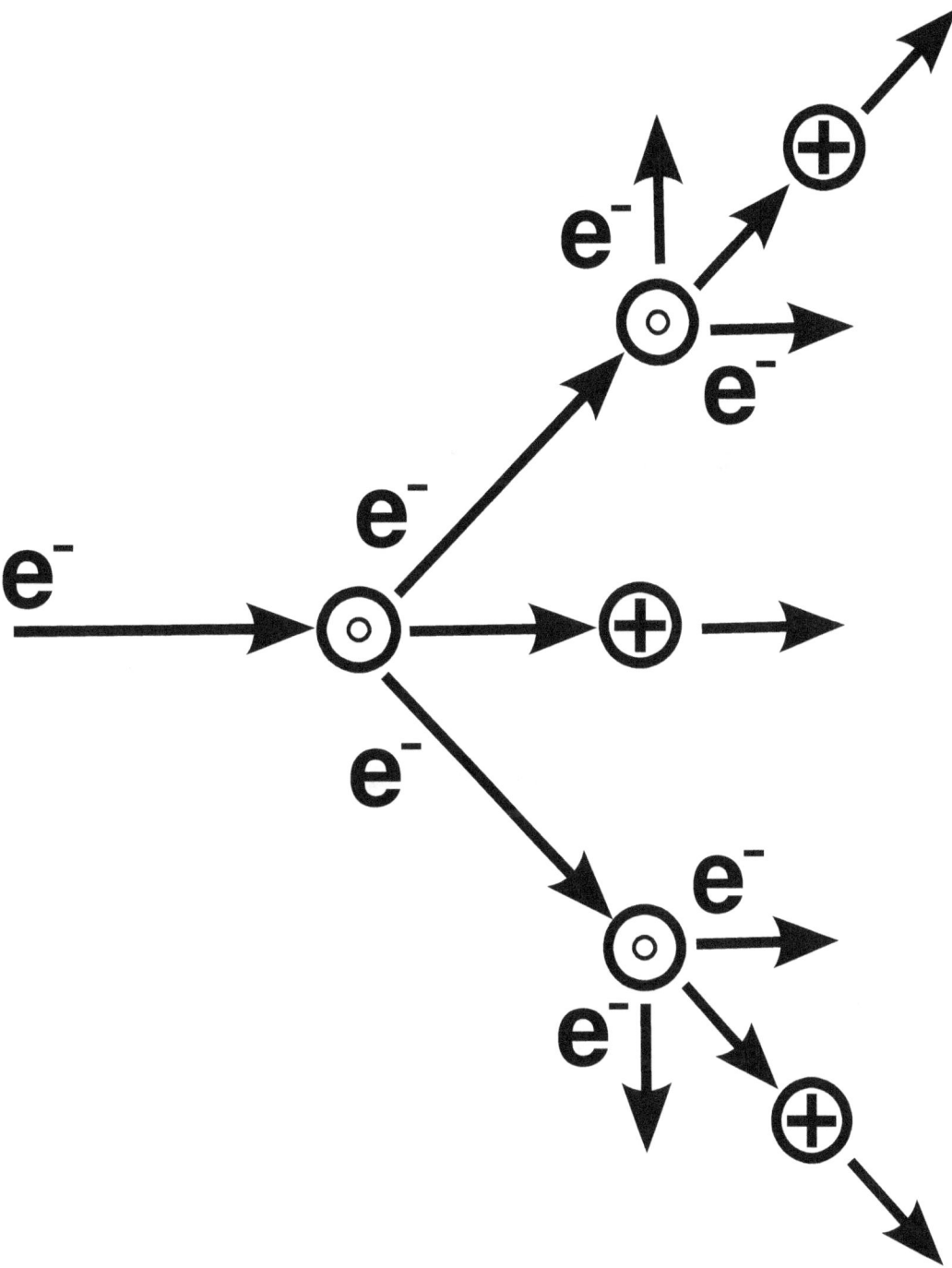

Cascade process of ionization. Electrons are 'e−', neutral atoms 'o', and cations '+'.

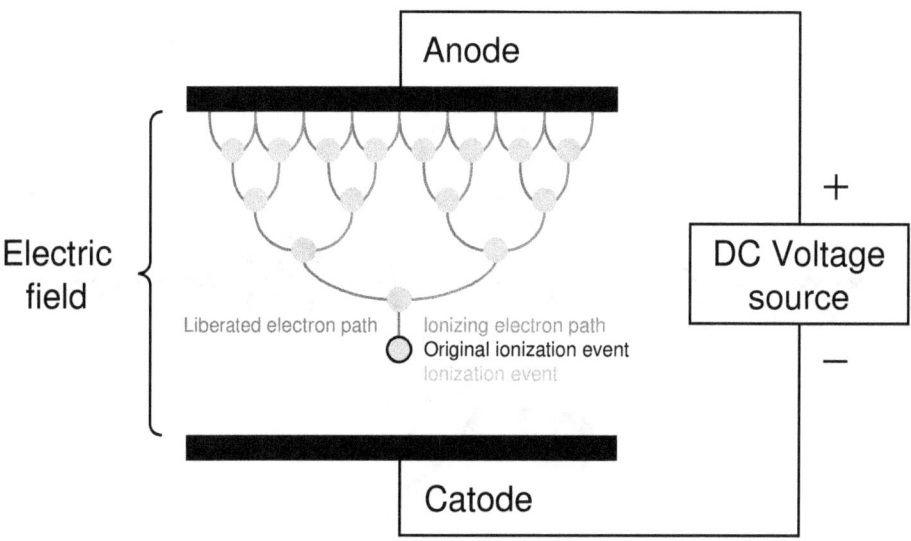

Avalanche effect between two electrodes. The original ionisation event liberates one electron, and each subsequent collision liberates a further electron, so two electrons emerge from each collision: the ionising electron and the liberated electron.

Hall effect thruster. The electric field in a plasma double layer is so effective at accelerating ions that electric fields are used in ion drives.

Chapter 13

Colloid

A **colloid**, in chemistry, is a substance in which one substance of microscopically dispersed insoluble particles is suspended throughout another substance. Sometimes the dispersed substance alone is called the colloid;[1] the term **colloidal suspension** refers unambiguously to the overall mixture (although a narrower sense of the word *suspension* is contradistinguished from colloids by larger particle size). Unlike a solution, whose solute and solvent constitute only one phase, a colloid has a dispersed phase (the suspended particles) and a continuous phase (the medium of suspension). To qualify as a colloid, the mixture must be one that does not settle or would take a very long time to settle appreciably.

The dispersed-phase particles have a diameter of between approximately 1 and 1000 nanometers.[2] Such particles are normally easily visible in an optical microscope, although at the smaller size range (r<250 nm), an ultramicroscope or an electron microscope may be required. Homogeneous mixtures with a dispersed phase in this size range may be called *colloidal aerosols*, *colloidal emulsions*, *colloidal foams*, *colloidal dispersions*, or *hydrosols*. The dispersed-phase particles or droplets are affected largely by the surface chemistry present in the colloid.

Some colloids are translucent because of the Tyndall effect, which is the scattering of light by particles in the colloid. Other colloids may be opaque or have a slight color.

Colloidal suspensions are the subject of interface and colloid science. This field of study was introduced in 1861 by Scottish scientist Thomas Graham.

IUPAC definition

Colloid: Short synonym for *colloidal* system.[3][4]

Colloidal: State of subdivision such that the molecules or polymolecular particles dispersed in a medium have at least one dimension between approximately 1 nm and 1 μm, or that in a system discontinuities are found at distances of that order.[3][4][5]

13.1 Classification

Because the size of the dispersed phase may be difficult to measure, and because colloids have the appearance of solutions, colloids are sometimes identified and characterized by their physico-chemical and transport properties. For example, if a colloid consists of a solid phase dispersed in a liquid, the solid particles will not diffuse through a membrane, whereas with a true solution the dissolved ions or molecules will diffuse through a membrane. Because of the size exclusion, the colloidal particles are unable to pass through the pores of an ultrafiltration membrane with a size smaller than their own dimension. The smaller the size of the pore of the ultrafiltration membrane, the lower the concentration of the dispersed colloidal particles remaining in the ultrafiltered liquid. The measured value of the concentration of a truly dissolved species will thus depend on the experimental conditions applied to separate it from the colloidal particles also dispersed in the liquid. This is particularly important for solubility studies of readily hydrolyzed species such as Al, Eu, Am, Cm, or organic matter complexing these species. Colloids can be classified as follows:

148

Milk is an emulsified colloid of liquid butterfat globules dispersed within a water-based solution.

Based on the nature of interaction between the dispersed phase and the dispersion medium, colloids can be classified as: **Hydrophilic colloids**: These are water-loving colloids. The colloid particles are attracted toward water. They are also called reversible sols. **Hydrophobic colloids**: These are opposite in nature to hydrophilic colloids. The colloid particles are repelled by water. They are also called irreversible sols.

In some cases, a colloid suspension can be considered a homogeneous mixture. This is because the distinction between "dissolved" and "particulate" matter can be sometimes a matter of approach, which affects whether or not it is homogeneous or heterogeneous.

13.2   Hydrocolloids

A *hydrocolloid* is defined as a colloid system wherein the colloid particles are hydrophilic polymers dispersed in water. A hydrocolloid has colloid particles spread throughout water, and depending on the quantity of water available that can take place in different states, e.g., gel or sol (liquid). Hydrocolloids can be either irreversible (single-state) or reversible. For example, agar, a reversible hydrocolloid of seaweed extract, can exist in a gel and solid state, and alternate between states with the addition or elimination of heat.

Many hydrocolloids are derived from natural sources. For example, agar-agar and carrageenan are extracted from seaweed, gelatin is produced by hydrolysis of proteins of mammalian and fish origins, and pectin is extracted from citrus peel and apple pomace.

Gelatin desserts like jelly or Jell-O are made from gelatin powder, another effective hydrocolloid. Hydrocolloids are employed in food mainly to influence texture or viscosity (e.g., a sauce). Hydrocolloid-based medical dressings are used for skin and wound treatment.

Other main hydrocolloids are xanthan gum, gum arabic, guar gum, locust bean gum, cellulose derivatives as carboxymethyl cellulose, alginate and starch.

13.3   Interaction between particles

The following forces play an important role in the interaction of colloid particles:

- Excluded volume repulsion: This refers to the impossibility of any overlap between hard particles.

- Electrostatic interaction: Colloidal particles often carry an electrical charge and therefore attract or repel each other. The charge of both the continuous and the dispersed phase, as well as the mobility of the phases are factors affecting this interaction.

- van der Waals forces: This is due to interaction between two dipoles that are either permanent or induced. Even if the particles do not have a permanent dipole, fluctuations of the electron density gives rise to a temporary dipole in a particle. This temporary dipole induces a dipole in particles nearby. The temporary dipole and the induced dipoles are then attracted to each other. This is known as van der Waals force, and is always present (unless the refractive indexes of the dispersed and continuous phases are matched), is short-range, and is attractive.

- Entropic forces: According to the second law of thermodynamics, a system progresses to a state in which entropy is maximized. This can result in effective forces even between hard spheres.

- Steric forces between polymer-covered surfaces or in solutions containing non-adsorbing polymer can modulate interparticle forces, producing an additional steric repulsive force (which is predominantly entropic in origin) or an attractive depletion force between them. Such an effect is specifically searched for with tailor-made superplasticizers developed to increase the workability of concrete and to reduce its water content.

13.4 Preparation

There are two principal ways of preparation of colloids:[6]

- Dispersion of large particles or droplets to the colloidal dimensions by milling, spraying, or application of shear (e.g., shaking, mixing, or high shear mixing).

- Condensation of small dissolved molecules into larger colloidal particles by precipitation, condensation, or redox reactions. Such processes are used in the preparation of colloidal silica or gold.

13.5 Stabilization (peptization)

The stability of a colloidal system is defined by particles remaining suspended in solution at equilibrium.

Stability is hindered by aggregation and sedimentation phenomena, which are driven by the colloid's tendency to reduce surface energy. Reducing the interfacial tension will stabilize the colloidal system by reducing this driving force.

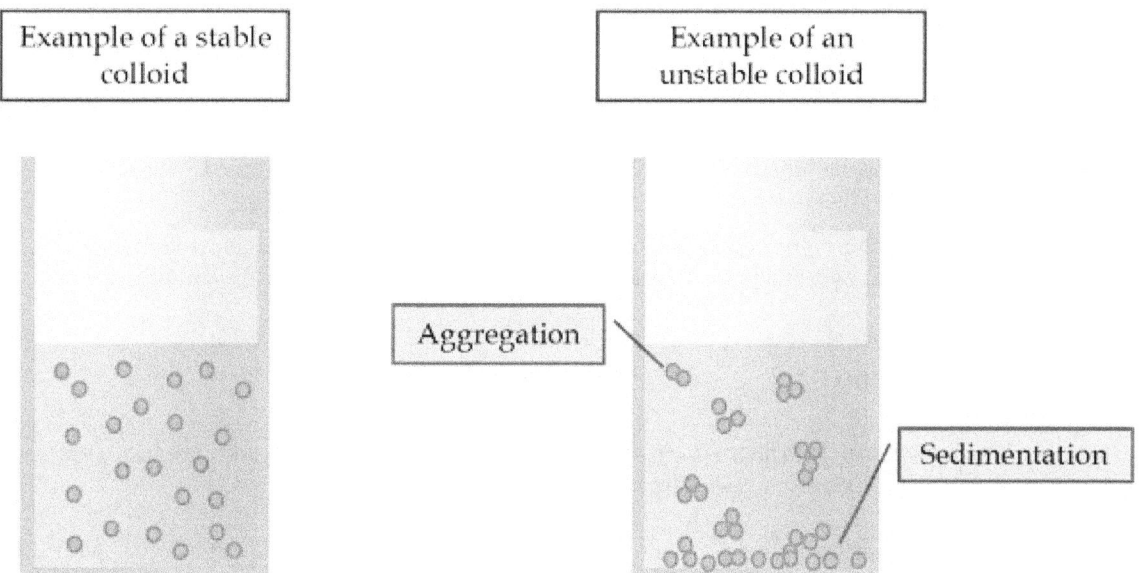

Examples of a stable and of an unstable colloidal dispersion.

Aggregation is due to the sum of the interaction forces between particles.[7][8] If attractive forces (such as van der Waals forces) prevail over the repulsive ones (such as the electrostatic ones) particles aggregate in clusters.

Electrostatic stabilization and steric stabilization are the two main mechanisms for stabilization against aggregation.

- Electrostatic stabilization is based on the mutual repulsion of like electrical charges. In general, different phases have different charge affinities, so that an electrical double layer forms at any interface. Small particle sizes lead to enormous surface areas, and this effect is greatly amplified in colloids. In a stable colloid, mass of a dispersed phase is so low that its buoyancy or kinetic energy is too weak to overcome the electrostatic repulsion between charged layers of the dispersing phase.

- Steric stabilization consists in covering the particles in polymers which prevents the particle to get close in the range of attractive forces.

A combination of the two mechanisms is also possible (electrosteric stabilization). All the above-mentioned mechanisms for minimizing particle aggregation rely on the enhancement of the repulsive interaction forces.

Electrostatic and steric stabilization do not directly address the sedimentation/floating problem.

Particle sedimentation (and also floating, although this phenomenon is less common) arises from a difference in the density of the dispersed and of the continuous phase. The higher the difference in densities, the faster the particle settling.

- The gel network stabilization represents the principal way to produce colloids stable to both aggregation and sedimentation.[9][10]

The method consists in adding to the colloidal suspension a polymer able to form a gel network and characterized by shear thinning properties. Examples of such substances are xanthan and guar gum.

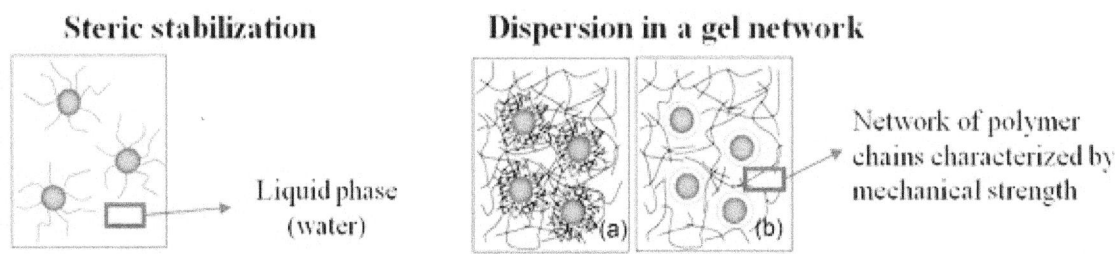

Steric and gel network stabilization.

Particle settling is hindered by the stiffness of the polymeric matrix where particles are trapped.[9] In addition, the long polymeric chains can provide a steric or electrosteric stabilization to dispersed particles.

The rheological shear thinning properties find beneficial in the preparation of the suspensions and in their use, as the reduced viscosity at high shear rates facilitates deagglomeration, mixing and in general the flow of the suspensions.

13.6   Destabilization

Unstable colloidal dispersions can form flocs as the particles aggregate due to interparticle attractions. In this way photonic glasses can be grown. This can be accomplished by a number of different methods:

- Removal of the electrostatic barrier that prevents aggregation of the particles. This can be accomplished by the addition of salt to a suspension or changing the pH of a suspension to effectively neutralize or "screen" the surface charge of the particles in suspension. This removes the repulsive forces that keep colloidal particles separate and allows for coagulation due to van der Waals forces.

- Addition of a charged polymer flocculant. Polymer flocculants can bridge individual colloidal particles by attractive electrostatic interactions. For example, negatively charged colloidal silica or clay particles can be flocculated by the addition of a positively charged polymer.

- Addition of non-adsorbed polymers called depletants that cause aggregation due to entropic effects.

- Physical deformation of the particle (e.g., stretching) may increase the van der Waals forces more than stabilization forces (such as electrostatic), resulting coagulation of colloids at certain orientations.

Unstable colloidal suspensions of low-volume fraction form clustered liquid suspensions, wherein individual clusters of particles fall to the bottom of the suspension (or float to the top if the particles are less dense than the suspending medium) once the clusters are of sufficient size for the Brownian forces that work to keep the particles in suspension to be overcome by gravitational forces. However, colloidal suspensions of higher-volume fraction form colloidal gels with viscoelastic properties. Viscoelastic colloidal gels, such as bentonite and toothpaste, flow like liquids under shear, but maintain their shape when shear is removed. It is for this reason that toothpaste can be squeezed from a toothpaste tube, but stays on the toothbrush after it is applied.

13.6.1 Monitoring stability

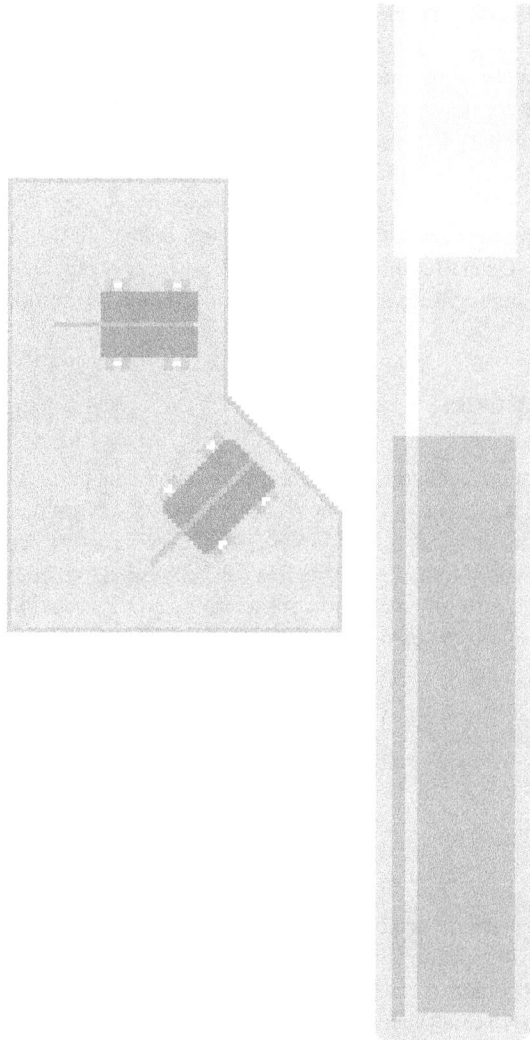
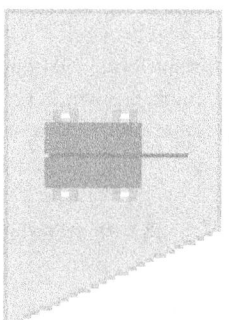

Measurement principle of multiple light scattering coupled with vertical scanning

Multiple light scattering coupled with vertical scanning is the most widely used technique to monitor the dispersion state of a product, hence identifying and quantifying destabilisation phenomena.[11][12][13][14] It works on concentrated dispersions without dilution. When light is sent through the sample, it is backscattered by the particles / droplets. The backscattering intensity is directly proportional to the size and volume fraction of the dispersed phase. Therefore, local changes in concentration (*e.g.*Creaming and Sedimentation) and global changes in size (*e.g.* flocculation, coalescence) are detected and monitored.

13.6.2   Accelerating methods for shelf life prediction

The kinetic process of destabilisation can be rather long (up to several months or even years for some products) and it is often required for the formulator to use further accelerating methods in order to reach reasonable development time for new product design. Thermal methods are the most commonly used and consists in increasing temperature to accelerate destabilisation (below critical temperatures of phase inversion or chemical degradation). Temperature affects not only the viscosity, but also interfacial tension in the case of non-ionic surfactants or more generally interactions forces inside the system. Storing a dispersion at high temperatures enables to simulate real life conditions for a product (e.g. tube of sunscreen cream in a car in the summer), but also to accelerate destabilisation processes up to 200 times. Mechanical acceleration including vibration, centrifugation and agitation are sometimes used. They subject the product to different forces that pushes the particles / droplets against one another, hence helping in the film drainage. However, some emulsions would never coalesce in normal gravity, while they do under artificial gravity.[15] Moreover, segregation of different populations of particles have been highlighted when using centrifugation and vibration.[16]

13.7   As a model system for atoms

In physics, colloids are an interesting model system for atoms. Micrometre-scale colloidal particles are large enough to be observed by optical techniques such as confocal microscopy. Many of the forces that govern the structure and behavior of matter, such as excluded volume interactions or electrostatic forces, govern the structure and behavior of colloidal suspensions. For example, the same techniques used to model ideal gases can be applied to model the behavior of a hard sphere colloidal suspension. In addition, phase transitions in colloidal suspensions can be studied in real time using optical techniques,[17] and are analogous to phase transitions in liquids. In many interesting cases optical fluidity is used to control colloid suspensions.[17][18]

13.8   Crystals

Main article: Colloidal crystal

A colloidal crystal is a highly ordered array of particles that can be formed over a very long range (typically on the order of a few millimeters to one centimeter) and that appear analogous to their atomic or molecular counterparts.[19] One of the finest natural examples of this ordering phenomenon can be found in precious opal, in which brilliant regions of pure spectral color result from close-packed domains of amorphous colloidal spheres of silicon dioxide (or silica, SiO_2).[20][21] These spherical particles precipitate in highly siliceous pools in Australia and elsewhere, and form these highly ordered arrays after years of sedimentation and compression under hydrostatic and gravitational forces. The periodic arrays of submicrometre spherical particles provide similar arrays of interstitial voids, which act as a natural diffraction grating for visible light waves, particularly when the interstitial spacing is of the same order of magnitude as the incident lightwave.[22][23]

Thus, it has been known for many years that, due to repulsive Coulombic interactions, electrically charged macromolecules in an aqueous environment can exhibit long-range crystal-like correlations with interparticle separation distances, often being considerably greater than the individual particle diameter. In all of these cases in nature, the same brilliant iridescence (or play of colors) can be attributed to the diffraction and constructive interference of visible lightwaves that satisfy Bragg's law, in a matter analogous to the scattering of X-rays in crystalline solids.

The large number of experiments exploring the physics and chemistry of these so-called "colloidal crystals" has emerged as a result of the relatively simple methods that have evolved in the last 20 years for preparing synthetic monodisperse colloids (both polymer and mineral) and, through various mechanisms, implementing and preserving their long-range order formation.

13.9 In biology

In the early 20th century, before enzymology was well understood, colloids were thought to be the key to the operation of enzymes; i.e., the addition of small quantities of an enzyme to a quantity of water would, in some fashion yet to be specified, subtly alter the properties of the water so that it would break down the enzyme's specific substrate, such as a solution of ATPase breaking down ATP. Furthermore, life itself was explainable in terms of the aggregate properties of all the colloidal substances that make up an organism. As more detailed knowledge of biology and biochemistry developed, the colloidal theory was replaced by the macromolecular theory, which explains an enzyme as a collection of identical huge molecules that act as very tiny machines, freely moving about between the water molecules of the solution and individually operating on the substrate, no more mysterious than a factory full of machinery. The properties of the water in the solution are not altered, other than the simple osmotic changes that would be caused by the presence of any solute. In humans, both the thyroid gland and the intermediate lobe (*pars intermedia*) of the pituitary gland contain colloid follicles.

13.10 In the environment

Colloidal particles can also serve as transport vector[24] of diverse contaminants in the surface water (sea water, lakes, rivers, fresh water bodies) and in underground water circulating in fissured rocks[25] (limestone, sandstone, granite, ...). Radionuclides and heavy metals easily sorb onto colloids suspended in water. Various types of colloids are recognised: inorganic colloids (clay particles, silicates, iron oxy-hydroxides, ...), organic colloids (humic and fulvic substances). When heavy metals or radionuclides form their own pure colloids, the term "*Eigencolloid*" is used to designate pure phases, e.g., $Tc(OH)_4$, $U(OH)_4$, $Am(OH)_3$. Colloids have been suspected for the long-range transport of plutonium on the Nevada Nuclear Test Site. They have been the subject of detailed studies for many years. However, the mobility of inorganic colloids is very low in compacted bentonites and in deep clay formations[26] because of the process of ultrafiltration occurring in dense clay membrane.[27] The question is less clear for small organic colloids often mixed in porewater with truly dissolved organic molecules.[28]

13.11 Intravenous therapy

Colloid solutions used in intravenous therapy belong to a major group of volume expanders, and can be used for intravenous fluid replacement. Colloids preserve a high colloid osmotic pressure in the blood,[29] and therefore, they should theoretically preferentially increases the intravascular volume, whereas other types of volume expanders called crystalloids also increases the interstitial volume and intracellular volume. However, there is still controversy to the actual difference in efficacy by this difference,[29] and much of the research related to this use of colloids is based on fraudulent research by Joachim Boldt.[30] Another difference is that crystalloids generally are much cheaper than colloids.[29]

13.12 References

[1] "Colloid". Britannica Online Encyclopedia. Retrieved 31 August 2009.

[2] Levine, Ira N. (2001). *Physical Chemistry* (5th ed.). Boston: McGraw-Hill. ISBN 0-07-231808-2., p. 955

[3] Richard G. Jones, Edward S. Wilks, W. Val Metanomski, Jaroslav Kahovec, Michael Hess, Robert Stepto, Tatsuki Kitayama, ed. (2009). *Compendium of Polymer Terminology and Nomenclature (IUPAC Recommendations 2008)* (2nd ed.). RSC Publ. p. 464. ISBN 978-0-85404-491-7.

[4] "Dispersity in polymer science (IUPAC Recommendations 2009)" (PDF). *Pure and Applied Chemistry* **81** (2): 351–353. 2009. doi:10.1351/PAC-REC-08-05-02.

[5] "Terminology of polymers
and polymerization processes in dispersed systems (IUPAC Recommendations 2011)" (PDF). *Pure and Applied Chemistry* **83** (12). 2011. doi:10.1351/PAC-REC-10-06-03.

[6] http://www.substech.com/dokuwiki/doku.php?id=preparation_of_colloids

[7] Israelachvili, Jacob N. (1991). *Intermolecular and Surface Forces.* Academic Press. ISBN 978-0-12-391927-4.

[8] Menachem Elimelech, John Gregory, Xiadong Jia, Richard Williams (1998). *Particle deposition and aggregation: measurement, modelling and simulation.* Butterworth-Heinemann. ISBN 0-7506-7024-X.

[9] Comba, Silvia; Sethi (August 2009). "Stabilization of highly concentrated suspensions of iron nanoparticles using shear-thinning gels of xanthan gum". *Water Research* **43** (15): 3717–3726. doi:10.1016/j.watres.2009.05.046.

[10] Cantrell, K.J.; Kaplan, D.I.; Gilmore, T.J. (1997). "Injection of colloidal Fe-0 particles in sand with shear-thinning fluids". *Journal of Environmental Engineering-Asce* **123** (8): 786–791. doi:10.1061/(ASCE)0733-9372(1997)123:8(786).

[11] Roland, I; Piel, G; Delattre, L; Evrard, B (2003). "Systematic characterization of oil-in-water emulsions for formulation design". *International Journal of Pharmaceutics* **263** (1–2): 85–94. doi:10.1016/S0378-5173(03)00364-8. PMID 12954183.

[12] Lemarchand, Caroline; Couvreur, Patrick; Besnard, Madeleine; Costantini, Dominique; Gref, Ruxandra (2003). "Novel polyester-polysaccharide nanoparticles". *Pharmaceutical Research* **20** (8): 1284–92. doi:10.1023/A:1025017502379. PMID 12948027.

[13] Mengual, O (1999). "Characterisation of instability of concentrated dispersions by a new optical analyser: the TURBISCAN MA 1000". *Colloids and Surfaces A: Physicochemical and Engineering Aspects* **152**: 111. doi:10.1016/S0927-7757(98)00680-3.

[14] P. Bru; et al. (2004). T. Provder and J. Texter, ed. *Particle sizing and characterisation.*

[15] J-L Salager (2000). Françoise Nielloud, Gilberte Marti-Mestres, ed. *Pharmaceutical emulsions and suspensions.* CRC press. p. 89. ISBN 0-8247-0304-9.

[16] Snabre, Patrick; Pouligny, Bernard (2008). "Size Segregation in a Fluid-like or Gel-like Suspension Settling under Gravity or in a Centrifuge". *Langmuir* **24** (23): 13338–47. doi:10.1021/la802459u. PMID 18986182.

[17] Greenfield, Elad; Nemirovsky, Jonathan; El-Ganainy, Ramy; Christodoulides, Demetri N; Segev, Mordechai (2013). "Shock-wave based nonlinear optical manipulation in densely scattering opaque suspensions". *Optics Express* **21** (20): 23785–23802. Bibcode:2013OExpr..2123785G. doi:10.1364/OE.21.023785. PMID 24104290.

[18] Greenfield, Elad; Rotschild, Carmel; Szameit, Alexander; Nemirovsky, Jonathan; El-Ganainy, Ramy; Christodoulides, Demetrios N; Saraf, Meirav; Lifshitz, Efrat; Segev, Mordechai (2011). "Light-induced self-synchronizing flow patterns" (PDF). *New Journal of Physics* **13** (5): 053021. Bibcode:2011NJPh...13e3021G. doi:10.1088/1367-2630/13/5/053021.

[19] Pieranski, P. (1983). "Colloidal Crystals".*Contemporary Physics***24**: 25.Bibcode:1983ConPh..24...25P.doi:10.1080/00107518

[20] Sanders, J.V.; Sanders, J. V.; Segnit, E. R. (1964). "Structure of Opal". *Nature* **204** (4962): 1151. Bibcode:1964Natur.204..990J. doi:10.1038/204990a0.

[21] Darragh, P.J.; et al. (1976). *Scientific American* **234**: 84. Missing or empty |title= (help)

[22] Luck, W.; et al. (1963). "Ber. Busenges". *Phys. Chem.* **67**: 84.

[23] Hiltner, P.A. and Krieger, I.M. (1969). "Diffraction of light by ordered suspensions".*J. Phys. Chem.***73**(7): 2306.doi:10.1021/

[24] Frimmel, Fritz H.; Frank von der Kammer; Hans-Curt Flemming (2007). *Colloidal transport in porous media* (1 ed.). Springer. p. 292. ISBN 3-540-71338-7.

[25] Alonso, U.; T. Missana; A. Patelli; V. Rigato (2007). "Bentonite colloid diffusion through the host rock of a deep geological repository". *Physics and Chemistry of the Earth, Parts A/B/C* **32** (1–7): 469–476. Bibcode:2007PCE....32..469A. doi:10.1016/j.pce.2006.04.021. ISSN 1474-7065.

[26] Voegelin, A.; Kretzschmar, R. (December 2002). *Stability and mobility of colloids in Opalinus Clay.* (PDF). Nagra Technical Report 02-14. Institute of Terrestrial Ecology, ETH Zürich. p. 47. ISSN 1015-2636.

[27] "Diffusion of colloids in compacted bentonite". Retrieved 12 February 2009.

[28] Wold, Susanna; Trygve Eriksen (2007). "Diffusion of humic colloids in compacted bentonite". *Physics and Chemistry of the Earth, Parts A/B/C* **32** (1–7): 477–484. Bibcode:2007PCE....32..477W. doi:10.1016/j.pce.2006.05.002. ISSN 1474-7065.

[29] An Update on Intravenous Fluids by Gregory S. Martin, MD, MSc

[30] Millions of surgery patients at risk in drug research fraud scandal, *The Telegraph*, 4 November 2011

13.13 Further reading

Lyklema, J. *Fundamentals of Interface and Colloid Science*, Vol. 2, p. 3208, 1995

Hunter, R.J. *Foundations of Colloid Science*, Oxford University Press, 1989

Dukhin, S.S. & Derjaguin, B.V. *Electrokinetic Phenomena*, J. Wiley and Sons, 1974

Russel, W.B., Saville, D.A. and Schowalter, W.R. *Colloidal Dispersions*, Cambridge, 1989 Cambridge University Press

Kruyt, H.R. *Colloid Science*, Volume 1, Irreversible systems, Elsevier, 1959

Dukhin, A.S. and Goetz, P.J. *Ultrasound for characterizing colloids*, Elsevier, 2002

Rodil, Ma. Lourdes C., *Chemistry The Central Science*, 7th Ed. ISBN 0-13-533480-2

Pieranski, P., Colloidal Crystals, *Contemp. Phys.*, Vol. 24, p. 25 (1983)

Sanders, J.V., *Structure of Opal*, Nature, Vol. 204, p. 1151, (1964);

Darragh, P.J., et al., Scientific American, Vol. 234, p. 84, (1976)

Luck, W. et al., Ber. Busenges Phys. Chem., Vol. 67, p. 84 (1963);

Hiltner, P.A. and Krieger, I.M., *Diffraction of Light by Ordered Suspensions*, J. Phys. Chem., Vol. 73, p. 2306 (1969)

Arora, A.K., Tata, B.V.R., Eds. *Ordering & Phase Transitions in Charged Colloids* Wiley, New York (1996)

Sood, A.K. in *Solid State Physics*, Eds. Ehrenreich, H., Turnbull, D., Vol. 45, p. 1 (1991)

Murray, C.A. and Grier, D.G., *Colloidal Crystals*, Amer. Scientist, Vol. 83, p. 238 (1995);

Video Microscopy of Monodisperse Colloidal Systems, Ann. Rev. Phys. Chem., Vol. 47, p. 421 (1996)

Tanaka, 1992, Phase Transition of Gel

Chapter 14

Degenerate matter

Degenerate matter[1][2] in physics is a collection of free, non-interacting particles with a pressure and other physical characteristics determined by quantum mechanical effects. It is the analogue of an ideal gas in classical mechanics. The degenerate state of matter, in the sense of deviant from an ideal gas, arises at extraordinarily high density (in compact stars) or at extremely low temperatures in laboratories.[3][4] It occurs for matter particles such as electrons, neutrons, protons, and fermions in general and is referred to as **electron-degenerate matter**, **neutron-degenerate matter**, etc. In a mixture of particles, such as ions and electrons in white dwarfs or metals, the electrons may be degenerate, while the ions are not.

In a quantum mechanical description, free particles limited to a finite volume may take only a discrete set of energies, called quantum states. The Pauli exclusion principle prevents identical fermions from occupying the same quantum state. At lowest total energy (when the thermal energy of the particles is negligible), all the lowest energy quantum states are filled. This state is referred to as full degeneracy. The pressure (called degeneracy pressure or Fermi pressure) remains nonzero even near absolute zero temperature.[3][4] Adding particles or reducing the volume forces the particles into higher-energy quantum states. This requires a compression force, and is made manifest as a resisting pressure. The key feature is that this degeneracy pressure does not depend on the temperature and only on the density of the fermions. It keeps dense stars in equilibrium independent of the thermal structure of the star.

Degenerate matter is also called a **Fermi gas** or a **degenerate gas**. A degenerate mass whose fermions have velocities close to the speed of light (particle energy larger than its rest mass energy) is called **relativistic degenerate matter**.

Degenerate matter was first described for a mixture of ions and electrons in 1926 by Ralph H. Fowler,[5] showing that at densities observed in white dwarfs the electrons (obeying Fermi–Dirac statistics, the term degenerate was not yet in use) have a pressure much higher than the partial pressure of the ions.

14.1   Concept

If a plasma is cooled and compressed repeatedly, it will eventually not be possible to compress the plasma any further. This is by the Pauli exclusion principle, which states that two fermions cannot share the same quantum state. When in this highly compressed state, since there is no extra space for any particles, a particle's location is extremely defined. This is due to the Heisenberg uncertainty principle, $\Delta p \Delta x \geq \hbar/2$, where Δp is the uncertainty in the particle's momentum and Δx is the uncertainty in position. This implies that the momentum of a highly compressed particle is extremely uncertain, since the particles are located in a very confined space. Therefore, *even though the plasma is cold*, such particles must be moving very fast on average. This leads to the conclusion that, in order to compress an object into a very small space, tremendous force is required to control its particles' momentum.

Unlike a classical ideal gas, whose pressure is proportional to its temperature ($P = nkT/V$, where P is pressure, V is the volume, n is the number of particles—typically atoms or molecules—k is Boltzmann's constant, and T is temperature), the pressure exerted by degenerate matter depends only weakly on its temperature. In particular, the pressure remains nonzero even at absolute zero temperature. At relatively low densities, the pressure of a fully degenerate gas is given by

$P = K(n/V)5/3$
, where K depends on the properties of the particles making up the gas. At very high densities, where most of the particles are forced into quantum states with relativistic energies, the pressure is given by $P = K'(n/V)4/3$
, where K' again depends on the properties of the particles making up the gas.[6]

All matter experiences both normal thermal pressure and degeneracy pressure, but in commonly encountered gases, thermal pressure dominates so much that degeneracy pressure can be ignored. Likewise, degenerate matter still has normal thermal pressure, but at extremely high densities, the degeneracy pressure usually dominates.

Exotic examples of degenerate matter include neutronium, strange matter, metallic hydrogen and white dwarf matter. Degeneracy pressure contributes to the pressure of conventional solids, but these are not usually considered to be degenerate matter because a significant contribution to their pressure is provided by electrical repulsion of atomic nuclei and the screening of nuclei from each other by electrons. In metals it is useful to treat the conduction electrons alone as a degenerate, free electron gas while the majority of the electrons are regarded as occupying bound quantum states. This contrasts with degenerate matter that forms the body of a white dwarf, where all the electrons would be treated as occupying free particle momentum states.

14.2 Degenerate gases

Degenerate gases are gases composed of fermions such as electrons, protons, and neutrons rather than molecules of ordinary matter. The electron gas in ordinary metals and in the interior of white dwarf stars are two examples. Following the Pauli exclusion principle, there can be only one fermion occupying each quantum state. In a degenerate gas, all quantum states are filled and therefore the volume of the gas cannot be compressed any further. Most stars are supported against their own gravitation by normal thermal gas pressure, while in white dwarf stars the supporting force comes from the degeneracy pressure of the electron gas in their interior. In neutron stars, the degenerate particles are neutrons.

A fermion gas in which all quantum states below a given energy level are filled is called a *fully degenerate* fermion gas. The difference between this energy level and the lowest energy level is known as the Fermi energy (see also Fermi–Dirac statistics).

14.2.1 Electron degeneracy

In an ordinary fermion gas in which thermal effects dominate, most of the available electron energy levels are unfilled and the electrons are free to move to these states. As particle density is increased, electrons progressively fill the lower energy states and additional electrons are forced to occupy states of higher energy even at low temperatures. Degenerate gases strongly resist further compression because the electrons cannot move to already filled lower energy levels due to the Pauli exclusion principle. Since electrons cannot give up energy by moving to lower energy states, no thermal energy can be extracted. The momentum of the fermions in the fermion gas nevertheless generates pressure, termed *degeneracy pressure*.

Under high densities the matter becomes a degenerate gas when the electrons are all stripped from their parent atoms. In the core of a star, once hydrogen burning in nuclear fusion reactions stops, it becomes a collection of positively charged ions, largely helium and carbon nuclei, floating in a sea of electrons, which have been stripped from the nuclei. Degenerate gas is an almost perfect conductor of heat and does not obey the ordinary gas laws. White dwarfs are luminous not because they are generating any energy but rather because they have trapped a large amount of heat which is gradually radiated away. Normal gas exerts higher pressure when it is heated and expands, but the pressure in a degenerate gas does not depend on the temperature. When gas becomes super-compressed, particles position right up against each other to produce degenerate gas that behaves more like a solid. In degenerate gases the kinetic energies of electrons are quite high and the rate of collision between electrons and other particles is quite low, therefore degenerate electrons can travel great distances at velocities that approach the speed of light. Instead of temperature, the pressure in a degenerate gas depends only on the speed of the degenerate particles; however, adding heat does not increase the speed. Pressure is only increased by the mass of the particles, which increases the gravitational force pulling the particles closer together. Therefore, the phenomenon is the opposite of that normally found in matter where if the mass of the matter is increased, the object becomes bigger. In degenerate gas, when the mass is increased, the pressure is increased, and the particles become spaced closer together, so the object becomes smaller. Degenerate gas can be compressed to very high densities, typical values being in the range of 10,000 kilograms per cubic centimeter.

There is an upper limit to the mass of an electron-degenerate object, the Chandrasekhar limit, beyond which electron degeneracy pressure cannot support the object against collapse. The limit is approximately 1.44[7] solar masses for objects with compositions similar to the Sun. The mass cutoff changes with the chemical composition of the object, as this affects the ratio of mass to number of electrons present. Celestial objects below this limit are white dwarf stars, formed by the gradual shrinking of the cores of stars that run out of fuel. During this shrinking, an electron-degenerate gas forms in the core, providing sufficient degeneracy pressure as it is compressed to resist further collapse. Above this mass limit, a neutron star (supported by neutron degeneracy pressure) or a black hole may be formed instead.

14.2.2   Proton degeneracy

Sufficiently dense matter containing protons experiences proton degeneracy pressure, in a manner similar to the electron degeneracy pressure in electron-degenerate matter: protons confined to a sufficiently small volume have a large uncertainty in their momentum due to the Heisenberg uncertainty principle. Because protons are much more massive than electrons, the same momentum represents a much smaller velocity for protons than for electrons. As a result, in matter with approximately equal numbers of protons and electrons, proton degeneracy pressure is much smaller than electron degeneracy pressure, and proton degeneracy is usually modeled as a correction to the equations of state of electron-degenerate matter.

14.2.3   Neutron degeneracy

Neutron degeneracy is analogous to electron degeneracy and is demonstrated in neutron stars, which are primarily supported by the pressure from a degenerate neutron gas.[8] This happens when the core of a white dwarf star above the vicinity of 1.4 solar masses, the Chandrasekhar limit, collapses and is not halted by the degenerate electrons. As the star collapses, the Fermi energy of the electrons increases to the point where it is energetically favorable for them to combine with protons to produce neutrons (via inverse beta decay, also termed electron capture and "neutralization"). The result of this collapse is an extremely compact star composed of nuclear matter, which is predominantly a degenerate neutron gas, sometimes called neutronium, with a small admixture of degenerate proton and electron gases.

Neutrons in a degenerate neutron gas are spaced much more closely than electrons in an electron-degenerate gas because the more massive neutron has a much shorter wavelength at a given energy. In the case of neutron stars and white dwarf stars, this is compounded by the fact that the pressures within neutron stars are much higher than those in white dwarfs. The pressure increase is caused by the fact that the compactness of a neutron star causes gravitational forces to be much higher than in a less compact body with similar mass. This results in a star with a diameter on the order of a thousandth that of a white dwarf.

There is an upper limit to the mass of a neutron-degenerate object, the Tolman–Oppenheimer–Volkoff limit, which is analogous to the Chandrasekhar limit for electron-degenerate objects. The precise limit is unknown, as it depends on the equations of state of nuclear matter, for which a highly accurate model is not yet available. Above this limit, a neutron star may collapse into a black hole, or into other, denser forms of degenerate matter (such as quark matter) if these forms exist and have suitable properties (mainly related to degree of compressibility, or "stiffness", described by the equations of state).

14.2.4   Quark degeneracy

At densities greater than those supported by neutron degeneracy, quark matter is expected to occur. Several variations of this have been proposed that represent quark-degenerate states. Strange matter is a degenerate gas of quarks that is often assumed to contain strange quarks in addition to the usual up and down quarks. Color superconductor materials are degenerate gases of quarks in which quarks pair up in a manner similar to Cooper pairing in electrical superconductors. The equations of state for the various proposed forms of quark-degenerate matter vary widely, and are usually also poorly defined, due to the difficulty of modeling strong force interactions.

Quark-degenerate matter may occur in the cores of neutron stars, depending on the equations of state of neutron-degenerate matter. It may also occur in hypothetical quark stars, formed by the collapse of objects above the Tolman–Oppenheimer–Volkoff mass limit for neutron-degenerate objects. Whether quark-degenerate matter forms at all in these situations depends on the equations of state of both neutron-degenerate matter and quark-degenerate matter, both of which are poorly known.

14.2.5 Preon degeneracy hypothesis

Preons are subatomic particles proposed to be the constituents of quarks, which become composite particles in preon-based models. If preons exist, preon-degenerate matter might occur at densities greater than that which can be supported by quark-degenerate matter. The expected properties of preon-degenerate matter depend very strongly on the model chosen to describe preons, and the existence of preons is not assumed by the majority of the scientific community, due to conflicts between the preon models originally proposed and experimental data from particle accelerators.

14.2.6 Singularity

At densities greater than those supported by any degeneracy, gravity overwhelms all other forces. To the best of our current understanding, the body collapses to form a black hole. At the same time, the material must be converted from fermions, subject to degeneracy pressure, to bosons, which are not. Physics cannot currently predict what sort of bosons those might be.

In the frame of reference that is co-moving with the collapsing matter, general relativity models without quantum mechanics have all the matter ending up in an infinitely dense singularity at the center of the event horizon. It is a general result of quantum mechanics that no object can be confined in a space smaller than its own wavelength, making such a singularity impossible, but we do not have a theory that combines GR and QM sufficiently to tell us what the structure inside a black hole might be.

In the frame of reference of an observer at infinity, the collapse viewed in electromagnetic radiation asymptotically appears to approach the event horizon. Observations in gravitational waves, which are planned at LIGO, would let distant observers detect effects from matter falling inside the event horizon.

As a consequence of relativity, the extreme gravitational field and orbital velocity experienced by infalling matter around a black hole would "slow" time for that matter relative to a distant observer. In the slowed proper time experienced by the infalling matter, the fall to the center of the black hole would be quite short in duration.

14.3 See also

- Compact star

- White dwarf

- Neutron star

- Quark star—QCD matter

- Preon star—preon

- Pauli exclusion principle

- Uncertainty principle

- Neutronium

- Electron degeneracy pressure

- Nuclear matter

- Gravitational time dilation

- Matter wave

- List of plasma (physics) articles

14.4 Notes

[1] H.S. Goldberg, M.D. Scadron (1987). *Physics of Stellar Evolution and Cosmology*. Taylor & Francis. p. 202. ISBN 0-677-05540-4.

[2] An Introduction to Modern Astrophysics §16.3 "The Physics of Degenerate Matter – Carroll & Ostlie, 2007, second edition. ISBN 0-8053-0402-9

[3] see http://apod.nasa.gov/apod/ap100228.html

[4] Andrew G. Truscott, Kevin E. Strecker, William I. McAlexander, Guthrie Partridge, and Randall G. Hulet, "Observation of Fermi Pressure in a Gas of Trapped Atoms", Science, 2 March 2001

[5] On Dense Matter, R. H. Fowler, *Monthly Notices of the Royal Astronomical Society* **87** (1926), pp. 114–122.

[6] *Stellar Structure and Evolution* section 15.3 – R Kippenhahn & A. Weigert, 1990, 3rd printing 1994. ISBN 0-387-58013-1

[7] ENCYCLOPAEDIA BRITANNICA

[8] Potekhin, A. Y. (2011). "The Physics of Neutron Stars".arXiv:1102.5735.Bibcode:2010PhyU...53.1235Y.doi:10.3367/UF

14.5 References

- Cohen-Tanoudji, Claude (2011). *Advances in Atomic Physics*. World Scientific. p. 791. ISBN 978-981-277-496-5.

14.6 External links

- Detailed mathematical explanation of degenerate gases

- Mass-radius diagram of degenerate star types

Chapter 15

Strange matter

This article is about physics. For the book series, see Strange Matter.

Strange matter is a particular form of quark matter, usually thought of as a "liquid" of up, down and strange quarks. It is to be contrasted with nuclear matter, which is a liquid of neutrons and protons (which themselves are built out of up and down quarks), and with non-strange quark matter, which is a quark liquid containing only up and down quarks. At high enough density, strange matter is expected to be color superconducting. Strange matter is hypothesized to occur in the core of neutron stars, or, more speculatively, as isolated droplets that may vary in size from femtometers (strangelets) to kilometers (quark stars).

15.1 Two meanings of the term "strange matter"

In particle physics and astrophysics, the term is used in two ways, one broader and the other more specific[1][2]

1. The broader meaning is simply quark matter that contains three flavors of quarks: up, down, and strange. In this definition, there is a critical pressure and an associated critical density, and when nuclear matter (made of protons and neutrons) is compressed beyond this density, the protons and neutrons dissociate into quarks, yielding quark matter (probably strange matter).

2. The narrower meaning is that quark matter is *more stable than nuclear matter*, i.e. that the true ground state of matter is quark matter. The idea that this could happen is the "strange matter hypothesis" of Bodmer[3] and Witten.[4] In this definition, the critical pressure is zero. The nuclei that we see in the matter around us, which are droplets of nuclear matter, are actually metastable, and given enough time (or the right external stimulus) would decay into droplets of strange matter, i.e. strangelets.

15.2 Strange matter that is only stable at high pressure

Under the broader definition, strange matter might occur inside neutron stars, if the pressure at their core is high enough (i.e. above the critical pressure). At the sort of densities we expect in the center of a neutron star, the quark matter would probably be strange matter. It could conceivably be non-strange quark matter, if the effective mass of the strange quark were too high. Charm and heavier quarks would only occur at much higher densities.

A neutron star with a quark matter core is often[1][2] called a hybrid star. However, it is hard to know whether hybrid stars really exist in nature because physicists currently have little idea of the likely value of the critical pressure or density. It seems plausible that the transition to quark matter will already have occurred when the separation between the nucleons becomes much smaller than their size, so the critical density must be less than about 100 times nuclear saturation density. But a more precise estimate is not yet available, because the strong interaction that governs the behavior of quarks is mathematically intractable, and numerical calculations using lattice QCD are currently blocked by the fermion sign problem.

One major area of activity in neutron star physics is the attempt to find observable signatures by which we could tell, from earth based observations of neutron stars, whether they have quark matter (probably strange matter) in their core.

15.3   Strange matter that is stable at zero pressure

If the "strange matter hypothesis" is true then nuclear matter is metastable against decaying into strange matter. The lifetime for spontaneous decay is very long, so we do not see this decay process happening around us.[4] However, under this hypothesis there should be strange matter in the universe:

1. Quark stars (often called "strange stars") consist of quark matter from their core to their surface. They would be several kilometers across, and may have a very thin crust of nuclear matter.[2]

2. Strangelets are small pieces of strange matter, perhaps as small as nuclei. They would be produced when strange stars are formed or collide, or when a nucleus decays.[1]

15.4   See also

- Exotic matter

- Negative matter

- Quark matter

- Quark star

- Strangeness production

- Strangelet

- Quark

- QCD matter

15.5   References

[1] J. Madsen, "Physics and astrophysics of strange quark matter" arXiv:astro-ph/9809032, Lect. Notes Phys. 516:162-203 (1999)

[2] F. Weber, "Strange quark matter and compact stars", arXiv:astro-ph/0407155, Prog.Part.Nucl.Phys.54:193-288,2005.

[3] A. Bodmer "Collapsed Nuclei" Phys. Rev. D4, 1601 (1971)

[4] Edward Witten, "Cosmic Separation Of Phases" Phys. Rev. D30, 272 (1984)

Chapter 16

Photonic molecule

Photonic molecules are a synthetic form of matter in which photons bind together to form "molecules". According to Mikhail Lukin, individual (massless) photons "interact with each other so strongly that they act as though they have mass". The effect is analogous to refraction. The light enters another medium, transferring part of its energy to the medium. Inside the medium, it exists as coupled light and matter, but it exits as light.[1]

Researchers drew analogies between the phenomenon and the fictional "lightsaber" from Star Wars.[1][2]

16.1 Construction

Gaseous rubidium atoms were pumped into a vacuum chamber. The cloud was cooled using lasers to just a few degrees above absolute zero. Using weak laser pulses, small numbers of photons were fired into the cloud.[1]

As the photons entered the cloud, their energy excited atoms along their path, causing them to lose speed. Inside the cloud medium the photons dispersively coupled to strongly interacting atoms in highly excited Rydberg states. This caused the photons to behave as massive particles with strong mutual attraction (photon molecules). Eventually the photons exited the cloud together as normal photons (often entangled in pairs).[1]

The effect is caused by a so-called Rydberg blockade, which, in the presence of one excited atom, prevents nearby atoms from being excited to the same degree. In this case, as two photons enter the atomic cloud, the first excites an atom, but must move forward before the second can excite nearby atoms. In effect the two photons push and pull each other through the cloud as their energy is passed from one atom to the next, forcing them to interact. This photonic interaction is mediated by the electromagnetic interaction between photons and atoms.[1]

16.2 Possible applications

The interaction of the photons suggests that the effect could be employed to build a system that can preserve quantum information, and process it using quantum logic operations.[1]

The system could also be useful in classical computing, given the much-lower power required to manipulate photons than electrons.[1]

It may be possible to arrange the photonic molecules in such a way within the medium that they form larger three-dimensional structures (similar to crystals).[1]

16.3   Interacting microcavities

The term photonic molecule has been also used since 1998 for an unrelated phenomenon involving electromagnetically-interacting optical microcavities. The properties of quantized confined photon states in optical micro- and nanocavities are very similar to those of confined electron states in atoms.[3] Owing to this similarity, optical microcavities can be termed 'photonic atoms'. Taking this analogy even further, a cluster of several mutually-coupled photonic atoms forms a photonic molecule.[4] When individual photonic atoms are brought into close proximity, their optical modes interact and give rise to a spectrum of hybridized super-modes of photonic molecules.[5]

"A micrometer-sized piece of semiconductor can trap photons inside it in such a way that they act like electrons in an atom. Now the 21 September PRL describes a way to link two of these "photonic atoms" together. The result of such a close relationship is a "photonic molecule," whose optical modes bear a strong resemblance to the electronic states of a diatomic molecule like hydrogen."[6]

"Photonic molecules, named by analogy with chemical molecules, are clusters of closely located electromagnetically interacting microcavities or "photonic atoms"."[7]

"Optically coupled microcavities have emerged as photonic structures with promising properties for investigation of fundamental science as well as for applications."[8]

The first demonstration of a lithographically-fabricated photonic molecule was inspired by an analogy with a simple diatomic molecule.[9] However, other nature-inspired PM structures (such as 'photonic benzene') have been proposed and shown to support confined optical modes closely analogous to the ground-state molecular orbitals of their chemical counterparts.[10]

Photonic molecules offer advantages over isolated photonic atoms in a variety of applications, including bio(chemical) sensing,[11][12] cavity optomechanics,[13][14] and microlasers,[15][16] ,[17][18] Photonic molecules can also be used as quantum simulators of many-body physics and as building blocks of future optical quantum information processing networks.[19]

In complete analogy, clusters of metal nanoparticles - which support confined surface plasmon states - have been termed 'plasmonic molecules.",[20][21][22][23][24]

Finally, hybrid photonic-plasmonic (or opto-plasmonic) molecules have also been proposed and demonstrated., [25][26][27][28]

16.4   References

[1] "Seeing light in a new light: Scientists create never-before-seen form of matter". Sciencedaily.com. Retrieved 2013-09-27.

[2] Firstenberg, O.; Peyronel, T.; Liang, Q. Y.; Gorshkov, A. V.; Lukin, M. D.; Vuletić, V. (2013). "Attractive photons in a quantum nonlinear medium". *Nature*. doi:10.1038/nature12512.

[3] Benson, T. M.; Boriskina, S. V.; Sewell, P.; Vukovic, A.; Greedy, S. C.; Nosich, A. I. (2006). "Micro-Optical Resonators for Microlasers and Integrated Optoelectronics". *Frontiers in Planar Lightwave Circuit Technology*. NATO Science Series II: Mathematics, Physics and Chemistry **216**. p. 39. doi:10.1007/1-4020-4167-5_02. ISBN 1-4020-4164-0.

[4] Boriskina, S. V. (2010). "Photonic Molecules and Spectral Engineering". *Photonic Microresonator Research and Applications*. Springer Series in Optical Sciences **156**. p. 393. doi:10.1007/978-1-4419-1744-7_16. ISBN 978-1-4419-1743-0.

[5] Rakovich, Y.; Donegan, J.; Gerlach, M.; Bradley, A.; Connolly, T.; Boland, J.; Gaponik, N.; Rogach, A. (2004). "Fine structure of coupled optical modes in photonic molecules". *Physical Review A* **70** (5). doi:10.1103/PhysRevA.70.051801.

[6] doi:10.1103/PhysRevFocus.2.14

[7] arXiv:0704.2154

[8] doi:10.1038/lsa.2013.38

[9] Bayer, M.; Gutbrod, T.; Reithmaier, J.; Forchel, A.; Reinecke, T.; Knipp, P.; Dremin, A.; Kulakovskii, V. (1998). "Optical Modes in Photonic Molecules". *Physical Review Letters* **81** (12): 2582. doi:10.1103/PhysRevLett.81.2582.

[10] Lin, B. (2003). "Variational analysis for photonic molecules: Application to photonic benzene waveguides". *Physical Review E* **68** (3). doi:10.1103/PhysRevE.68.036611.

[11] Boriskina, S. V. (2006). "Spectrally engineered photonic molecules as optical sensors with enhanced sensitivity: A proposal and numerical analysis". *Journal of the Optical Society of America B* **23** (8): 1565. doi:10.1364/JOSAB.23.001565.

[12] Boriskina, S. V.; Dal Negro, L. (2010). "Self-referenced photonic molecule bio(chemical)sensor". *Optics Letters* **35** (14): 2496–8. doi:10.1364/OL.35.002496. PMID 20634875.

[13] Jiang, X.; Lin, Q.; Rosenberg, J.; Vahala, K.; Painter, O. (2009). "High-Q double-disk microcavities for cavity optomechanics". *Optics Express* **17** (23): 20911–9. doi:10.1364/OE.17.020911. PMID 19997328.

[14] Hu, Y. W.; Xiao, Y. F.; Liu, Y. C.; Gong, Q. (2013). "Optomechanical sensing with on-chip microcavities". *Frontiers of Physics* **8** (5): 475. doi:10.1007/s11467-013-0384-y.

[15] Hara, Y.; Mukaiyama, T.; Takeda, K.; Kuwata-Gonokami, M. (2003). "Photonic molecule lasing". *Optics Letters* **28** (24): 2437–9. doi:10.1364/OL.28.002437. PMID 14690107.

[16] Nakagawa, A.; Ishii, S.; Baba, T. (2005). "Photonic molecule laser composed of GaInAsP microdisks". *Applied Physics Letters* **86** (4): 041112. doi:10.1063/1.1855388.

[17] Boriskina, S. V. (2006). "Theoretical prediction of a dramatic Q-factor enhancement and degeneracy removal of whispering gallery modes in symmetrical photonic molecules". *Optics Letters* **31** (3): 338–40. doi:10.1364/OL.31.000338. PMID 16480201.

[18] Smotrova, E. I.; Nosich, A. I.; Benson, T. M.; Sewell, P. (2006). "Threshold reduction in a cyclic photonic molecule laser composed of identical microdisks with whispering-gallery modes". *Optics Letters* **31** (7): 921–3. doi:10.1364/OL.31.000921. PMID 16599212.

[19] Hartmann, M.; Brandão, F.; Plenio, M. (2007). "Effective Spin Systems in Coupled Microcavities". *Physical Review Letters* **99** (16). doi:10.1103/PhysRevLett.99.160501.

[20] Nordlander, P.; Oubre, C.; Prodan, E.; Li, K.; Stockman, M. I. (2004). "Plasmon Hybridization in Nanoparticle Dimers". *Nano Letters* **4** (5): 899. doi:10.1021/nl049681c.

[21] Fan, J. A.; Bao, K.; Wu, C.; Bao, J.; Bardhan, R.; Halas, N. J.; Manoharan, V. N.; Shvets, G.; Nordlander, P.; Capasso, F. (2010). "Fano-like Interference in Self-Assembled Plasmonic Quadrumer Clusters". *Nano Letters* **10** (11): 4680–5. doi:10.1021/nl1029 732.PMID20923179.

[22] Liu, N.; Mukherjee, S.; Bao, K.; Brown, L. V.; Dorfmüller, J.; Nordlander, P.; Halas, N. J. (2012). "Magnetic Plasmon Formation and Propagation in Artificial Aromatic Molecules". *Nano Letters* **12** (1): 364–9. doi:10.1021/nl203641z. PMID 22122612.

[23] Yan, B.; Boriskina, S. V.; Reinhard, B. R. M. (2011). "Optimizing Gold Nanoparticle Cluster Configurations (n≤ 7) for Array Applications". *The Journal of Physical Chemistry C* **115** (11): 4578. doi:10.1021/jp112146d.

[24] Yan, B.; Boriskina, S. V.; Reinhard, B. R. M. (2011). "Design and Implementation of Noble Metal Nanoparticle Cluster Arrays for Plasmon Enhanced Biosensing". *The Journal of Physical Chemistry C* **115** (50): 24437. doi:10.1021/jp207821t.

[25] Boriskina, S. V.; Reinhard, B. M. (2011). "Spectrally and spatially configurable superlenses for optoplasmonic nanocircuits". *Proceedings of the National Academy of Sciences* **108** (8): 3147. doi:10.1073/pnas.1016181108.

[26] Boriskina, S. V.; Reinhard, B. R. M. (2011). "Adaptive on-chip control of nano-optical fields with optoplasmonic vortex nanogates". *Optics Express* **19** (22): 22305–15. doi:10.1364/OE.19.022305. PMC 3298770. PMID 22109072.

[27] Hong, Y.; Pourmand, M.; Boriskina, S. V.; Reinhard, B. R. M. (2013). "Enhanced Light Focusing in Self-Assembled Optoplasmonic Clusters with Subwavelength Dimensions". *Advanced Materials* **25**: 115. doi:10.1002/adma.201202830.

[28] Ahn, W.; Boriskina, S. V.; Hong, Y.; Reinhard, B. R. M. (2012). "Photonic–Plasmonic Mode Coupling in On-Chip Integrated Optoplasmonic Molecules". *ACS Nano* **6** (1): 951–60. doi:10.1021/nn204577v. PMID 22148502.

16.5 Notes

- http://prl.aps.org/abstract/PRL/v81/i12/p2582_1

Chapter 17

Quantum Hall effect

The **quantum Hall effect** (or **integer quantum Hall effect**) is a quantum-mechanical version of the Hall effect, observed in two-dimensional electron systems subjected to low temperatures and strong magnetic fields, in which the Hall conductance σ undergoes certain quantum Hall transitions to take on the quantized values

$$\sigma = \frac{I_{\text{channel}}}{V_{\text{Hall}}} = \nu\,\frac{e^2}{h},$$

where I_{channel} is the channel current, V_{Hall} is the Hall voltage, e is the elementary charge and h is Planck's constant. The prefactor ν is known as the "filling factor", and can take on either integer ($\nu = 1, 2, 3, ...$) or fractional ($\nu = 1/3, 2/5,$ $3/7, 2/3, 3/5, 1/5, 2/9, 3/13, 5/2, 12/5, ...$) values. The quantum Hall effect is referred to as the integer or fractional quantum Hall effect depending on whether ν is an integer or fraction, respectively. The integer quantum Hall effect is very well understood, and can be simply explained in terms of single-particle orbitals of an electron in a magnetic field (see Landau quantization). The fractional quantum Hall effect is more complicated, as its existence relies fundamentally on electron–electron interactions. Although the microscopic origins of the fractional quantum Hall effect are unknown, there are several phenomenological approaches that provide accurate approximations. For example the effect can be thought of as an integer quantum Hall effect, not of electrons but of charge-flux composites known as composite fermions. In 1988, it was proposed that there was quantum Hall effect without Landau levels. This quantum Hall effect is referred to as the quantum anomalous Hall (QAH) effect. There is also a new concept of the quantum spin Hall effect which is an analogue of the quantum Hall effect, where spin currents flow instead of charge currents.[1]

17.1   Applications

The quantization of the Hall conductance has the important property of being incredibly precise. Actual measurements of the Hall conductance have been found to be integer or fractional multiples of e^2/h to nearly one part in a billion. This phenomenon, referred to as "exact quantization", has been shown to be a subtle manifestation of the principle of gauge invariance.[2] It has allowed for the definition of a new practical standard for electrical resistance, based on the resistance quantum given by the von Klitzing constant $RK = h/e^2 = 25812.807557(18)\ \Omega$.[3] This is named after Klaus von Klitzing, the discoverer of exact quantization. Since 1990, a fixed conventional value $RK\text{-}_{90}$ is used in resistance calibrations worldwide.[4] The quantum Hall effect also provides an extremely precise independent determination of the fine structure constant, a quantity of fundamental importance in quantum electrodynamics.

17.2   History

The integer quantization of the Hall conductance was originally predicted by Ando, Matsumoto, and Uemura in 1975, on the basis of an approximate calculation which they themselves did not believe to be true. Several workers subsequently observed the effect in experiments carried out on the inversion layer of MOSFETs. It was only in 1980 that Klaus von Klitzing, working at the high magnetic field laboratory in Grenoble with silicon-based samples developed by Michael Pepper and Gerhard Dorda, made the unexpected discovery that the Hall conductivity was *exactly* quantized. For this finding, von Klitzing was awarded the 1985 Nobel Prize in Physics. The link between exact quantization and gauge invariance was subsequently found by Robert Laughlin. Most integer quantum Hall experiments are now performed on gallium arsenide heterostructures, although many other semiconductor materials can be used. In 2007, the integer quantum Hall effect was reported in graphene at temperatures as high as room temperature,[5] and in the oxide ZnO-$Mg_xZn_{1-x}O$.[6]

17.3 Integer quantum Hall effect – Landau levels

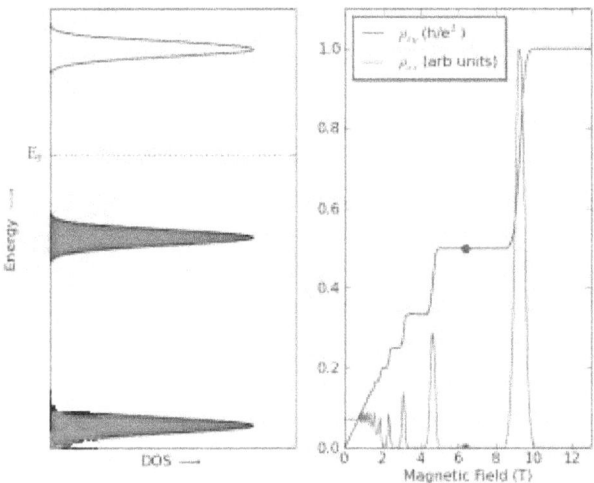

In two dimensions, when classical electrons are subjected to a magnetic field they follow circular cyclotron orbits. When the system is treated quantum mechanically, these orbits are quantized. The energy levels of these quantized orbitals take on discrete values:

$$E_n = \hbar\omega_c(n + 1/2),$$

where $\omega c = eB/m$ is the cyclotron frequency. These orbitals are known as Landau levels, and at weak magnetic fields, their existence gives rise to many interesting "quantum oscillations" such as the Shubnikov–de Haas oscillations and the de Haas–van Alphen effect (which is often used to map the Fermi surface of metals). For strong magnetic fields, each Landau level is highly degenerate (i.e. there are many single particle states which have the same energy En). Specifically, for a sample of area A, in magnetic field B, the degeneracy of each Landau level is

$$N = g_sBA/\phi_0,$$

where gs represents a factor of 2 for spin degeneracy, and $\phi_0 = 2\cdot10^{-15}$ Wb is the magnetic flux quantum. For sufficiently strong B-fields, each Landau level may have so many states that all of the free electrons in the system sit in only a few Landau levels; it is in this regime where one observes the quantum Hall effect.

17.4 Mathematics

The integers that appear in the Hall effect are examples of topological quantum numbers. They are known in mathematics as the first Chern numbers and are closely related to Berry's phase. A striking model of much interest in this context

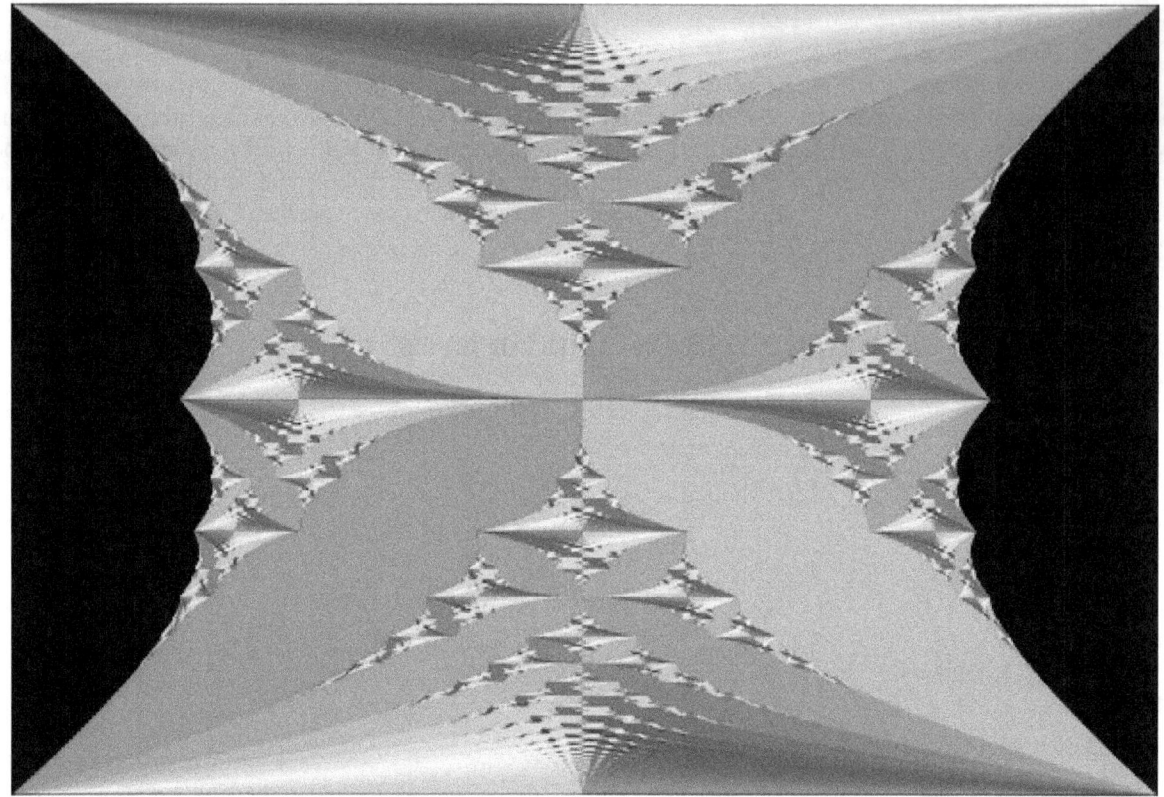

Hofstadter's butterfly

is the Azbel-Harper-Hofstadter model whose quantum phase diagram is the Hofstadter butterfly shown in the figure. The vertical axis is the strength of the magnetic field and the horizontal axis is the chemical potential, which fixes the electron density. The colors represent the integer Hall conductances. Warm colors represent positive integers and cold colors negative integers. The phase diagram is fractal and has structure on all scales. In the figure there is an obvious self-similarity.

Concerning physical mechanisms, impurities and/or particular states (e.g., edge currents) are important for both the 'integer' and 'fractional' effects. In addition, Coulomb interaction is also essential in the fractional quantum Hall effect. The observed strong similarity between integer and fractional quantum Hall effects is explained by the tendency of electrons to form bound states with an even number of magnetic flux quanta, called *composite fermions*.

17.5   See also

- Quantum Hall transitions

- Fractional quantum Hall effect

- Composite fermions

- Hall effect

- Hall probe

- Graphene

- Quantum spin Hall effect

- Coulomb potential between two current loops embedded in a magnetic field

17.6 References

[1] Ezawa, Zyun F. (2013). *Quantum Hall Effects: Recent Theoretical and Experimental Developments* (3rd ed.). World Scientific. ISBN 978-981-4360-75-3.

[2] Laughlin, R. (1981)."Quantized Hall conductivity in two dimensions".*Physical Review B***23**(10): 5632–5633.Bibcode:1981P doi:10.1103/PhysRevB.23.5632. Retrieved 8 May 2012.

[3] Tzalenchuk, Alexander; Lara-Avila, Samuel; Kalaboukhov, Alexei; Paolillo, Sara; Syväjärvi, Mikael; Yakimova, Rositza; Kazakova, Olga; Janssen, T. J. B. M.; Fal'ko, Vladimir; Kubatkin, Sergey (2010). "Towards a quantum resistance standard based on epitaxial graphene". *Nature Nanotechnology* **5** (3): 186–189. arXiv:0909.1220. Bibcode:2010NatNa...5..186T. doi:10.1038/nnano.2009.474. PMID 20081845.

[4] "conventional value of von Klitzing constant". *NIST*.

[5] Novoselov, K. S.; Jiang, Z.; Zhang, Y.; Morozov, S. V.; Stormer, H. L.; Zeitler, U.; Maan, J. C.; Boebinger, G. S.; Kim, P.; Geim, A. K. (2007). "Room-Temperature Quantum Hall Effect in Graphene". *Science* **315** (5817): 1379. arXiv:cond-mat/0702408. Bibcode:2007Sci...315.1379N. doi:10.1126/science.1137201. PMID 17303717.

[6] Tsukazaki, A.; Ohtomo, A.; Kita, T.; Ohno, Y.; Ohno, H.; Kawasaki, M. (2007). "Quantum Hall Effect in Polar Oxide Heterostructures". *Science* **315** (5817): 1388–91. Bibcode:2007Sci...315.1388T. doi:10.1126/science.1137430. PMID 17255474.

17.7 Further reading

- Ando, Tsuneya; Matsumoto, Yukio; Uemura, Yasutada (1975). "Theory of Hall Effect in a Two-Dimensional Electron System". *J. Phys. Soc. Jpn.* **39** (2): 279–288. Bibcode:1975JPSJ...39..279A. doi:10.1143/JPSJ.39.279.

- Klitzing, K.; Dorda, G.; Pepper, M. (1980). "New Method for High-Accuracy Determination of the Fine-Structure Constant Based on Quantized Hall Resistance". *Phys. Rev. Lett.* **45** (6): 494–497. Bibcode:1980PhRvL..45..494K. doi:10.1103/PhysRevLett.45.494.

- Laughlin, R. B. (1981). "Quantized Hall conductivity in two dimensions". *Phys. Rev. B.* **23** (10): 5632–5633. Bibcode:1981PhRvB..23.5632L. doi:10.1103/PhysRevB.23.5632.

- Yennie, D. R. (1987). "Integral quantum Hall effect for nonspecialists". *Rev. Mod. Phys.* **59** (3): 781–824. Bibcode:1987RvMP...59..781Y. doi:10.1103/RevModPhys.59.781.

- Hsieh, D.; Qian, D.; Wray, L.; Xia, Y.; Hor, Y. S.; Cava, R. J.; Hasan, M. Z. (2008). "A topological Dirac insulator in a quantum spin Hall phase". *Nature* **452** (7190): 970–974. arXiv:0902.1356. Bibcode:2008Natur.452..970H. doi:10.1038/nature06843. PMID 18432240.

- *25 years of Quantum Hall Effect*, K. von Klitzing, Poincaré Seminar (Paris-2004). Postscript. Pdf.

- Magnet Lab Press Release Quantum Hall Effect Observed at Room Temperature

- Avron, Joseph E.; Osadchy, Daniel; Seiler, Ruedi (2003). "A Topological Look at the Quantum Hall Effect". *Physics Today* **56** (8): 38. Bibcode:2003PhT....56h..38A. doi:10.1063/1.1611351. Retrieved 8 May 2012.

- Zyun F. Ezawa: *Quantum Hall Effects - Field Theoretical Approach and Related Topics.* World Scientific, Singapore 2008, ISBN 978-981-270-032-2

- Sankar D. Sarma, Aron Pinczuk: *Perspectives in Quantum Hall Effects.* Wiley-VCH, Weinheim 2004, ISBN 978-0-471-11216-7

- Baumgartner, A.; Ihn, T.; Ensslin, K.; Maranowski, K.; Gossard, A. (2007). "Quantum Hall effect transition in scanning gate experiments".*Physical Review B***76**(8).Bibcode:2007PhRvB..76h5316B.doi:10.1103/PhysRevB.7

- E. I. Rashba and V. B. Timofeev, Quantum Hall Effect, Sov. Phys. - Semiconductors v. 20, pp. 617–647 (1986).

Chapter 18

Quantum spin Hall effect

The quantum spin Hall state is a state of matter proposed to exist in special, two-dimensional, semiconductors that have a quantized spin-Hall conductance and a vanishing charge-Hall conductance. The quantum spin Hall state of matter is the cousin of the integer quantum Hall state, and both states can be realized on lattice that does not require the application of a large magnetic field. The quantum spin Hall state does not break charge conservation symmetry and spin- S_z conservation symmetry (in order to have well defined Hall conductances). The first proposal for the existence of a quantum spin Hall state was developed by Kane and Mele[1] who adapted an earlier model for graphene by F. Duncan M. Haldane[2] which exhibits an integer quantum Hall effect. The Kane and Mele model is two copies of the Haldane model such that the spin up electron exhibits a chiral integer quantum Hall Effect while the spin down electron exhibits an anti-chiral integer quantum Hall effect.

Overall the Kane-Mele model has a charge-Hall conductance of exactly zero but a spin-Hall conductance of exactly $\sigma_{xy}^{spin} = 2$ (in units of $\frac{e}{4\pi}$). Independently, a quantum spin Hall model was proposed by Bernevig and Zhang[3] in an intricate strain architecture which engineers, due to spin-orbit coupling, a magnetic field pointing upwards for spin-up electrons and a magnetic field pointing downwards for spin-down electrons. The main ingredient is the existence of spin-orbit coupling, which can be understood as a momentum-dependent magnetic field coupling to the spin of the electron.

Real experimental systems, however, are far from the idealized picture presented above in which spin-up and spin-down electrons are not coupled. A very important achievement was the realization that the quantum spin Hall state remain to be non-trivial even after the introduction of spin-up spin-down scattering,[4] which destroy the quantum spin Hall effect. In a separate paper, Kane and Mele introduced a topological Z_2 invariant who characterizes a state as trivial or non-trivial band insulator (regardless if the state exhibits or does not exhibit a quantum spin Hall effect). Further stability studies of the edge liquid through which conduction takes place in the quantum spin Hall state proved, both analytically and numerically that the non-trivial state is robust to both interactions and extra spin-orbit coupling terms that mix spin-up and spin-down electrons. Such a non-trivial state (exhibiting or not exhibiting a quantum spin Hall effect) is called a topological insulator, which is an example of symmetry protected topological order protected by charge conservation symmetry and time reversal symmetry. (Note that the quantum spin Hall state is also a symmetry protected topological state protected by charge conservation symmetry and spin- S_z conservation symmetry. We do not need time reversal symmetry to protect quantum spin Hall state. Topological insulator and quantum spin Hall state are different symmetry protected topological states. So Topological insulator and quantum spin Hall state are different states of matter.)

18.1   In HgTe quantum wells

Since graphene has extremely weak spin-orbit coupling, it is very unlikely to support a quantum spin Hall state at temperatures achievable with today's technologies. A very realistic theoretical proposal for the existence of the quantum spin Hall state has been put forward in 1987 by Pankratov, Pakhomov and Volkov in Cadmium Telluride/Mercury Telluride/Cadmium Telluride (CdTe/HgTe/CdTe) quantum wells in which a thin (5-7 nanometers) sheet of HgTe is sandwiched between two sheets of CdTe, [5] and subsequently experimentally realized [6] (In fact, the proposed quantum wells do not exhibit the quantum spin Hall effect and do not demonstrate the existence of the quantum spin Hall state. It should be regarded as topological insulators. So far quantum spin Hall effect has not been observed in experiments.) Different quantum wells of varying HgTe thickness can be built. When the sheet of HgTe in between the CdTe is thin, the system behaves like an ordinary insulator and does not conduct when the Fermi level resides in the band-gap. When the sheet of HgTe is varied and made thicker (this requires the fabrication of separate quantum wells), an interesting phenomenon happens. Due to the inverted band structure of HgTe, at some critical HgTe thickness, a Lifshitz transition occurs in which the system closes the bulk band gap to become a semi-metal, and then re-opens it to become a quantum spin Hall insulator.

In the gap closing and re-opening process, two edge states are brought out from the bulk and cross the bulk-gap. As such, when the Fermi level resides in the bulk gap, the conduction is dominated by the edge channels that cross the gap. The two-terminal conductance is $G_{xx} = 2\frac{e^2}{h}$ in the quantum spin Hall state and zero in the normal insulating state. As the conduction is dominated by the edge channels, the value of the conductance should be insensitive to how wide the sample is. A magnetic field should destroy the quantum spin Hall state by breaking time-reversal invariance and allowing spin-up spin-down electron scattering processes at the edge. All these predictions have been experimentally verified in an experiment [6] performed in the Molenkamp labs at Universitat Würzburg in Germany. (In fact, a magnetic field in z-direction does not destroy the quantum spin Hall state which has conserved S_z spins. The fact that the quantization of the two-terminal conductance is destroyed by magnetic field suggests that the quantum well is not a quantum spin Hall state, but a topological insulator.)

18.2   References

[1] C.L. Kane and E.J. Mele, *Quantum Spin Hall Effect in Graphene*, Physical Review Letters 95, 226801 (2005).

[2] F.D.M. Haldane, *Model for a Quantum Hall Effect without Landau Levels: Condensed-Matter Realization of the "Parity Anomaly"*, Physical Review Letters 61, 2015 (1988).

[3] B.A. Bernevig and S.C. Zhang, *Quantum Spin Hall Effect*, Physical Review Letters 96, 106802 (2006)

[4] C.L. Kane and E.J. Mele, *Z2 Topological Order and the Quantum Spin Hall Effect*, Physical Review Letters 95, 146802 (2005)

[5] Pankratov, O.A.; Pakhomov, S.V.; Volkov, B.A. (January 1987). "Supersymmetry in heterojunctions: Band-inverting contact on the basis of Pb1-xSnxTe and Hg1-xCdxTe". *Solid State Communications* **61** (2): 93–96. doi:10.1016/0038-1098(87)90934-3.

[6] Markus König, Steffen Wiedmann, Christoph Brüne, Andreas Roth, Hartmut Buhmann, Laurens W. Molenkamp, Xiao-Liang Qi, and Shou-Cheng Zhang, *Quantum Spin Hall Insulator State in HgTe Quantum Wells*, Published online September 20, 2007; 10.1126/science.1148047 (Science Express Research Articles)

18.3   Further reading

- Maciejko, J.; Hughes, T. L.; Zhang, S. C. (2011). "The Quantum Spin Hall Effect". *Annual Review of Condensed Matter Physics* **2**: 31. doi:10.1146/annurev-conmatphys-062910-140538.

Chapter 19

Rydberg matter

Rydberg matter is a phase of matter formed by Rydberg atoms; it was predicted around 1980 by É. A. Manykin, M. I. Ozhovan and P. P. Poluéktov.[1][2] It has been formed from various elements like caesium,[3] potassium,[4] hydrogen[5][6] and nitrogen;[7] studies have been conducted on theoretical possibilities like sodium, beryllium, magnesium and calcium.[8] It has been suggested to be a material that diffuse interstellar bands may arise from;[9] circular[10] Rydberg states, where the outermost electron is found in a planar circular orbit, are the most long-lived with lifetimes of up to several hours[11] and are the most common.[12][13][14]

19.1   Physical

Rydberg matter consists of usually[16] hexagonal ɪplanar clusters; these cannot be very big because of the retardation eff ect caused by thefinite velocity of the speed of light.[18]Hence,they are not gases or plasmas;nor are they solids or liquids;they are most similar todusty plasmaswith small clusters in a gas.Though Rydberg matter can be studied in the laboratory bylaser probing, [19] the largest cluster reported consists of only 91 atoms,[6] but it has been shown to be behind extended clouds in space[9][20] and the upper atmosphere of planets.[21] Bonding in Rydberg matter is caused by delocalisation of the high-energy electrons to form an overall lower energy state.[2] The way in which the electrons delocalise is to form standing waves on loops surrounding nuclei, creating quantised angular momentum and the defining characteristics of Rydberg matter. It is a generalised metal by way of the quantum numbers influencing loop size but restricted by the bonding requirement for strong electron correlation;[18] it shows exchange-correlation properties similar to covalent bonding.[22] Electronic excitation and vibrational motion of these bonds can be studied by Raman spectroscopy.[23]

19.2   Lifetime

Due to reasons still debated by the physics community because of the lack of methods to observe clusters,[26] Rydberg matter is highly stable against disintegration by emission of radiation; the characteristic lifetime of a cluster at n = 12 is 25 seconds.[25][27] Reasons given include the lack of overlap between excited and ground states, the forbidding of transitions between them and exchange-correlation effects hindering emission through necessitating tunnelling[22] that causes a long delay in excitation decay.[24] Excitation plays a role in determining lifetimes, with a higher excitation giving a longer lifetime;[25] n = 80 gives a lifetime comparable to the age of the Universe.[28]

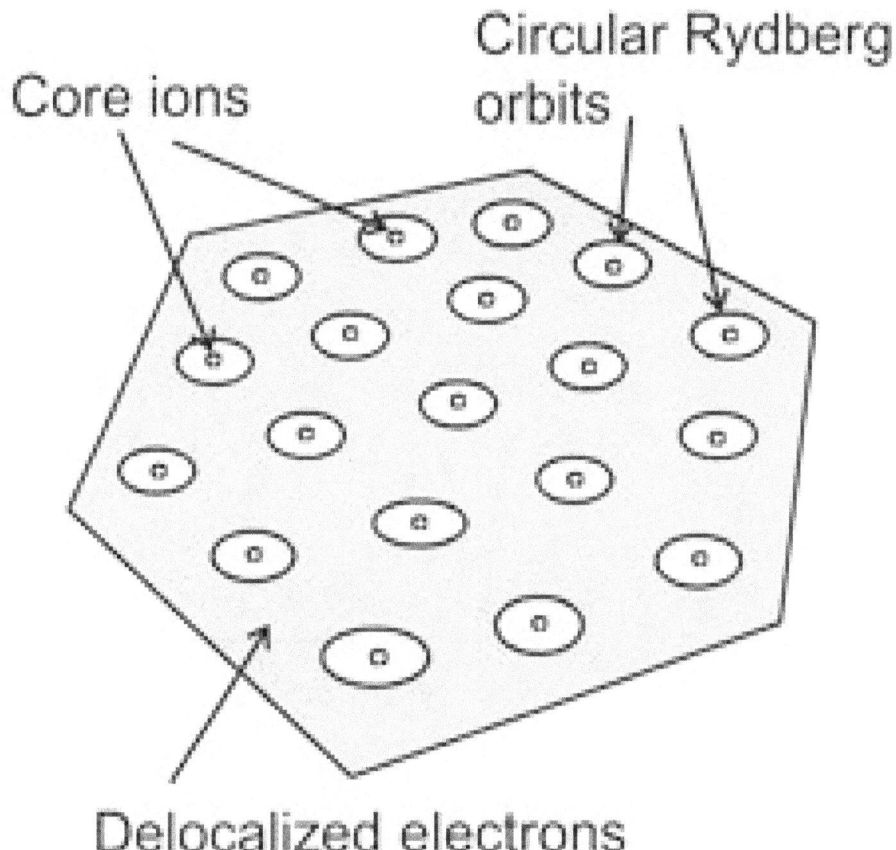

A 19-atom planar Rydberg matter cluster. At the seventh excitation level, spectroscopy on K_{19} clusters showed the bond distance to be 5.525 nm.[15]

19.3 Excitations

In ordinary metals, interatomic distances are nearly constant through a wide range of temperatures and pressures; this is not the case with Rydberg matter, whose distances and thus properties vary greatly with excitations. A key variable in determining these properties is the principal quantum number n that can be any integer greater than 1; the highest values reported for it are around 100.[28][29] Bond distance d in Rydberg matter is given by

$$d = 2.9n^2a_0$$

where a_0 is the Bohr radius. The approximate factor 2.9 was first experimentally determined, then measured with rotational spectroscopy in different clusters.[15] Examples of d calculated this way, along with selected values of the density D, are given in the table to the right.

19.4 Condensation

Like bosons that can be condensed to form Bose–Einstein condensates, Rydberg matter can be condensed, but not in the same way as bosons. The reason for this is that Rydberg matter behaves similarly to a gas, meaning that it cannot be

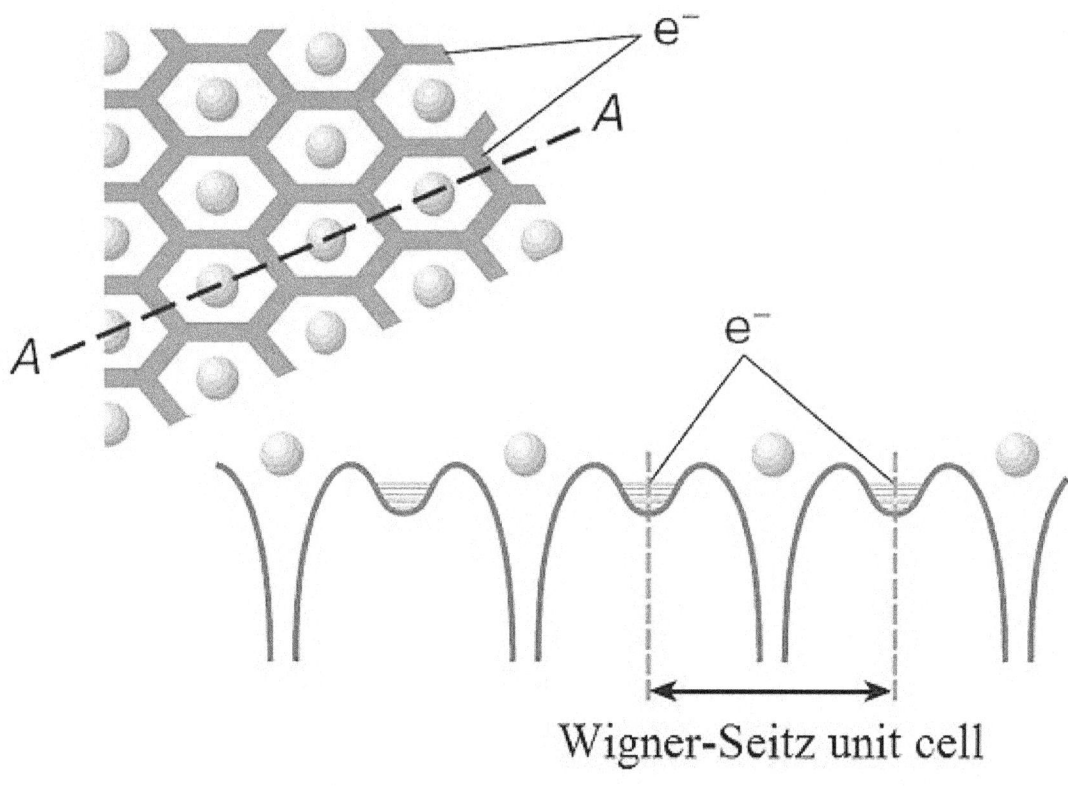

Schematic of valence electron distribution in a Rydberg matter made of excited (n=10) Cs atoms.

condensed without removing the condensation energy; ionisation occurs if this is not done. All solutions to this problem so far involve using an adjacent surface in some way, the best being evaporating the atoms of which the Rydberg matter is to be formed from and leaving the condensation energy on the surface.[30] Using caesium atoms, graphite-covered surfaces and thermionic converters as containment, the work function of the surface has been measured to be 0.5eV,[31] indicating that the cluster is between the ninth and fourteenth excitation levels.[24]

19.5   See also

- State of matter

- The Journal of Cluster Science has dedicated in 2012 a special issue (Volume 23, Issue 1) to "Rydberg-Matter and Excited-State Clusters".[32]

- The President of Russian Academy of Sciences Vladimir Fortov has recently noted the works on Rydberg Matter as a great scientific event [33]

19.6   References

[1] É.A. Manykin, M.I. Ozhovan, P.P. Poluéktov (1980). "Transition of an excited gas to a metallic state". *Sov. Phys. Tech. Phys. Lett.* **6**: 95.

[2] É.A. Manykin, M.I. Ozhovan, P.P. Poluéktov; Ozhovan; Poluéktov (1981). "On the collective electronic state in a system of strongly excited atoms". *Sov. Phys. Dokl.* **26**: 974–975. Bibcode:1981SPhD...26..974M.

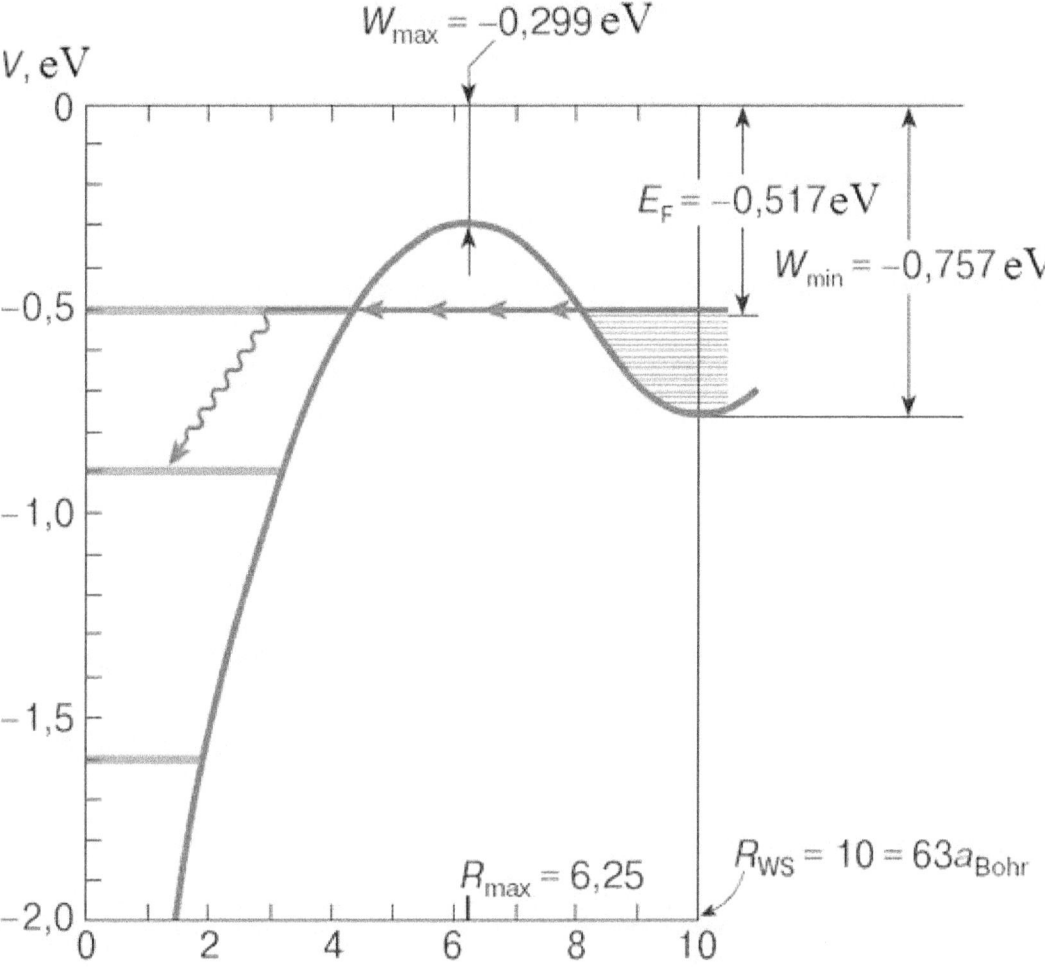

Schematic of an effective potential within a Wigner-Seitz cell of a Rydberg matter made of excited (n=10) Cs atoms.[24][25]

[3] V.I. Yarygin, V.N. Sidel'nikov, I.I. Kasikov, V.S. Mironov, and S.M. Tulin (2003). "Experimental Study on the Possibility of Formation of a Condensate of Excited States in a Substance (Rydberg Matter)". *JETP Letters* **77** (6): 280. Bibcode:2003JETPL.. doi:10.1134/1.1577757.

[4] S. Badiei and L. Holmlid (2002). "Neutral Rydberg Matter clusters from K: Extreme cooling of translational degrees of freedom observed by neutral time-of-flight". *Chemical Physics* **282**: 137–146. Bibcode:2002CP....282..137B. doi:10.1016/S0301-0104(02)00601-8.

[5] S. Badiei and L. Holmlid (2006). "Experimental studies of fast fragments of H Rydberg matter". *Journal of Physics B* **39** (20): 4191–4212. Bibcode:2006JPhB...39.4191B. doi:10.1088/0953-4075/39/20/017.

[6] J. Wang; Holmlid, Leif (2002). "Rydberg Matter clusters of hydrogen $(H_2)N^*$ with well-defined kinetic energy release observed by neutral time-of-flight". *Chemical Physics* **277** (2): 201. Bibcode:2002CP....277..201W. doi:10.1016/S0301-0104(02)00303-8.

[7] S. Badiei and L. Holmlid (2002). "Rydberg Matter of K and N_2: Angular dependence of the time-of-flight for neutral and ionized clusters formed in Coulomb explosions". *International Journal of Mass Spectrometry* **220** (2): 127. Bibcode:2002IJMSp.220 ..127B.doi:10.1016/S1387-3806(02)00689-9.

[8] A.V. Popov (2006). "Search for Rydberg matter: Beryllium, magnesium and calcium". *Czechoslovak Journal of Physics* **56**: B1294. Bibcode:2006CzJPh..56B1294P. doi:10.1007/s10582-006-0365-2.

[9] L. Holmlid (2008). "The diffuse interstellar band carriers in interstellar space: All intense bands calculated from He doubly excited states embedded in Rydberg Matter". *Monthly Notices of the Royal Astronomical Society* **384** (2): 764–774. Bibcode:2008MNRAS.384..764H. doi:10.1111/j.1365-2966.2007.12753.x.

[10] J. Liang, M. Gross, P. Goy, S. Haroche (1986). "Circular Rydberg-state spectroscopy". *Physical Review A* **33** (6): 4437–4439. Bibcode:1986PhRvA..33.4437L. doi:10.1103/PhysRevA.33.4437. PMID 9897204.

[11] R.L. Sorochenko (1990). "Postulation, detection and observations of radio recombination lines". In M.A. Gordon, R.L. Sorochenko. *Radio recombination lines: 25 years of investigation*. Kluwer. p. 1. ISBN 0-7923-0804-2.

[12] L. Holmlid (2007). "Direct observation of circular Rydberg electrons in a Rydberg Matter surface layer by electronic circular dichroism". *Journal of Physics: Condensed Matter* **19** (27): 276206. Bibcode:2007JPCM...19A6206H. doi:10.1088/0953-8984/19/27/276206.

[13] L. Holmlid (2007). "Stimulated emission spectroscopy of Rydberg Matter: observation of Rydberg orbits in the core ions". *Applied Physics B* **87** (2): 273–281. Bibcode:2007ApPhB..87..273H. doi:10.1007/s00340-007-2579-9.

[14] L. Holmlid (2009). "Nuclear spin transitions in the kHz range in Rydberg Matter clusters give precise values of the internal magnetic field from orbiting Rydberg electrons".*Chemical Physics***358**: 61–67.Bibcode:2009CP....358...61H.doi:10.1016/j.chem

[15] L. Holmlid, "Rotational spectra of large Rydberg Matter clusters K_{37}, K_{61} and K_{91} give trends in K-K bond distances relative to electron orbit radius". J. Mol. Struct. 885 (2008) 122–130.

[16] L. Holmlid, "Clusters HN^+ (N = 4, 6, 12) from condensed atomic hydrogen and deuterium indicating close-packed structures in the desorbed phase at an active catalyst surface". Surf. Sci. 602 (2008) 3381–3387.

[17] L. Holmlid, "Precision bond lengths for Rydberg Matter clusters K_{19} in excitation levels n = 4, 5 and 6 from rotational radio-frequency emission spectra". Mol. Phys. 105 (2007) 933–939.

[18] L. Holmlid, "Classical energy calculations with electron correlation of condensed excited states – Rydberg Matter". Chem. Phys. 237 (1998) 11–19. doi:10.1016/S0301-0104(98)00259-6

[19] H. Åkesson, S. Badiei and L. Holmlid, "Angular variation of time-of-flight of neutral clusters released from Rydberg Matter: primary and secondary Coulomb explosion processes". Chem. Phys. 321 (2006) 215–222.

[20] L. Holmlid, "Amplification by stimulated emission in Rydberg Matter clusters as the source of intense maser lines in interstellar space". Astrophys. Space Sci. 305 (2006) 91–98.

[21] L. Holmlid, "The alkali metal atmospheres on the Moon and Mercury: explaining the stable exospheres by heavy Rydberg Matter clusters". Planet. Space Sci. 54 (2006) 101–112.

[22] E.A. Manykin, M.I. Ojovan, P.P. Poluektov. "Theory of the condensed state in a system of excited atoms". Sov. Phys. JETP 57 (1983) 256–262.

[23] L. Holmlid, "Vibrational transitions in Rydberg Matter clusters from stimulated Raman and Rabi-flopping phase-delay in the infrared". J. Raman Spectr. 39 (2008) 1364–1374.

[24] É. A. Manykin, M. I. Ozhovan, P. P. Poluéktov, "Decay of a condensate consisting of excited cesium atoms". Zh. Éksp. Teor. Fiz. 102, 1109 (1992) [Sov. Phys. JETP 75, 602 (1992)].

[25] E.A. Manykin, M.I. Ojovan, P.P. Poluektov. "Impurity recombination of Rydberg matter". JETP 78 (1994) 27–32.

[26] Holmlid, Leif (2002). "Conditions for forming Rydberg matter: condensation of Rydberg states in the gas phase versus at surfaces". *Journal of Physics: Condensed Matter* **14** (49): 13469. doi:10.1088/0953-8984/14/49/305.

[27] I. L. Beigman and V. S. Lebedev, "Collision theory of Rydberg atoms with neutral and charged particles". Phys. Rep. 250, 95 (1995).

[28] L. Holmlid, "Redshifts in space caused by stimulated Raman scattering in cold intergalactic Rydberg Matter with experimental verification". J. Exp. Theor. Phys. JETP 100 (2005) 637–644.

[29] Badiei, Shahriar; Holmlid, Leif (2002). "Magnetic field in the intracluster medium: Rydberg matter with almost free electrons". *Monthly Notices of the Royal Astronomical Society* **335** (4): L94. Bibcode:2002MNRAS.335L..94B. doi:10.1046/j.1365-8711.2002.05911.x.

[30] J. Wang, K. Engvall and L. Holmlid, "Cluster KN formation by Rydberg collision complex stabilization during scattering of a K beam off zirconia surfaces". J. Chem. Phys. 110 (1999) 1212–1220.

[31] R. Svensson and L. Holmlid, "Very low work function surfaces from condensed excited states: Rydberg matter of cesium". Surface Sci. 269/270 (1992) 695–699.

[32] http://rd.springer.com/article/10.1007%2Fs10876-012-0449-z

[33] http://itar-tass.com/opinions/interviews/1853

Chapter 20

Bose–Einstein condensate

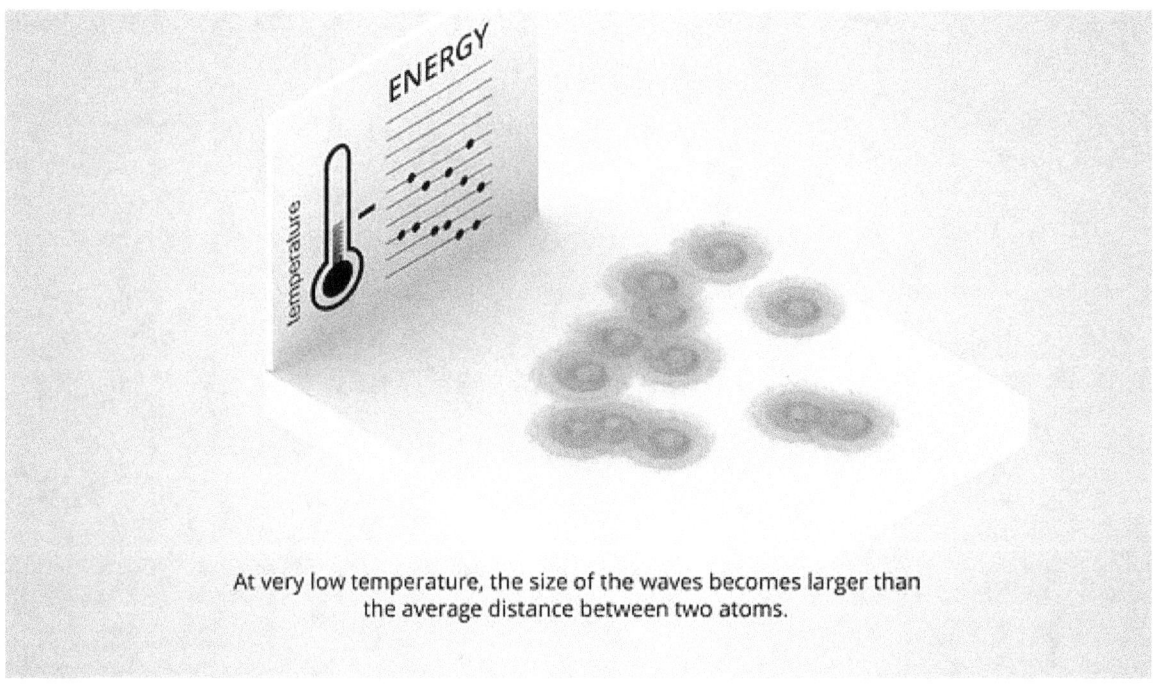

At very low temperature, the size of the waves becomes larger than
the average distance between two atoms.

Schematic Bose-Einstein Condensation versus temperature and the energy diagram

A **Bose–Einstein condensate** (**BEC**) is a state of matter of a dilute gas of bosons cooled to temperatures very close to absolute zero (that is, very near 0 K or −273.15 °C). Under such conditions, a large fraction of bosons occupy the lowest quantum state, at which point macroscopic quantum phenomena become apparent.

This state was first predicted, generally, in 1924–25 by Satyendra Nath Bose and Albert Einstein.

20.1 History

Satyendra Nath Bose first sent a paper to Einstein on the quantum statistics of light quanta (now called photons), deriving Planck's quantum radiation law without any reference to classical physics, and Einstein was impressed, translated the paper himself from English to German and submitted it for Bose to the *Zeitschrift für Physik*, which published it. (The Einstein manuscript, once believed to be lost, was found in a library at Leiden University in 2005.[1]). Einstein then extended Bose's ideas to matter in two other papers.[2] The result of their efforts is the concept of a Bose gas, governed

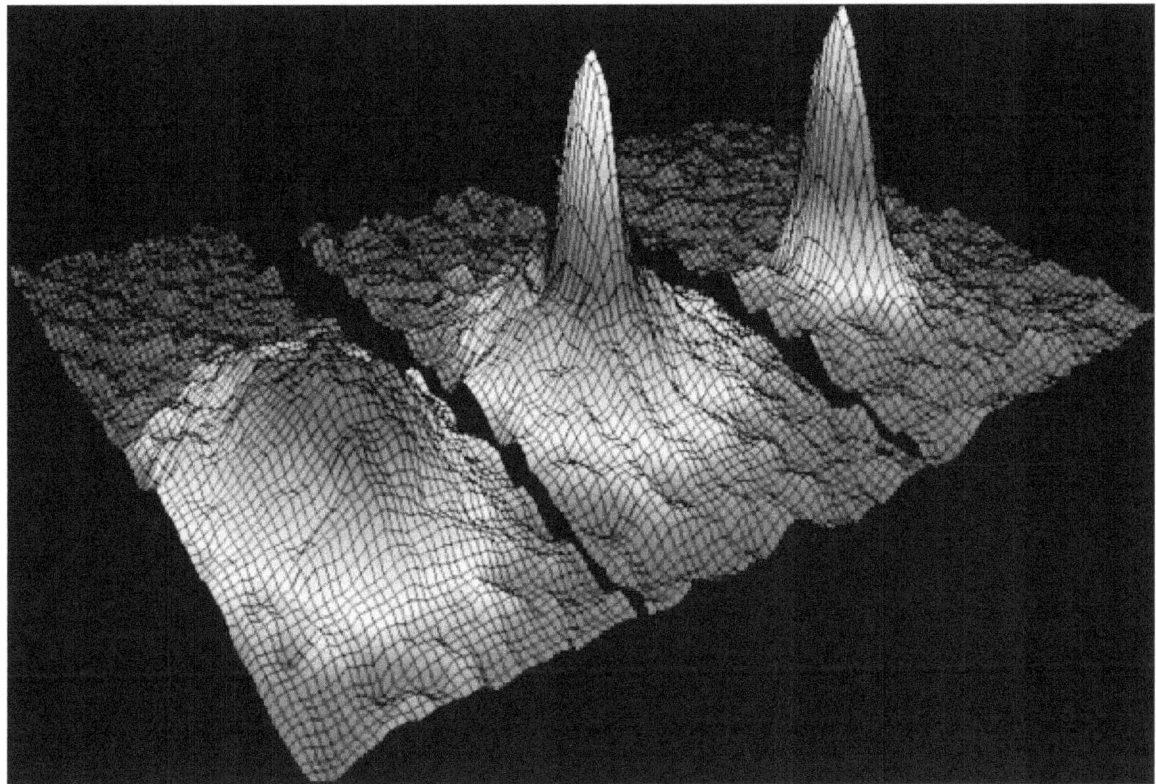

Velocity-distribution data (3 views) for a gas of rubidium atoms, confirming the discovery of a new phase of matter, the Bose–Einstein condensate. Left: just before the appearance of a Bose–Einstein condensate. Center: just after the appearance of the condensate. Right: after further evaporation, leaving a sample of nearly pure condensate.

by Bose–Einstein statistics, which describes the statistical distribution of identical particles with integer spin, now called bosons. Bosons, which include the photon as well as atoms such as helium-4 (^{4}He), are allowed to share a quantum state. Einstein proposed that cooling bosonic atoms to a very low temperature would cause them to fall (or "condense") into the lowest accessible quantum state, resulting in a new form of matter.

In 1938 Fritz London proposed BEC as a mechanism for superfluidity in ^{4}He and superconductivity.[3][4]

On June 5, 1995 the first gaseous condensate was produced by Eric Cornell and Carl Wieman at the University of Colorado at Boulder NIST–JILA lab, in a gas of rubidium atoms cooled to 170 nanokelvin (nK).[5] Shortly thereafter, Wolfgang Ketterle at MIT demonstrated important BEC properties. For their achievements Cornell, Wieman, and Ketterle received the 2001 Nobel Prize in Physics.[6]

Many isotopes were soon condensed, then molecules, quasi-particles, and photons in 2010.[7]

20.2   Critical temperature

This transition to BEC occurs below a critical temperature, which for a uniform three-dimensional gas consisting of non-interacting particles with no apparent internal degrees of freedom is given by:

$$T_c = \left(\frac{n}{\zeta(3/2)} \right)^{2/3} \frac{2\pi\hbar^2}{mk_B} \approx 3.3125 \, \frac{\hbar^2 n^{2/3}}{mk_B}$$

where:

Interactions shift the value and the corrections can be calculated by mean-field theory.

20.3 Models

20.3.1 Einstein's non-interacting gas

Consider a collection of N noninteracting particles, which can each be in one of two quantum states, $|0\rangle$ and $|1\rangle$. If the two states are equal in energy, each different configuration is equally likely.

If we can tell which particle is which, there are 2^N different configurations, since each particle can be in $|0\rangle$ or $|1\rangle$ independently. In almost all of the configurations, about half the particles are in $|0\rangle$ and the other half in $|1\rangle$. The balance is a statistical effect: the number of configurations is largest when the particles are divided equally.

If the particles are indistinguishable, however, there are only $N+1$ different configurations. If there are K particles in state $|1\rangle$, there are $N - K$ particles in state $|0\rangle$. Whether any particular particle is in state $|0\rangle$ or in state $|1\rangle$ cannot be determined, so each value of K determines a unique quantum state for the whole system.

Suppose now that the energy of state $|1\rangle$ is slightly greater than the energy of state $|0\rangle$ by an amount E. At temperature T, a particle will have a lesser probability to be in state $|1\rangle$ by $e^{-E/kT}$. In the distinguishable case, the particle distribution will be biased slightly towards state $|0\rangle$. But in the indistinguishable case, since there is no statistical pressure toward equal numbers, the most-likely outcome is that most of the particles will collapse into state $|0\rangle$.

In the distinguishable case, for large N, the fraction in state $|0\rangle$ can be computed. It is the same as flipping a coin with probability proportional to $p = \exp(-E/T)$ to land tails.

In the indistinguishable case, each value of K is a single state, which has its own separate Boltzmann probability. So the probability distribution is exponential:

$$P(K) = Ce^{-KE/T} = Cp^K.$$

For large N, the normalization constant C is $(1 - p)$. The expected total number of particles not in the lowest energy state, in the limit that $N\to\infty$, is equal to $\sum_{n>0} Cnp^n = p/(1-p)$. It does not grow when N is large; it just approaches a constant. This will be a negligible fraction of the total number of particles. So a collection of enough Bose particles in thermal equilibrium will mostly be in the ground state, with only a few in any excited state, no matter how small the energy difference.

Consider now a gas of particles, which can be in different momentum states labeled $|k\rangle$. If the number of particles is less than the number of thermally accessible states, for high temperatures and low densities, the particles will all be in different states. In this limit, the gas is classical. As the density increases or the temperature decreases, the number of accessible states per particle becomes smaller, and at some point, more particles will be forced into a single state than the maximum allowed for that state by statistical weighting. From this point on, any extra particle added will go into the ground state.

To calculate the transition temperature at any density, integrate, over all momentum states, the expression for maximum number of excited particles, $p/(1 - p)$:

$$N = V \int \frac{d^3k}{(2\pi)^3} \frac{p(k)}{1 - p(k)} = V \int \frac{d^3k}{(2\pi)^3} \frac{1}{e^{\frac{k^2}{2mT}} - 1}$$

$$p(k) = e^{\frac{-k^2}{2mT}}.$$

When the integral is evaluated with factors of k_B and $\hbar$ restored by dimensional analysis, it gives the critical temperature formula of the preceding section. Therefore, this integral defines the critical temperature and particle number corresponding to the conditions of negligible chemical potential. In Bose–Einstein statistics distribution, μ is actually still nonzero for BECs; however, μ is less than the ground state energy. Except when specifically talking about the ground state, μ can be approximated for most energy or momentum states as $\mu \approx 0$.

20.3.2 Bogoliubov theory for weakly interacting gas

Bogoliubov considered perturbations on the limit of dilute gas,[9] finding a finite pressure at zero temperature and positive chemical potential. This leads to corrections for the ground state. The Bogoliubov state has pressure(T=0): $P = g/2n^2$.

The original interacting system can be converted to a system of non-interacting particles with a dispersion law.

20.3.3 Gross–Pitaevskii equation

Main article: Gross–Pitaevskii equation

In some simplest cases, the state of condensed particles can be described with a nonlinear Schrödinger equation, also known as Gross-Pitaevskii or Ginzburg-Landau equation. The validity of this approach is actually limited to the case of ultracold temperatures, which fits well for the most alkali atoms experiments.

This approach originates from the assumption that the state of the BEC can be described by the unique wavefunction of the condensate $\psi(\vec{r})$. For a system of this nature, $|\psi(\vec{r})|^2$ is interpreted as the particle density, so the total number of atoms is $N = \int d\vec{r} |\psi(\vec{r})|^2$

Provided essentially all atoms are in the condensate (that is, have condensed to the ground state), and treating the bosons using mean field theory, the energy (E) associated with the state $\psi(\vec{r})$ is:

$$E = \int d\vec{r} \left[\frac{\hbar^2}{2m} |\nabla \psi(\vec{r})|^2 + V(\vec{r})|\psi(\vec{r})|^2 + \frac{1}{2} U_0 |\psi(\vec{r})|^4 \right]$$

Minimizing this energy with respect to infinitesimal variations in $\psi(\vec{r})$, and holding the number of atoms constant, yields the Gross–Pitaevski equation (GPE) (also a non-linear Schrödinger equation):

$$i\hbar \frac{\partial \psi(\vec{r})}{\partial t} = \left(-\frac{\hbar^2 \nabla^2}{2m} + V(\vec{r}) + U_0 |\psi(\vec{r})|^2 \right) \psi(\vec{r})$$

where:

In the case of zero external potential, the dispersion law of interacting Bose-Einstein-condensed particles is given by so-called Bogoliubov spectrum (for $T = 0$):

$$\omega_p = \sqrt{\frac{p^2}{2m} \left(\frac{p^2}{2m} + 2U_0 n_0 \right)}$$

The Gross-Pitaevskii equation (GPE) provides a relatively good description of the behavior of atomic BEC's. However, GPE does not take into account the temperature dependence of dynamical variables, and is therefore valid only for $T = 0$. It is not applicable, for example, for the condensates of excitons, magnons and photons, where the critical temperature is up to room one.

Weaknesses of Gross–Pitaevskii model

The Gross–Pitaevskii model of BEC is a physical approximation valid for certain classes of BECs. By construction, the GPE uses the following simplifications: it assumes that interactions between condensate particles are of the contact two-body type and also neglects anomalous contributions to self-energy.[10] These assumptions are suitable mostly for the dilute three-dimensional condensates. If one relaxes any of these assumptions, the equation for the condensate wavefunction acquires the terms containing higher-order powers of the wavefunction. Moreover, for some physical systems the amount of such terms turns out to be infinite, therefore, the equation becomes essentially non-polynomial. The examples where this could happen are the Bose–Fermi composite condensates,[11][12][13][14] effectively lower-dimensional condensates,[15] and dense condensates and superfluid clusters and droplets.[16]

20.3.4   Other

However, it is clear that in a general case the behaviour of Bose–Einstein condensate can be described by coupled evolution equations for condensate density, superfluid velocity and distribution function of elementary excitations. This problem was in 1977 by Peletminskii et al. in microscopical approach. The Peletminskii equations are valid for any finite temperatures below the critical point. Years after, in 1985, Kirkpatrick and Dorfman obtained similar equations using another microscopical approach. The Peletminskii equations also reproduce Khalatnikov hydrodynamical equations for superfluid as a limiting case.

20.3.5   Superfluidity of BEC and Landau criterion

The phenomena of superfluidity of a Bose gas and superconductivity of a strongly-correlated Fermi gas (a gas of Cooper pairs) are tightly connected to Bose-Einstein condensation. Under corresponding conditions, below the temperature of phase transition, these phenomena were observed in helium-4 and different classes of superconductors. In this sense, the superconductivity is often called the superfluidity of Fermi gas. In the simplest form, the origin of superfluidity can be seen from the weakly interacting bosons model.

20.4   Experimental observation

20.4.1   Superfluid He-4

In 1938, Pyotr Kapitsa, John Allen and Don Misener discovered that helium-4 became a new kind of fluid, now known as a superfluid, at temperatures less than 2.17 K (the lambda point). Superfluid helium has many unusual properties, including zero viscosity (the ability to flow without dissipating energy) and the existence of quantized vortices. It was quickly believed that the superfluidity was due to partial Bose–Einstein condensation of the liquid. In fact, many properties of superfluid helium also appear in gaseous condensates created by Cornell, Wieman and Ketterle (see below). Superfluid helium-4 is a liquid rather than a gas, which means that the interactions between the atoms are relatively strong; the original theory of Bose–Einstein condensation must be heavily modified in order to describe it. Bose–Einstein condensation remains, however, fundamental to the superfluid properties of helium-4. Note that helium-3, a fermion, also enters a superfluid phase at low temperature, which can be explained by the formation of bosonic Cooper pairs of two atoms (see also fermionic condensate).

20.4.2   Gaseous

The first "pure" Bose–Einstein condensate was created by Eric Cornell, Carl Wieman, and co-workers at JILA on 5 June 1995. They cooled a dilute vapor of approximately two thousand rubidium-87 atoms to below 170 nK using a combination of laser cooling (a technique that won its inventors Steven Chu, Claude Cohen-Tannoudji, and William D. Phillips the 1997 Nobel Prize in Physics) and magnetic evaporative cooling. About four months later, an independent effort led by Wolfgang Ketterle at MIT condensed sodium-23. Ketterle's condensate had a hundred times more atoms, allowing important results such as the observation of quantum mechanical interference between two different condensates. Cornell, Wieman and Ketterle won the 2001 Nobel Prize in Physics for their achievements.[17]

A group led by Randall Hulet at Rice University announced a condensate of lithium atoms only one month following the JILA work.[18] Lithium has attractive interactions, causing the condensate to be unstable and collapse for all but a few atoms. Hulet's team subsequently showed the condensate could be stabilized by confinement quantum pressure for up to about 1000 atoms. Various isotopes have since been condensed.

Velocity-distribution data graph

In the image accompanying this article, the velocity-distribution data indicates the formation of a Bose–Einstein condensate out of a gas of rubidium atoms. The false colors indicate the number of atoms at each velocity, with red being the fewest and white being the most. The areas appearing white and light blue are at the lowest velocities. The peak is not infinitely narrow because of the Heisenberg uncertainty principle: spatially confined atoms have a minimum width velocity distribution. This width is given by the curvature of the magnetic potential in the given direction. More tightly confined directions have bigger widths in the ballistic velocity distribution. This anisotropy of the peak on the right is a purely quantum-mechanical effect and does not exist in the thermal distribution on the left. This graph served as the cover design for the 1999 textbook *Thermal Physics* by Ralph Baierlein.[19]

20.4.3 Quasiparticles

Main article: Bose-Einstein condensation of quasiparticles

Bose–Einstein condensation also applies to quasiparticles in solids. Magnons, Excitons, and Polaritons have integer spin and form condensates.

Magnons, electron spin waves, can be controlled by a magnetic field. Densities from the limit of a dilute gas to a strongly interacting Bose liquid are possible. Magnetic ordering is the analog of superfluidity. In 1999 condensation was demonstrated in antiferromagnetic $TlCuCl_3$,[20] at temperatures as large as 14 K. The high transition temperature (relative to atomic gases) is due to the magnons small mass (near an electron) and greater achievable density. In 2006, condensation in a ferromagnetic Yttrium-iron-garnet thin film was seen even at room temperature,[21][22] with optical pumping.

Excitons, electron-hole pairs, were predicted to condense at low temperature and high density by Boer et al. in 1961. Bilayer system experiments first demonstrated condensation in 2003, by Hall voltage disappearance. Fast optical exciton creation was used to form condensates in sub-Kelvin Cu_2O in 2005 on.

Polariton condensation was detected in a 5 K quantum well microcavity.

20.5 Peculiar properties

20.5.1 Vortices

As in many other systems, vortices can exist in BECs. These can be created, for example, by 'stirring' the condensate with lasers, or rotating the confining trap. The vortex created will be a quantum vortex. These phenomena are allowed for by the non-linear $|\psi(\vec{r})|^2$ term in the GPE. As the vortices must have quantized angular momentum the wavefunction may have the form $\psi(\vec{r}) = \phi(\rho, z)e^{i\ell\theta}$ where ρ, z and θ are as in the cylindrical coordinate system, and ℓ is the angular number. This is particularly likely for an axially symmetric (for instance, harmonic) confining potential, which is commonly used. The notion is easily generalized. To determine $\phi(\rho, z)$, the energy of $\psi(\vec{r})$ must be minimized, according to the constraint $\psi(\vec{r}) = \phi(\rho, z)e^{i\ell\theta}$. This is usually done computationally, however in a uniform medium the analytic form

$$\phi = \frac{nx}{\sqrt{2 + x^2}}$$

demonstrates the correct behavior, and is a good approximation.

A singly charged vortex ($\ell = 1$) is in the ground state, with its energy ϵ_v given by

$$\epsilon_v = \pi n \frac{\hbar^2}{m} \ln\left(1.464 \frac{b}{\xi}\right)$$

where b is the farthest distance from the vortex considered.(To obtain an energy which is well defined it is necessary to include this boundary b .)

For multiply charged vortices ($\ell > 1$) the energy is approximated by

$$\epsilon_v \approx \ell^2 \pi n \frac{\hbar^2}{m} \ln\left(\frac{b}{\xi}\right)$$

which is greater than that of ℓ singly charged vortices, indicating that these multiply charged vortices are unstable to decay. Research has, however, indicated they are metastable states, so may have relatively long lifetimes.

Closely related to the creation of vortices in BECs is the generation of so-called dark solitons in one-dimensional BECs. These topological objects feature a phase gradient across their nodal plane, which stabilizes their shape even in propagation and interaction. Although solitons carry no charge and are thus prone to decay, relatively long-lived dark solitons have been produced and studied extensively.[23]

20.5.2   Attractive interactions

Experiments led by Randall Hulet at Rice University from 1995 through 2000 showed that lithium condensates with attractive interactions could stably exist up to a critical atom number. Quench cooling the gas, they observed the condensate to grow, then subsequently collapse as the attraction overwhelmed the zero-point energy of the confining potential, in a burst reminiscent of a supernova, with an explosion preceded by an implosion.

Further work on attractive condensates was performed in 2000 by the JILA team, of Cornell, Wieman and coworkers. Their instrumentation now had better control so they used naturally *attracting* atoms of rubidium-85 (having negative atom–atom scattering length). Through Feshbach resonance involving a sweep of the magnetic field causing spin flip collisions, they lowered the characteristic, discrete energies at which rubidium bonds, making their Rb-85 atoms repulsive and creating a stable condensate. The reversible flip from attraction to repulsion stems from quantum interference among wave-like condensate atoms.

When the JILA team raised the magnetic field strength further, the condensate suddenly reverted to attraction, imploded and shrank beyond detection, then exploded, expelling about two-thirds of its 10,000 atoms. About half of the atoms in the condensate seemed to have disappeared from the experiment altogether, not seen in the cold remnant or expanding gas cloud.[17] Carl Wieman explained that under current atomic theory this characteristic of Bose–Einstein condensate could not be explained because the energy state of an atom near absolute zero should not be enough to cause an implosion; however, subsequent mean field theories have been proposed to explain it. Most likely they formed molecules of two rubidium atoms,[24] energy gained by this bond imparts velocity sufficient to leave the trap without being detected.

20.6   Current research

Compared to more commonly encountered states of matter, Bose–Einstein condensates are extremely fragile. The slightest interaction with the external environment can be enough to warm them past the condensation threshold, eliminating their interesting properties and forming a normal gas.

Nevertheless, they have proven useful in exploring a wide range of questions in fundamental physics, and the years since the initial discoveries by the JILA and MIT groups have seen an increase in experimental and theoretical activity. Examples include experiments that have demonstrated interference between condensates due to wave–particle duality,[25] the study of superfluidity and quantized vortices, the creation of bright matter wave solitons from Bose condensates confined to one dimension, and the slowing of light pulses to very low speeds using electromagnetically induced transparency.[26] Vortices in Bose–Einstein condensates are also currently the subject of analogue gravity research, studying the possibility of modeling black holes and their related phenomena in such environments in the laboratory. Experimenters have also realized "optical lattices", where the interference pattern from overlapping lasers provides a periodic potential. These have been used to explore the transition between a superfluid and a Mott insulator,[27] and may be useful in studying Bose–Einstein condensation in fewer than three dimensions, for example the Tonks–Girardeau gas.

Bose–Einstein condensates composed of a wide range of isotopes have been produced.[28]

Cooling fermions to extremely low temperatures has created degenerate gases, subject to the Pauli exclusion principle. To exhibit Bose–Einstein condensation, the fermions must "pair up" to form bosonic compound particles (e.g. molecules or Cooper pairs). The first molecular condensates were created in November 2003 by the groups of Rudolf Grimm at the University of Innsbruck, Deborah S. Jin at the University of Colorado at Boulder and Wolfgang Ketterle at MIT. Jin quickly went on to create the first fermionic condensate composed of Cooper pairs.[29]

In 1999, Danish physicist Lene Hau led a team from Harvard University which slowed a beam of light to about 17 meters per second., using a superfluid.[30] Hau and her associates have since made a group of condensate atoms recoil from a light pulse such that they recorded the light's phase and amplitude, recovered by a second nearby condensate, in what they term "slow-light-mediated atomic matter-wave amplification" using Bose–Einstein condensates: details are discussed in *Nature*.[31]

Researchers in the new field of atomtronics use the properties of Bose–Einstein condensates when manipulating groups of identical cold atoms using lasers.[32] Further, BECs have been proposed by Emmanuel David Tannenbaum for anti-stealth technology.[33]

20.6.1 Isotopes

The effect has mainly been observed on alkaline atoms which have nuclear properties particularly suitable for working with traps. As of 2012, using ultra-low temperatures of 10^{-7} K or below, Bose–Einstein condensates had been obtained for a multitude of isotopes, mainly of alkaline, alkaline earth, and lanthanoid atoms (^{7}Li, ^{23}Na, ^{39}K, ^{41}K, ^{85}Rb, ^{87}Rb, ^{133}Cs, ^{52}Cr, ^{40}Ca, ^{84}Sr, ^{86}Sr, ^{88}Sr, ^{174}Yb, ^{164}Dy, and ^{168}Er). Research was finally successful in hydrogen with aid of special methods. In contrast, the superfluid state of ^{4}He below 2.17 K is not a good example, because the interaction between the atoms is too strong. Only 8% of atoms are in the ground state near absolute zero, rather than the 100% of a true condensate.

The bosonic behavior of some of these alkaline gases appears odd at first sight, because their nuclei have half-integer total spin. It arises from a subtle interplay of electronic and nuclear spins: at ultra-low temperatures and corresponding excitation energies, the half-integer total spin of the electronic shell and half-integer total spin of the nucleus are coupled by a very weak hyperfine interaction. The total spin of the atom, arising from this coupling, is an integer lower value. The chemistry of systems at room temperature is determined by the electronic properties, which is essentially fermionic, since room temperature thermal excitations have typical energies much higher than the hyperfine values.

20.7 See also

- Atom laser

- Atomic coherence

- Bose–Einstein correlations

- Bose–Einstein condensation: a network theory approach

- Bose-Einstein condensation of excitons

- Cold Atom Laboratory

- Electromagnetically induced transparency

- Fermionic condensate

- Gas in a box

- Gross–Pitaevskii equation

- Macroscopic quantum phenomena

- Macroscopic quantum self-trapping

- Slow light

- Superconductivity

- Superfluid film

- Superfluid helium-4

- Supersolid

- Tachyon condensation

- Timeline of low-temperature technology

- Super-heavy atom

- Ultracold atom

- Wiener sausage

20.8 References

[1] "Leiden University Einstein archive". Lorentz.leidenuniv.nl. 27 October 1920. Retrieved 23 March 2011.

[2] Clark, Ronald W. (1971). *Einstein: The Life and Times*. Avon Books. pp. 408–409. ISBN 0-380-01159-X.

[3] London, F. (1938). "The λ-Phenomenon of Liquid Helium and the Bose–Einstein Degeneracy". *Nature* **141** (3571): 643–644. Bibcode:1938Natur.141..643L. doi:10.1038/141643a0.

[4] London, F. *Superfluids* Vol.I and II, (reprinted New York: Dover 1964)

[5] http://www.nist.gov/public_affairs/releases/bec_background.cfm

[6] Levi, Barbara Goss (2001). "Cornell, Ketterle, and Wieman Share Nobel Prize for Bose–Einstein Condensates". *Search & Discovery*. Physics Today online. Archived from the original on 24 October 2007. Retrieved 26 January 2008.

[7] Klaers, Jan; Schmitt, Julian; Vewinger, Frank; Weitz, Martin (2010). "Bose–Einstein condensation of photons in an optical microcavity". *Nature* **468** (7323): 545–548. arXiv:1007.4088. Bibcode:2010Natur.468..545K. doi:10.1038/nature09567. PMID 21107426.

[8] (sequence A078434 in OEIS)

[9] N. N. Bogoliubov (1947). "On the theory of superfluidity.". *J. Phys. (USSR), 11:23.*

[10] Beliaev, S. T. Zh. Eksp. Teor. Fiz. 34, 418–432 (1958); ibid. 433–446 [Soviet Phys. JETP 3, 299 (1957)].

[11] Schick, M. (1971). "Two-Dimensional System of Hard-Core Bosons". *Physical Review A* **3**(3): 1067. Bibcode:1971PhRvA...3. doi:10.1103/PhysRevA.3.1067.

[12] Kolomeisky, E.; Straley, J. (1992). "Renormalization-group analysis of the ground-state properties of dilute Bose systems in d spatial dimensions". *Physical Review B* **46** (18): 11749. Bibcode:1992PhRvB..4611749K. doi:10.1103/PhysRevB.46.11749.

[13] Kolomeisky, E. B.; Newman, T. J.; Straley, J. P.; Qi, X. (2000). "Low-Dimensional Bose Liquids: Beyond the Gross-Pitaevskii Approximation". *Physical Review Letters* **85** (6): 1146–1149. arXiv:cond-mat/0002282. Bibcode:2000PhRvL..85.1146K. doi:10.1103/PhysRevLett.85.1146. PMID 10991498.

[14] Chui, S.; Ryzhov, V. (2004). "Collapse transition in mixtures of bosons and fermions". *Physical Review A* **69** (4): 043607. Bibcode:2004PhRvA..69d3607C. doi:10.1103/PhysRevA.69.043607.

[15] Salasnich, L.; Parola, A.; Reatto, L. (2002). "Effective wave equations for the dynamics of cigar-shaped and disk-shaped Bose condensates".*Phys. Rev. A***65**(4): 043614.arXiv:cond-mat/0201395.Bibcode:2002PhRvA..65d3614S.doi:10.1103/PhysRev

[16] Avdeenkov, A. V.; Zloshchastiev, K. G. (2011). "Quantum Bose liquids with logarithmic nonlinearity: Self-sustainability and emergence of spatial extent". *J. Phys. B: At. Mol. Opt. Phys.* **44** (19): 195303. arXiv:1108.0847. Bibcode:2011JPhB...44s5303A. doi:10.1088/0953-4075/44/19/195303.

[17] "Eric A. Cornell and Carl E. Wieman — Nobel Lecture" (PDF). nobelprize.org.

[18] Bradley, C. C.; Sackett, C. A.; Tollett, J. J.; Hulet, R. G. (1995). "Evidence of Bose-Einstein Condensation in an Atomic Gas with Attractive Interactions" (PDF). *Physical review letters* **75** (9): 1687–1690. Bibcode:1995PhRvL..75.1687B. doi:10.1103/P hysRevLett.75.1687.PMID10060366.

[19] Baierlein, Ralph (1999). *Thermal Physics*. Cambridge University Press. ISBN 0-521-65838-1.

[20] Nikuni, T.; Oshikawa, M.; Oosawa, A.; Tanaka, H. (1999). "Bose–Einstein Condensation of Dilute Magnons in TlCuCl₃". *Physical Review Letters* **84** (25): 5868–71. arXiv:cond-mat/9908118. Bibcode:2000PhRvL..84.5868N. doi:10.1103/PhysRev Lett.84.5868.PMID10991075.

[21] Demokritov, S.O.; Demidov, VE; Dzyapko, O; Melkov, GA; Serga, AA; Hillebrands, B; Slavin, AN (2006). "Bose–Einstein condensation of quasi-equilibrium magnons at room temperature under pumping". *Nature* **443** (7110): 430–433. Bibcode:2006 Natur.443..430D.doi:10.1038/nature05117.PMID17006509.

[22] *Magnon Bose Einstein Condensation* made simple. Website of the "Westfählische Wilhelms Universität Münster" Prof.Demokritov. Retrieved 25 June 2012.

[23] Becker, Christoph; Stellmer, Simon; Soltan-Panahi, Parvis; Dörscher, Sören; Baumert, Mathis; Richter, Eva-Maria; Kronjäger, Jochen; Bongs, Kai; Sengstock, Klaus (2008). "Oscillations and interactions of dark and dark–bright solitons in Bose–Einstein condensates". *Nature Physics* **4** (6): 496–501. arXiv:0804.0544. Bibcode:2008NatPh...4..496B. doi:10.1038/nphys962.

[24]van Putten, M.H.P.M. (2010). "Pair condensates produced in bosenovae".*Physics Letters A***374**(33): 3346.Bibcode:2010PhL doi:10.1016/j.physleta.2010.06.020.

[25] Gorlitz, Axel. "Interference of Condensates (BEC@MIT)". Cua.mit.edu. Retrieved 13 October 2009.

[26] Dutton, Zachary; Ginsberg, Naomi S.; Slowe, Christopher and Hau, Lene Vestergaard (2004). "The art of taming light: ultra-slow and stopped light" (PDF). *Europhysics News* **35** (2): 33. Bibcode:2004ENews..35...33D. doi:10.1051/epn:2004201.

[27] "From Superfluid to Insulator: Bose–Einstein Condensate Undergoes a Quantum Phase Transition". Qpt.physics.harvard.edu. Retrieved 13 October 2009.

[28] "Ten of the best for BEC". Physicsweb.org. 1 June 2005.

[29] "Fermionic condensate makes its debut". Physicsweb.org. 28 January 2004.

[30] Cromie, William J. (18 February 1999). "Physicists Slow Speed of Light". The Harvard University Gazette. Retrieved 26 January 2008.

[31] Ginsberg, N. S.; Garner, S. R.; Hau, L. V. (2007). "Coherent control of optical information with matter wave dynamics". *Nature* **445** (7128): 623–626. doi:10.1038/nature05493. PMID 17287804.

[32] Weiss, P. (12 February 2000). "Atomtronics may be the new electronics". *Science News Online* **157** (7): 104. doi:10.2307/4012185. JSTOR 4012185. Retrieved 12 February 2011.

[33] Tannenbaum, Emmanuel David (1970). "Gravimetric Radar: Gravity-based detection of a point-mass moving in a static background". arXiv:1208.2377 [physics.ins-det].

20.9   Further reading

• Bose, S. N. (1924). "Plancks Gesetz und Lichtquantenhypothese".*Zeitschrift für Physik***26**: 178.Bibcode:1924ZPhy doi:10.1007/BF01327326.

• Einstein, A. (1925). "Quantentheorie des einatomigen idealen Gases". *Sitzungsberichte der Preussischen Akademie der Wissenschaften* **1**: 3.,

• Landau, L. D. (1941). "The theory of Superfluity of Helium 111". *J. Phys. USSR* **5**: 71–90.

• L. Landau(1941). "Theory of the Superfluidity of Helium II".*Physical Review***60**(4): 356–358.Bibcode:1941PhR doi:10.1103/PhysRev.60.356.

• M.H. Anderson, J.R. Ensher, M.R. Matthews, C.E. Wieman, and E.A. Cornell (1995). "Observation of Bose–Einstein Condensation in a Dilute Atomic Vapor". *Science* **269** (5221): 198–201. Bibcode:1995Sci...269..198A. doi:10.1126/science.269.5221.198. JSTOR 2888436. PMID 17789847.

• C. Barcelo, S. Liberati and M. Visser (2001). "Analogue gravity from Bose–Einstein condensates". *Classical and Quantum Gravity* **18** (6): 1137–1156. arXiv:gr-qc/0011026. Bibcode:2001CQGra..18.1137B. doi:10.1088/0264-9381/18/6/312.

• P.G. Kevrekidis, R. Carretero-Gonzlaez, D.J. Frantzeskakis and I.G. Kevrekidis (2006). "Vortices in Bose–Einstein Condensates: Some Recent Developments". *Modern Physics Letters B* **5** (33).

• K.B. Davis, M.-O. Mewes, M.R. Andrews, N.J. van Druten, D.S. Durfee, D.M. Kurn, and W. Ketterle (1995). "Bose–Einstein condensation in a gas of sodium atoms". *Physical Review Letters* **75** (22): 3969–3973. Bibcode:1995 PhRvL..75.3969D.doi:10.1103/PhysRevLett.75.3969.PMID10059782..

• D. S. Jin, J. R. Ensher, M. R. Matthews, C. E. Wieman, and E. A. Cornell (1996). "Collective Excitations of a Bose–Einstein Condensate in a Dilute Gas". *Physical Review Letters* **77** (3): 420–423. Bibcode:1996PhRvL..77..420J. doi:10.1103/PhysRevLett.77.420. PMID 10062808.

• M. R. Andrews, C. G. Townsend, H.-J. Miesner, D. S. Durfee, D. M. Kurn, and W. Ketterle (1997). "Observation of interference between two Bose condensates". *Science* **275** (5300): 637–641. doi:10.1126/science.275.5300.637. PMID 9005843..

• Eric A. Cornell and Carl E. Wieman (1998). "The Bose–Einstein Condensate". *Scientific American* **278** (3): 40–45. doi:10.1038/scientificamerican0398-40.

• M. R. Matthews, B. P. Anderson, P. C. Haljan, D. S. Hall, C. E. Wieman, and E. A. Cornell (1999). "Vortices in a Bose–Einstein Condensate". *Physical Review Letters* **83** (13): 2498–2501. arXiv:cond-mat/9908209. Bibcode:1999PhRvL..83.2498M. doi:10.1103/PhysRevLett.83.2498.

• E.A. Donley, N.R. Claussen, S.L. Cornish, J.L. Roberts, E.A. Cornell, and C.E. Wieman (2001). "Dynamics of collapsing and exploding Bose–Einstein condensates". *Nature* **412** (6844): 295–299. arXiv:cond-mat/0105019. Bibcode:2001Natur.412..295D. doi:10.1038/35085500. PMID 11460153.

• A. G. Truscott, K. E. Strecker, W. I. McAlexander, G. B. Partridge, and R. G. Hulet (2001). "Observation of Fermi Pressure in a Gas of Trapped Atoms". *Science* **291** (5513): 2570–2572. Bibcode:2001Sci...291.2570T. doi:10.1126/science.1059318. PMID 11283362.

• M. Greiner, O. Mandel, T. Esslinger, T. W. Hänsch, I. Bloch (2002). "Quantum phase transition from a super-fluid to a Mott insulator in a gas of ultracold atoms". *Nature* **415** (6867): 39–44. Bibcode:2002Natur.415...39G. doi:10.1038/415039a. PMID 11780110..

• S. Jochim, M. Bartenstein, A. Altmeyer, G. Hendl, S. Riedl, C. Chin, J. Hecker Denschlag, and R. Grimm (2003). "Bose–Einstein Condensation of Molecules". *Science* **302** (5653): 2101–2103. Bibcode:2003Sci...302.2101J. doi:10.1126/science.1093280. PMID 14615548.

- Markus Greiner, Cindy A. Regal and Deborah S. Jin (2003). "Emergence of a molecular Bose–Einstein condensate from a Fermi gas". *Nature* **426** (6966): 537–540. Bibcode:2003Natur.426..537G. doi:10.1038/nature02199. PMID 14647340.

- M. W. Zwierlein, C. A. Stan, C. H. Schunck, S. M. F. Raupach, S. Gupta, Z. Hadzibabic, and W. Ketterle (2003). "Observation of Bose–Einstein Condensation of Molecules". *Physical Review Letters* **91** (25): 250401. arXiv:cond-mat/0311617. Bibcode:2003PhRvL..91y0401Z. doi:10.1103/PhysRevLett.91.250401. PMID 14754098.

- C. A. Regal, M. Greiner, and D. S. Jin (2004). "Observation of Resonance Condensation of Fermionic Atom Pairs". *Physical Review Letters* **92** (4): 040403. arXiv:cond-mat/0401554. Bibcode:2004PhRvL..92d0403R. doi:10.1103/PhysRevLett.92.040403. PMID 14995356.

- C. J. Pethick and H. Smith, *Bose–Einstein Condensation in Dilute Gases*, Cambridge University Press, Cambridge, 2001.

- Lev P. Pitaevskii and S. Stringari, *Bose–Einstein Condensation*, Clarendon Press, Oxford, 2003.

- Mackie M, Suominen KA, Javanainen J., "Mean-field theory of Feshbach-resonant interactions in 85Rb condensates." Phys Rev Lett. 2002 Oct 28;89(18):180403.

20.10 External links

- Bose–Einstein Condensation 2009 Conference Bose–Einstein Condensation 2009 – Frontiers in Quantum Gases

- BEC Homepage General introduction to Bose–Einstein condensation

- Nobel Prize in Physics 2001 – for the achievement of Bose–Einstein condensation in dilute gases of alkali atoms, and for early fundamental studies of the properties of the condensates

- Physics Today: Cornell, Ketterle, and Wieman Share Nobel Prize for Bose–Einstein Condensates

- Bose–Einstein Condensates at JILA

- Atomcool at Rice University

- Alkali Quantum Gases at MIT

- Atom Optics at UQ

- Einstein's manuscript on the Bose–Einstein condensate discovered at Leiden University

- Bose–Einstein condensate on arxiv.org

- Bosons – The Birds That Flock and Sing Together

- Easy BEC machine – information on constructing a Bose–Einstein condensate machine.

- Verging on absolute zero – Cosmos Online

- Lecture by W Ketterle at MIT in 2001

- Bose–Einstein Condensation at NIST – NIST resource on BEC

Chapter 21

Fermionic condensate

A **fermionic condensate** is a superfluid phase formed by fermionic particles at low temperatures. It is closely related to the Bose–Einstein condensate, a superfluid phase formed by bosonic atoms under similar conditions. Unlike the Bose–Einstein condensates, fermionic condensates are formed using fermions instead of bosons. The earliest recognized fermionic condensate described the state of electrons in a superconductor; the physics of other examples including recent work with fermionic atoms is analogous. The first atomic fermionic condensate was created by a team led by Deborah S. Jin in 2003. A **chiral condensate** is an example of a fermionic condensate that appears in theories of massless fermions with chiral symmetry breaking.

21.1   Background

21.1.1   Superfluidity

Fermionic condensates are attained at temperatures lower than Bose–Einstein condensates. Fermionic condensates are a type of superfluid. As the name suggests, a superfluid possesses fluid properties similar to those possessed by ordinary liquids and gases, such as the lack of a definite shape and the ability to flow in response to applied forces. However, superfluids possess some properties that do not appear in ordinary matter. For instance, they can flow at low velocities without dissipating any energy—i.e. zero viscosity. At higher velocities, energy is dissipated by the formation of quantized vortices, which act as "holes" in the medium where superfluidity breaks down.

Superfluidity was originally discovered in liquid helium-4, in 1938, by Pyotr Kapitsa, John Allen and Don Misener. Superfluidity in helium-4, which occurs at temperatures below 2.17 kelvins (K), has long been understood to result from Bose condensation, the same mechanism that produces the Bose–Einstein condensates. The primary difference between superfluid helium and a Bose–Einstein condensate is that the former is condensed from a liquid while the latter is condensed from a gas.

21.1.2   Fermionic superfluids

It is far more difficult to produce a fermionic superfluid than a bosonic one, because the Pauli exclusion principle prohibits fermions from occupying the same quantum state. However, there is a well-known mechanism by which a superfluid may be formed from fermions. This is the BCS transition, discovered in 1957 by John Bardeen, Leon Cooper and Robert Schrieffer for describing superconductivity. These authors showed that, below a certain temperature, electrons (which are fermions) can pair up to form bound pairs now known as Cooper pairs. As long as collisions with the ionic lattice of the solid do not supply enough energy to break the Cooper pairs, the electron fluid will be able to flow without dissipation. As a result, it becomes a superfluid, and the material through which it flows a superconductor.

The BCS theory was phenomenally successful in describing superconductors. Soon after the publication of the BCS paper, several theorists proposed that a similar phenomenon could occur in fluids made up of fermions other than electrons, such as helium-3 atoms. These speculations were confirmed in 1971, when experiments performed by Douglas D. Osheroff showed that helium-3 becomes a superfluid below 0.0025 K. It was soon verified that the superfluidity of helium-3 arises from a BCS-like mechanism. (The theory of superfluid helium-3 is a little more complicated than the BCS theory of superconductivity. These complications arise because helium atoms repel each other much more strongly than electrons, but the basic idea is the same.)

21.1.3 Creation of the first fermionic condensates

When Eric Cornell and Carl Wieman produced a Bose–Einstein condensate from rubidium atoms in 1995, there naturally arose the prospect of creating a similar sort of condensate made from fermionic atoms, which would form a superfluid by the BCS mechanism. However, early calculations indicated that the temperature required for producing Cooper pairing in atoms would be too cold to achieve. In 2001, Murray Holland at JILA suggested a way of bypassing this difficulty. He speculated that fermionic atoms could be coaxed into pairing up by subjecting them to a strong magnetic field.

In 2003, working on Holland's suggestion, Deborah Jin at JILA, Rudolf Grimm at the University of Innsbruck, and Wolfgang Ketterle at MIT managed to coax fermionic atoms into forming molecular bosons, which then underwent Bose–Einstein condensation. However, this was not a true fermionic condensate. On December 16, 2003, Jin managed to produce a condensate out of fermionic atoms for the first time. The experiment involved 500,000 potassium−40 atoms cooled to a temperature of 5×10^{-8} K, subjected to a time-varying magnetic field. The findings were published in the online edition of *Physical Review Letters* on January 24, 2004.

21.2 Examples

21.2.1 BCS theory

The BCS theory of superconductivity has a fermion condensate. A pair of electrons in a metal, with opposite spins can form a scalar bound state called a Cooper pair. Then, the bound states themselves form a condensate. Since the Cooper pair has electric charge, this fermion condensate breaks the electromagnetic gauge symmetry of a superconductor, giving rise to the wonderful electromagnetic properties of such states.

21.2.2 QCD

In quantum chromodynamics (QCD) the chiral condensate is also called the **quark condensate**. This property of the QCD vacuum is partly responsible for giving masses to hadrons (along with other condensates like the gluon condensate).

In an approximate version of QCD, which has vanishing quark masses for N quark flavours, there is an exact chiral SU(N) × SU(N) symmetry of the theory. The QCD vacuum breaks this symmetry to SU(N) by forming a quark condensate. The existence of such a fermion condensate was first shown explicitly in the lattice formulation of QCD. The quark condensate is therefore an order parameter of transitions between several phases of quark matter in this limit.

This is very similar to the BCS theory of superconductivity. The Cooper pairs are analogous to the pseudoscalar mesons. However, the vacuum carries no charge. Hence all the gauge symmetries are unbroken. Corrections for the masses of the quarks can be incorporated using chiral perturbation theory.

21.2.3 Helium-3 superfluid

A helium-3 atom is a fermion and at very low temperatures, they form two-atom Cooper pairs which are bosonic and condense into a superfluid. These Cooper pairs are substantially larger than the interatomic separation.

21.3   References

- Guenault, Tony (2003). *Basic superfluids*. Taylor & Francis. ISBN 0-7484-0892-4.

- University of Colorado (January 28, 2004). *NIST/University of Colorado Scientists Create New Form of Matter: A Fermionic Condensate*. Press Release.

- Rodgers, Peter & Dumé, Bell (January 28, 2004). *Fermionic condensate makes its debut*. PhysicWeb.

- Haegler, Philipp, "Hadron Structure from Lattice Quantum Chromodynamics", Physics Reports 490, 49-175 (2010) [DOI 10.1016/j.physrep.2009.12.008]

Chapter 22

Superconductivity

A magnet levitating above a high-temperature superconductor, cooled with liquid nitrogen. Persistent electric current flows on the surface of the superconductor, acting to exclude the magnetic field of the magnet (Faraday's law of induction). This current effectively forms an electromagnet that repels the magnet.

Superconductivity is a phenomenon of exactly zero electrical resistance and expulsion of magnetic fields occurring in certain materials when cooled below a characteristic critical temperature. It was discovered by Dutch physicist Heike Kamerlingh Onnes on April 8, 1911 in Leiden. Like ferromagnetism and atomic spectral lines, superconductivity is a quantum mechanical phenomenon. It is characterized by the Meissner effect, the complete ejection of magnetic field lines from the interior of the superconductor as it transitions into the superconducting state. The occurrence of the Meissner effect indicates that superconductivity cannot be understood simply as the idealization of *perfect conductivity* in classical

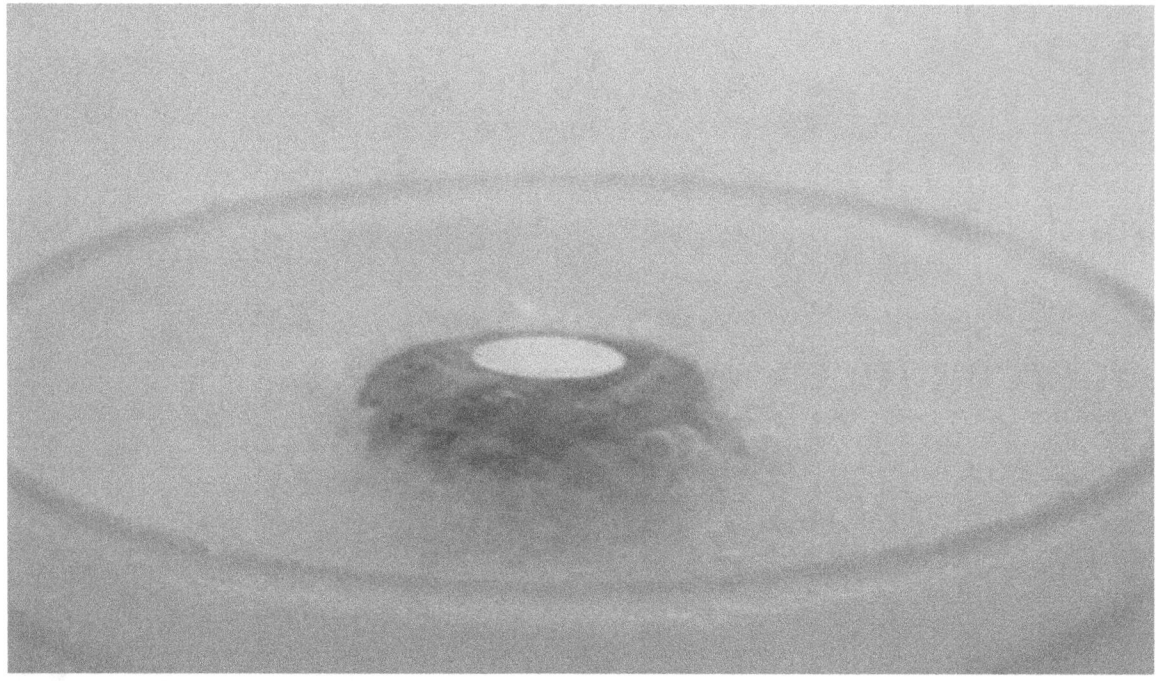

Video of a Meissner effect in a high temperature superconductor (black pellet) with a NdFeB magnet (metallic)

A high-temperature superconductor levitating above a magnet

physics.

The electrical resistivity of a metallic conductor decreases gradually as temperature is lowered. In ordinary conductors, such as copper or silver, this decrease is limited by impurities and other defects. Even near absolute zero, a real sample of a normal conductor shows some resistance. In a superconductor, the resistance drops abruptly to zero when the material is cooled below its critical temperature. An electric current flowing through a loop of superconducting wire can persist indefinitely with no power source.[1][2][3][4]

In 1986, it was discovered that some cuprate-perovskite ceramic materials have a critical temperature above 90 K (−183 °C).[5] Such a high transition temperature is theoretically impossible for a conventional superconductor, leading the materials to be termed high-temperature superconductors. Liquid nitrogen boils at 77 K, and superconduction at higher temperatures than this facilitates many experiments and applications that are less practical at lower temperatures.

22.1 Classification

Main article: Superconductor classification

There are many criteria by which superconductors are classified. The most common are:

- **Response to a magnetic field**: A superconductor can be *Type I*, meaning it has a single critical field, above which all superconductivity is lost; or *Type II*, meaning it has two critical fields, between which it allows partial penetration of the magnetic field.

- **By theory of operation**: It is *conventional* if it can be explained by the BCS theory or its derivatives, or *unconventional*, otherwise.

- **By critical temperature**: A superconductor is generally considered *high temperature* if it reaches a superconducting state when cooled using liquid nitrogen – that is, at only $T_c > 77$ K) – or *low temperature* if more aggressive cooling techniques are required to reach its critical temperature.

- **By material**: Superconductor material classes include chemical elements (e.g. mercury or lead), alloys (such as niobium-titanium, germanium-niobium, and niobium nitride), ceramics (YBCO and magnesium diboride), or organic superconductors (fullerenes and carbon nanotubes; though perhaps these examples should be included among the chemical elements, as they are composed entirely of carbon).

22.2 Elementary properties of superconductors

Most of the physical properties of superconductors vary from material to material, such as the heat capacity and the critical temperature, critical field, and critical current density at which superconductivity is destroyed.

On the other hand, there is a class of properties that are independent of the underlying material. For instance, all superconductors have *exactly* zero resistivity to low applied currents when there is no magnetic field present or if the applied field does not exceed a critical value. The existence of these "universal" properties implies that superconductivity is a thermodynamic phase, and thus possesses certain distinguishing properties which are largely independent of microscopic details.

22.2.1 Zero electrical DC resistance

The simplest method to measure the electrical resistance of a sample of some material is to place it in an electrical circuit in series with a current source I and measure the resulting voltage V across the sample. The resistance of the sample is given by Ohm's law as $R = V / I$. If the voltage is zero, this means that the resistance is zero.

Superconductors are also able to maintain a current with no applied voltage whatsoever, a property exploited in superconducting electromagnets such as those found in MRI machines. Experiments have demonstrated that currents in superconducting

Electric cables for accelerators at CERN. Both the massive and slim cables are rated for 12,500 A. Top: *conventional cables for LEP;* bottom: *superconductor-based cables for the LHC*

coils can persist for years without any measurable degradation. Experimental evidence points to a current lifetime of at least 100,000 years. Theoretical estimates for the lifetime of a persistent current can exceed the estimated lifetime of the universe, depending on the wire geometry and the temperature.[3]

In a normal conductor, an electric current may be visualized as a fluid of electrons moving across a heavy ionic lattice. The electrons are constantly colliding with the ions in the lattice, and during each collision some of the energy carried by the current is absorbed by the lattice and converted into heat, which is essentially the vibrational kinetic energy of the lattice ions. As a result, the energy carried by the current is constantly being dissipated. This is the phenomenon of electrical resistance and Joule heating.

The situation is different in a superconductor. In a conventional superconductor, the electronic fluid cannot be resolved into individual electrons. Instead, it consists of bound *pairs* of electrons known as Cooper pairs. This pairing is caused by an attractive force between electrons from the exchange of phonons. Due to quantum mechanics, the energy spectrum of this Cooper pair fluid possesses an *energy gap*, meaning there is a minimum amount of energy ΔE that must be supplied in order to excite the fluid. Therefore, if ΔE is larger than the thermal energy of the lattice, given by kT, where k is Boltzmann's constant and T is the temperature, the fluid will not be scattered by the lattice. The Cooper pair fluid is thus a superfluid, meaning it can flow without energy dissipation.

In a class of superconductors known as type II superconductors, including all known high-temperature superconductors, an extremely small amount of resistivity appears at temperatures not too far below the nominal superconducting transition when an electric current is applied in conjunction with a strong magnetic field, which may be caused by the electric current. This is due to the motion of magnetic vortices in the electronic superfluid, which dissipates some of the energy carried by the current. If the current is sufficiently small, the vortices are stationary, and the resistivity vanishes. The resistance due to this effect is tiny compared with that of non-superconducting materials, but must be taken into account in sensitive experiments. However, as the temperature decreases far enough below the nominal superconducting transition, these vortices can become frozen into a disordered but stationary phase known as a "vortex glass". Below this vortex glass transition temperature, the resistance of the material becomes truly zero.

22.2.2 Superconducting phase transition

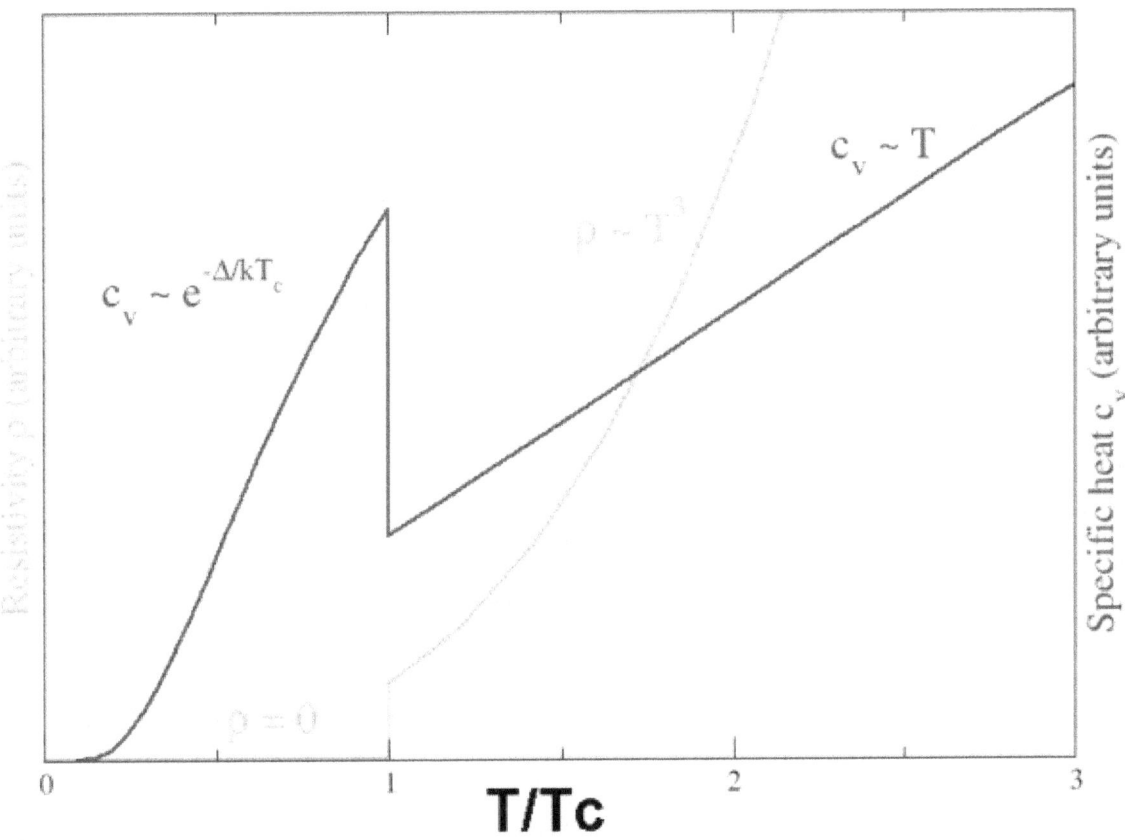

Behavior of heat capacity (cv, blue) and resistivity (ϱ, green) at the superconducting phase transition

In superconducting materials, the characteristics of superconductivity appear when the temperature T is lowered below a **critical temperature** Tc. The value of this critical temperature varies from material to material. Conventional superconductors usually have critical temperatures ranging from around 20 K to less than 1 K. Solid mercury, for example, has a critical temperature of 4.2 K. As of 2009, the highest critical temperature found for a conventional superconductor is 39 K for magnesium diboride (MgB_2),[6][7] although this material displays enough exotic properties that there is some doubt about classifying it as a "conventional" superconductor.[8] Cuprate superconductors can have much higher critical temperatures: $YBa_2Cu_3O_7$, one of the first cuprate superconductors to be discovered, has a critical temperature of 92 K, and mercury-based cuprates have been found with critical temperatures in excess of 130 K. The explanation for these high critical temperatures remains unknown. Electron pairing due to phonon exchanges explains superconductivity in conventional superconductors, but it does not explain superconductivity in the newer superconductors that have a very high critical temperature.

Similarly, at a fixed temperature below the critical temperature, superconducting materials cease to superconduct when an external magnetic field is applied which is greater than the *critical magnetic field*. This is because the Gibbs free energy of the superconducting phase increases quadratically with the magnetic field while the free energy of the normal phase is roughly independent of the magnetic field. If the material superconducts in the absence of a field, then the superconducting phase free energy is lower than that of the normal phase and so for some finite value of the magnetic field (proportional to the square root of the difference of the free energies at zero magnetic field) the two free energies will be equal and a phase transition to the normal phase will occur. More generally, a higher temperature and a stronger magnetic field lead to a smaller fraction of the electrons in the superconducting band and consequently a longer London penetration depth of external magnetic fields and currents. The penetration depth becomes infinite at the phase transition.

The onset of superconductivity is accompanied by abrupt changes in various physical properties, which is the hallmark of a phase transition. For example, the electronic heat capacity is proportional to the temperature in the normal (non-superconducting) regime. At the superconducting transition, it suffers a discontinuous jump and thereafter ceases to be linear. At low temperatures, it varies instead as $e^{-\alpha/T}$ for some constant, α. This exponential behavior is one of the pieces of evidence for the existence of the energy gap.

The order of the superconducting phase transition was long a matter of debate. Experiments indicate that the transition is second-order, meaning there is no latent heat. However, in the presence of an external magnetic field there is latent heat, because the superconducting phase has a lower entropy below the critical temperature than the normal phase. It has been experimentally demonstrated[9] that, as a consequence, when the magnetic field is increased beyond the critical field, the resulting phase transition leads to a decrease in the temperature of the superconducting material.

Calculations in the 1970s suggested that it may actually be weakly first-order due to the effect of long-range fluctuations in the electromagnetic field. In the 1980s it was shown theoretically with the help of a disorder field theory, in which the vortex lines of the superconductor play a major role, that the transition is of second order within the type II regime and of first order (i.e., latent heat) within the type I regime, and that the two regions are separated by a tricritical point.[10] The results were strongly supported by Monte Carlo computer simulations.[11]

22.2.3 Meissner effect

Main article: Meissner effect

When a superconductor is placed in a weak external magnetic field **H**, and cooled below its transition temperature, the magnetic field is ejected. The Meissner effect does not cause the field to be completely ejected but instead the field penetrates the superconductor but only to a very small distance, characterized by a parameter λ, called the London penetration depth, decaying exponentially to zero within the bulk of the material. The Meissner effect is a defining characteristic of superconductivity. For most superconductors, the London penetration depth is on the order of 100 nm.

The Meissner effect is sometimes confused with the kind of diamagnetism one would expect in a perfect electrical conductor: according to Lenz's law, when a *changing* magnetic field is applied to a conductor, it will induce an electric current in the conductor that creates an opposing magnetic field. In a perfect conductor, an arbitrarily large current can be induced, and the resulting magnetic field exactly cancels the applied field.

The Meissner effect is distinct from this—it is the spontaneous expulsion which occurs during transition to superconductivity. Suppose we have a material in its normal state, containing a constant internal magnetic field. When the material is cooled below the critical temperature, we would observe the abrupt expulsion of the internal magnetic field, which we would not expect based on Lenz's law.

The Meissner effect was given a phenomenological explanation by the brothers Fritz and Heinz London, who showed that the electromagnetic free energy in a superconductor is minimized provided

$$\nabla^2 \mathbf{H} = \lambda^{-2} \mathbf{H}$$

where **H** is the magnetic field and λ is the London penetration depth.

This equation, which is known as the London equation, predicts that the magnetic field in a superconductor decays exponentially from whatever value it possesses at the surface.

A superconductor with little or no magnetic field within it is said to be in the Meissner state. The Meissner state breaks down when the applied magnetic field is too large. Superconductors can be divided into two classes according to how this breakdown occurs. In Type I superconductors, superconductivity is abruptly destroyed when the strength of the applied field rises above a critical value *Hc*. Depending on the geometry of the sample, one may obtain an intermediate state[12] consisting of a baroque pattern[13] of regions of normal material carrying a magnetic field mixed with regions of superconducting material containing no field. In Type II superconductors, raising the applied field past a critical value Hc_1 leads to a mixed state (also known as the vortex state) in which an increasing amount of magnetic flux penetrates the material, but there remains no resistance to the flow of electric current as long as the current is not too large. At a second critical field strength Hc_2, superconductivity is destroyed. The mixed state is actually caused by vortices in the electronic superfluid, sometimes called fluxons because the flux carried by these vortices is quantized. Most pure elemental superconductors, except niobium and carbon nanotubes, are Type I, while almost all impure and compound superconductors are Type II.

22.2.4 London moment

Conversely, a spinning superconductor generates a magnetic field, precisely aligned with the spin axis. The effect, the London moment, was put to good use in Gravity Probe B. This experiment measured the magnetic fields of four superconducting gyroscopes to determine their spin axes. This was critical to the experiment since it is one of the few ways to accurately determine the spin axis of an otherwise featureless sphere.

22.3 History of superconductivity

Heike Kamerlingh Onnes (right), the discoverer of superconductivity. Paul Ehrenfest, Hendrik Lorentz, Niels Bohr stand to his left.

Main article: History of superconductivity

Superconductivity was discovered on April 8, 1911 by Heike Kamerlingh Onnes, who was studying the resistance of solid mercury at cryogenic temperatures using the recently produced liquid helium as a refrigerant. At the temperature of

4.2 K, he observed that the resistance abruptly disappeared.[14] In the same experiment, he also observed the superfluid transition of helium at 2.2 K, without recognizing its significance. The precise date and circumstances of the discovery were only reconstructed a century later, when Onnes's notebook was found.[15] In subsequent decades, superconductivity was observed in several other materials. In 1913, lead was found to superconduct at 7 K, and in 1941 niobium nitride was found to superconduct at 16 K.

Great efforts have been devoted to finding out how and why superconductivity works; the important step occurred in 1933, when Meissner and Ochsenfeld discovered that superconductors expelled applied magnetic fields, a phenomenon which has come to be known as the Meissner effect.[16] In 1935, Fritz and Heinz London showed that the Meissner effect was a consequence of the minimization of the electromagnetic free energy carried by superconducting current.[17]

22.3.1   London theory

The first phenomenological theory of superconductivity was London theory. It was put forward by the brothers Fritz and Heinz London in 1935, shortly after the discovery that magnetic fields are expelled from superconductors. A major triumph of the equations of this theory is their ability to explain the Meissner effect,[18] wherein a material exponentially expels all internal magnetic fields as it crosses the superconducting threshold. By using the London equation, one can obtain the dependence of the magnetic field inside the superconductor on the distance to the surface.[19]

There are two London equations:

$$\frac{\partial \mathbf{j}_s}{\partial t} = \frac{n_s e^2}{m} \mathbf{E}, \qquad \nabla \times \mathbf{j}_s = -\frac{n_s e^2}{m} \mathbf{B}.$$

The first equation follows from Newton's second law for superconducting electrons.

22.3.2   Conventional theories (1950s)

During the 1950s, theoretical condensed matter physicists arrived at a solid understanding of "conventional" superconductivity, through a pair of remarkable and important theories: the phenomenological Ginzburg-Landau theory (1950) and the microscopic BCS theory (1957).[20][21]

In 1950, the phenomenological Ginzburg-Landau theory of superconductivity was devised by Landau and Ginzburg.[22] This theory, which combined Landau's theory of second-order phase transitions with a Schrödinger-like wave equation, had great success in explaining the macroscopic properties of superconductors. In particular, Abrikosov showed that Ginzburg-Landau theory predicts the division of superconductors into the two categories now referred to as Type I and Type II. Abrikosov and Ginzburg were awarded the 2003 Nobel Prize for their work (Landau had received the 1962 Nobel Prize for other work, and died in 1968). The four-dimensional extension of the Ginzburg-Landau theory, the Coleman-Weinberg model, is important in quantum field theory and cosmology.

Also in 1950, Maxwell and Reynolds *et al.* found that the critical temperature of a superconductor depends on the isotopic mass of the constituent element.[23][24] This important discovery pointed to the electron-phonon interaction as the microscopic mechanism responsible for superconductivity.

The complete microscopic theory of superconductivity was finally proposed in 1957 by Bardeen, Cooper and Schrieffer.[21] This BCS theory explained the superconducting current as a superfluid of Cooper pairs, pairs of electrons interacting through the exchange of phonons. For this work, the authors were awarded the Nobel Prize in 1972.

The BCS theory was set on a firmer footing in 1958, when N. N. Bogolyubov showed that the BCS wavefunction, which had originally been derived from a variational argument, could be obtained using a canonical transformation of the electronic Hamiltonian.[25] In 1959, Lev Gor'kov showed that the BCS theory reduced to the Ginzburg-Landau theory close to the critical temperature.[26][27]

Generalizations of BCS theory for conventional superconductors form the basis for understanding of the phenomenon of superfluidity, because they fall into the lambda transition universality class. The extent to which such generalizations can be applied to unconventional superconductors is still controversial.

22.3.3 Further history

The first practical application of superconductivity was developed in 1954 with Dudley Allen Buck's invention of the cryotron.[28] Two superconductors with greatly different values of critical magnetic field are combined to produce a fast, simple, switch for computer elements.

Soon after discovering superconductivity in 1911, Kamerlingh Onnes attempted to make an electromagnet with superconducting windings but found that relatively low magnetic fields destroyed superconductivity in the materials he investigated. Much later, in 1955, G.B. Yntema [29] succeeded in constructing a small 0.7-tesla iron-core electromagnet with superconducting niobium wire windings. Then, in 1961, J.E. Kunzler, E. Buehler, F.S.L. Hsu, and J.H. Wernick [30] made the startling discovery that, at 4.2 degrees kelvin, a compound consisting of three parts niobium and one part tin, was capable of supporting a current density of more than 100,000 amperes per square centimeter in a magnetic field of 8.8 tesla. Despite being brittle and difficult to fabricate, niobium-tin has since proved extremely useful in supermagnets generating magnetic fields as high as 20 tesla. In 1962 T.G. Berlincourt and R.R. Hake [31][32] discovered that alloys of niobium and titanium are suitable for applications up to 10 tesla. Promptly thereafter, commercial production of niobium-titanium supermagnet wire commenced at Westinghouse Electric Corporation and at Wah Chang Corporation. Although niobium-titanium boasts less-impressive superconducting properties than those of niobium-tin, niobium-titanium has, nevertheless, become the most widely-used "workhorse" supermagnet material, in large measure a consequence of its very-high ductility and ease of fabrication. However, both niobium-tin and niobium-titanium find wide application in MRI medical imagers, bending and focusing magnets for enormous high-energy-particle accelerators, and a host of other applications. Conectus, a European superconductivity consortium, estimated that in 2014, global economic activity for which superconductivity was indispensable amounted to about five billion euros, with MRI systems accounting for about 80% of that total.

In 1962, Josephson made the important theoretical prediction that a supercurrent can flow between two pieces of superconductor separated by a thin layer of insulator.[33] This phenomenon, now called the Josephson effect, is exploited by superconducting devices such as SQUIDs. It is used in the most accurate available measurements of the magnetic flux quantum $\Phi_0 = h/(2e)$, where h is the Planck constant. Coupled with the quantum Hall resistivity, this leads to a precise measurement of the Planck constant. Josephson was awarded the Nobel Prize for this work in 1973.

In 2008, it was proposed that the same mechanism that produces superconductivity could produce a superinsulator state in some materials, with almost infinite electrical resistance.[34]

22.4 High-temperature superconductivity

Main article: High-temperature superconductivity

Until 1986, physicists had believed that BCS theory forbade superconductivity at temperatures above about 30 K. In that year, Bednorz and Müller discovered superconductivity in a lanthanum-based cuprate perovskite material, which had a transition temperature of 35 K (Nobel Prize in Physics, 1987).[5] It was soon found that replacing the lanthanum with yttrium (i.e., making YBCO) raised the critical temperature to 92 K.[35]

This temperature jump is particularly significant, since it allows liquid nitrogen as a refrigerant, replacing liquid helium.[35] This can be important commercially because liquid nitrogen can be produced relatively cheaply, even on-site. Also, the higher temperatures help avoid some of the problems that arise at liquid helium temperatures, such as the formation of plugs of frozen air that can block cryogenic lines and cause unanticipated and potentially hazardous pressure buildup.[36][37]

Many other cuprate superconductors have since been discovered, and the theory of superconductivity in these materials is one of the major outstanding challenges of theoretical condensed matter physics.[38] There are currently two main hypotheses – the resonating-valence-bond theory, and spin fluctuation which has the most support in the research community.[39] The second hypothesis proposed that electron pairing in high-temperature superconductors is mediated by short-range spin waves known as paramagnons.[40][41]

Since about 1993, the highest temperature superconductor was a ceramic material consisting of mercury, barium, calcium, copper and oxygen ($HgBa_2Ca_2Cu_3O_{8+\delta}$) with T_c = 133–138 K.[42][43] The latter experiment (138 K) still awaits experimental confirmation, however.

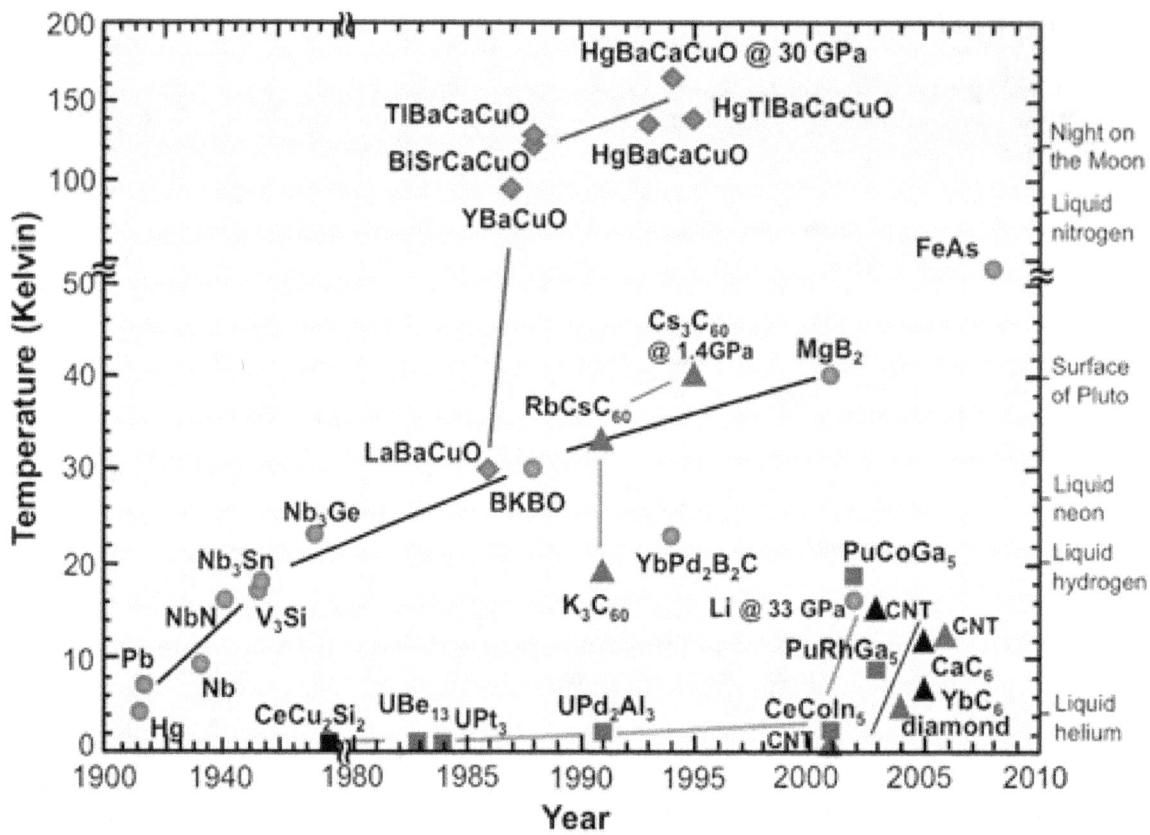

Timeline of superconducting materials

In February 2008, an iron-based family of high-temperature superconductors was discovered.[44][45] Hideo Hosono, of the Tokyo Institute of Technology, and colleagues found lanthanum oxygen fluorine iron arsenide ($LaO_{1-x}F_xFeAs$), an oxypnictide that superconducts below 26 K. Replacing the lanthanum in $LaO_{1-x}FxFeAs$ with samarium leads to superconductors that work at 55 K.[46]

In May 2014, hydrogen sulfide (H
2S) was predicted to be a high-temperature superconductor with a transition temperate of 80 at 160 gigapascals.[47] In 2015, H
2S has been observed to exhibit superconductivity at below 203 K but at extremely high pressures — around 150 gigapascals.[48]

22.5   Applications

Main article: Technological applications of superconductivity
Superconducting magnets are some of the most powerful electromagnets known. They are used in MRI/NMR machines, mass spectrometers, and the beam-steering magnets used in particle accelerators. They can also be used for magnetic separation, where weakly magnetic particles are extracted from a background of less or non-magnetic particles, as in the pigment industries.

In the 1950s and 1960s, superconductors were used to build experimental digital computers using cryotron switches. More recently, superconductors have been used to make digital circuits based on rapid single flux quantum technology and RF and microwave filters for mobile phone base stations.

Superconductors are used to build Josephson junctions which are the building blocks of SQUIDs (superconducting quantum interference devices), the most sensitive magnetometers known. SQUIDs are used in scanning SQUID microscopes

Video of superconducting levitation of YBCO

and magnetoencephalography. Series of Josephson devices are used to realize the SI volt. Depending on the particular mode of operation, a superconductor-insulator-superconductor Josephson junction can be used as a photon detector or as a mixer. The large resistance change at the transition from the normal- to the superconducting state is used to build thermometers in cryogenic micro-calorimeter photon detectors. The same effect is used in ultrasensitive bolometers made from superconducting materials.

Other early markets are arising where the relative efficiency, size and weight advantages of devices based on high-temperature superconductivity outweigh the additional costs involved. For example in wind turbines the lower weight and volume of superconducting generators could lead to savings in construction and tower costs, offsetting the higher costs for the generator and lowering the total LCOE.[49]

Promising future applications include high-performance smart grid, electric power transmission, transformers, power storage devices, electric motors (e.g. for vehicle propulsion, as in vactrains or maglev trains), magnetic levitation devices, fault current limiters, enhancing spintronic devices with superconducting materials,[50] and superconducting magnetic refrigeration. However, superconductivity is sensitive to moving magnetic fields so applications that use alternating current (e.g. transformers) will be more difficult to develop than those that rely upon direct current. Compared to to traditional power lines superconducting transmission lines are more efficient and require only a fraction of the space, which would not only lead to a better environmental performance but could also improve public acceptance for expansion of the electric grid.[51]

22.6 Nobel Prizes for superconductivity

- Heike Kamerlingh Onnes (1913), "for his investigations on the properties of matter at low temperatures which led, inter alia, to the production of liquid helium"

- John Bardeen, Leon N. Cooper, and J. Robert Schrieffer (1972), "for their jointly developed theory of superconductivity, usually called the BCS-theory"

- Leo Esaki, Ivar Giaever, and Brian D. Josephson (1973), "for their experimental discoveries regarding tunneling phenomena in semiconductors and superconductors, respectively," and "for his theoretical predictions of the properties of a supercurrent through a tunnel barrier, in particular those phenomena which are generally known as the Josephson effects"

- Georg Bednorz and K. Alex Müller (1987), "for their important break-through in the discovery of superconductivity in ceramic materials"

- Alexei A. Abrikosov, Vitaly L. Ginzburg, and Anthony J. Leggett (2003), "for pioneering contributions to the theory of superconductors and superfluids"[52]

22.7   See also

- Andreev reflection

- Charge transfer complex

- Color superconductivity in quarks

- Composite Reaction Texturing

- Conventional superconductor

- Covalent superconductors

- Flux pumping

- High-temperature superconductivity

- Homes's law

- Iron-based superconductor

- Kondo effect

- List of superconductors

- Little-Parks effect

- Magnetic levitation

- Macroscopic quantum phenomena

- Magnetic sail

- National Superconducting Cyclotron Laboratory

- Oxypnictide

- Persistent current

- Proximity effect

- Room-temperature superconductor

- Rutherford cable

- Spallation Neutron Source

- Superconducting RF

- Superconductor classification

- Superfluid film

- Superfluidity

- Superstripes

- Technological applications of superconductivity

- Timeline of low-temperature technology

- Type-I superconductor

- Type-II superconductor

- Unconventional superconductor

- BCS theory

- Bean's critical state model

22.8 References

[1] John Bardeen; Leon Cooper; J. R. Schriffer (December 1, 1957). "Theory of Superconductivity". *Physical Review* **8** (5): 1178. Bibcode:1957PhRv..108.1175B. doi:10.1103/physrev.108.1175. ISBN 9780677000800. Retrieved June 6, 2014. reprinted in Nikolaĭ Nikolaevich Bogoliubov (1963) *The Theory of Superconductivity, Vol. 4*, CRC Press, ISBN 0677000804, p. 73

[2] John Daintith (2009). *The Facts on File Dictionary of Physics* (4th ed.). Infobase Publishing. p. 238. ISBN 1438109490.

[3] John C. Gallop (1990). *SQUIDS, the Josephson Effects and Superconducting Electronics*. CRC Press. pp. 3, 20. ISBN 0-7503-0051-5.

[4] Durrant, Alan (2000). *Quantum Physics of Matter*. CRC Press. pp. 102–103. ISBN 0750307218.

[5] J. G. Bednorz & K. A. Müller (1986). "Possible high T_c superconductivity in the Ba–La–Cu–O system". *Z. Physik, B* **64** (1): 189–193. Bibcode:1986ZPhyB..64..189B. doi:10.1007/BF01303701.

[6] Jun Nagamatsu; Norimasa Nakagawa; Takahiro Muranaka; Yuji Zenitani; et al. (2001). "Superconductivity at 39 K in magnesium diboride". *Nature* **410** (6824): 63–4. Bibcode:2001Natur.410...63N. doi:10.1038/35065039. PMID 11242039.

[7] Paul Preuss (14 August 2002). "A most unusual superconductor and how it works: first-principles calculation explains the strange behavior of magnesium diboride". *Research News* (Lawrence Berkeley National Laboratory). Retrieved 2009-10-28.

[8] Hamish Johnston (17 February 2009). "Type-1.5 superconductor shows its stripes". *Physics World* (Institute of Physics). Retrieved 2009-10-28.

[9] R. L. Dolecek (1954). "Adiabatic Magnetization of a Superconducting Sphere". *Physical Review* **96** (1): 25–28. Bibcode:1954Ph doi:10.1103/PhysRev.96.25.

[10] H. Kleinert (1982). "Disorder Version of the Abelian Higgs Model and the Order of the Superconductive Phase Transition" (PDF). *Lettere al Nuovo Cimento* **35** (13): 405–412. doi:10.1007/BF02754760.

[11] J. Hove; S. Mo; A. Sudbo (2002). "Vortex interactions and thermally induced crossover from type-I to type-II superconductivity" (PDF). *Physical Review B* **66** (6): 064524. arXiv:cond-mat/0202215. Bibcode:2002PhRvB..66f4524H. doi:10.1103/PhysRev

[12] Lev D. Landau; Evgeny M. Lifschitz (1984). *Electrodynamics of Continuous Media*. Course of Theoretical Physics **8**. Oxford: Butterworth-Heinemann. ISBN 0-7506-2634-8.

[13] David J. E. Callaway (1990). "On the remarkable structure of the superconducting intermediate state". *Nuclear Physics B* **344** (3): 627–645. Bibcode:1990NuPhB.344..627C. doi:10.1016/0550-3213(90)90672-Z.

[14] H. K. Onnes (1911). "The resistance of pure mercury at helium temperatures". *Commun. Phys. Lab. Univ. Leiden* **12**: 120.

[15] Dirk vanDelft & Peter Kes (September 2010). "The Discovery of Superconductivity" (PDF). *Physics Today* (American Institute of Physics).

[16] W. Meissner & R. Ochsenfeld (1933). "Ein neuer Effekt bei Eintritt der Supraleitfähigkeit". *Naturwissenschaften* **21** (44): 787–788. Bibcode:1933NW.....21..787M. doi:10.1007/BF01504252.

[17] F. London & H. London (1935). "The Electromagnetic Equations of the Supraconductor". *Proceedings of the Royal Society of London A* **149** (866): 71–88. Bibcode:1935RSPSA.149...71L. doi:10.1098/rspa.1935.0048. JSTOR 96265.

[18] Meissner, W.; R. Ochsenfeld (1933). "Ein neuer Effekt bei Eintritt der Supraleitfähigkeit". *Naturwissenschaften* **21** (44): 787–788. Bibcode:1933NW.....21..787M. doi:10.1007/BF01504252.

[19] "The London equations". The Open University. Retrieved 2011-10-16.

[20] J. Bardeen; L. N. Cooper & J. R. Schrieffer (1957). "Microscopic Theory of Superconductivity". *Physical Review* **106** (1): 162–164. Bibcode:1957PhRv..106..162B. doi:10.1103/PhysRev.106.162.

[21] J. Bardeen; L. N. Cooper & J. R. Schrieffer (1957). "Theory of Superconductivity". *Physical Review* **108** (5): 1175–1205. Bibcode:1957PhRv..108.1175B. doi:10.1103/PhysRev.108.1175.

[22] V. L. Ginzburg & L.D. Landau (1950). "On the theory of superconductivity". *Zhurnal Eksperimental'noi i Teoreticheskoi Fiziki* **20**: 1064.

[23] E. Maxwell (1950). "Isotope Effect in the Superconductivity of Mercury". *Physical Review* **78**(4): 477. Bibcode:1950PhRv...78 doi:10.1103/PhysRev.78.477.

[24] C. A. Reynolds; B. Serin; W. H. Wright & L. B. Nesbitt (1950). "Superconductivity of Isotopes of Mercury". *Physical Review* **78** (4): 487. Bibcode:1950PhRv...78..487R. doi:10.1103/PhysRev.78.487.

[25] N. N. Bogoliubov (1958). "A new method in the theory of superconductivity". *Zhurnal Eksperimental'noi i Teoreticheskoi Fiziki* **34**: 58.

[26] L. P. Gor'kov (1959). "Microscopic derivation of the Ginzburg—Landau equations in the theory of superconductivity". *Zhurnal Eksperimental'noi i Teoreticheskoi Fiziki* **36**: 1364.

[27] M. Combescot; W.V. Pogosov and O. Betbeder-Matibet (2013). "BCS ansatz for superconductivity in the light of the Bogoliubov approach and the Richardson–Gaudin exact wave function". *Physica C: Superconductivity* **485**: 47–57. arXiv:1111.4781. Bibcode:2013PhyC..485...47C. doi:10.1016/j.physc.2012.10.011. Retrieved 11 August 2014.

[28] Buck, Dudley A. "The Cryotron - A Superconductive Computer Component" (PDF). Lincoln Laboratory, Massachusetts Institute of Technology. Retrieved 10 August 2014.

[29] G.B.Yntema (1955). "Superconducting Winding for Electromagnet". *Physical Review* **98**(4): 1197. Bibcode:1955PhRv...98.11. doi:10.1103/PhysRev.98.1144.

[30] J.E. Kunzler, E. Buehler, F.L.S. Hsu, and J.H. Wernick (1961). "Superconductivity in Nb3Sn at High Current Density in a Magnetic Field of 88 kgauss". *Physical Review Letters* **6** (3): 89–91. Bibcode:1961PhRvL...6...89K. doi:10.1103/PhysRevLett.6.89. line feed character in |title= at position 65 (help)

[31] T.G. Berlincourt and R.R. Hake (1962). "Pulsed-Magnetic-Field Studies of Superconducting Transition Metal Alloys at High and Low Current Densities". *Bulletin of the American Physical Society* **II–7**: 408. line feed character in |title= at position 60 (help)

[32] T.G. Berlincourt (1987). "Emergence of Nb-Ti as Supermagnet Material". *Cryogenics* **27**(6): 283–289. Bibcode:1987Cryo...27 doi:10.1016/0011-2275(87)90057-9.

[33] B. D. Josephson (1962). "Possible new effects in superconductive tunnelling". *Physics Letters* **1**(7): 251–253. Bibcode:1962PhL doi:10.1016/0031-9163(62)91369-0.

[34] "Newly discovered fundamental state of matter, a superinsulator, has been created.". Science Daily. April 9, 2008. Retrieved 2008-10-23.

[35] M. K. Wu; et al. (1987). "Superconductivity at 93 K in a New Mixed-Phase Y-Ba-Cu-O Compound System at Ambient Pressure". *Physical Review Letters* **58** (9): 908–910. Bibcode:1987PhRvL..58..908W. doi:10.1103/PhysRevLett.58.908. PMID 10035069.

[36] "Introduction to Liquid Helium". *"Cryogenics and Fluid Branch"*. Goddard Space Flight Center, NASA.

[37] "Section 4.1 "Air plug in the fill line"". *"Superconducting Rock Magnetometer Cryogenic System Manual"*. 2G Enterprises. Archived from the original on May 6, 2009. Retrieved 9 October 2012.

[38] Alexei A. Abrikosov (8 December 2003). "type II Superconductors and the Vortex Lattice". *Nobel Lecture.*

[39] Adam Mann (Jul 20, 2011). "High-temperature superconductivity at 25: Still in suspense". *Nature* **475** (7356): 280–2. Bibcode:2011Natur.475..280M. doi:10.1038/475280a. PMID 21776057.

[40] Pines, D. (2002), "The Spin Fluctuation Model for High Temperature Superconductivity: Progress and Prospects", *The Gap Symmetry and Fluctuations in High-Tc Superconductors*, NATO Science Series: B: **371**, New York: Kluwer Academic, pp. 111–142, doi:10.1007/0-306-47081-0_7, ISBN 0-306-45934-5

[41] P. Monthoux; A. V. Balatsky & D. Pines (1991). "Toward a theory of high-temperature superconductivity in the antiferromagnetically correlated cuprate oxides". *Phys. Rev. Lett.* **67** (24): 3448–3451. Bibcode:1991PhRvL..67.3448M. doi:10.1103/PhysRevLett.67.3448.PMID10044736.

[42] A. Schilling; et al. (1993). "Superconductivity above 130 K in the Hg–Ba–Ca–Cu–O system". *Nature* **363** (6424): 56–58. Bibcode:1993Natur.363..56C. doi:10.1038/363056a0.

[43] P. Dai; B. C. Chakoumakos; G. F. Sun; K. W. Wong; et al. (1995). "Synthesis and neutron powder diffraction study of the superconductor $HgBa_2Ca_2Cu_3O_{8+}\delta$ by Tl substitution". *Physica C* **243** (3–4): 201–206. Bibcode:1995PhyC..243..201D. doi:10.1016/0921-4534(94)02461-8.

[44] Hiroki Takahashi; Kazumi Igawa; Kazunobu Arii; Yoichi Kamihara; et al. (2008). "Superconductivity at 43 K in an iron-based layered compound $LaO_{1-x}F_xFeAs$". *Nature* **453** (7193): 376–378. Bibcode:2008Natur.453..376T. doi:10.1038/nature06972. PMID 18432191.

[45] Adrian Cho. "Second Family of High-Temperature Superconductors Discovered". ScienceNOW Daily News.

[46] Zhi-An Ren; et al. (2008). "Superconductivity and phase diagram in iron-based arsenic-oxides ReFeAsO1-d (Re = rare-earth metal) without fluorine doping". *EPL* **83**: 17002. arXiv:0804.2582. Bibcode:2008EL.....8317002R. doi:10.1209/0295-5075/83/17002.

[47] Li, Yinwei; Hao, Jian; Liu, Hanyu; Li, Yanling; Ma, Yanming (2014-05-07). "The metallization and superconductivity of dense hydrogen sulfide". *The Journal of Chemical Physics* **140** (17): 174712. doi:10.1063/1.4874158. ISSN 0021-9606.

[48] Drozdov, A. P.; Eremets, M. I.; Troyan, I. A.; Ksenofontov, V.; Shylin, S. I. (2015). "Conventional superconductivity at 203 kelvin at high pressures in the sulfur hydride system". *Nature* **525** (7567): 73–6. doi:10.1038/nature14964. ISSN 0028-0836. PMID 26280333.

[49] Islam et al, *A review of offshore wind turbine nacelle: Technical challenges, and research and developmental trends.* In: *Renewable and Sustainable Energy Reviews* 33, (2014), 161–176, doi:10.1016/j.rser.2014.01.085

[50] Linder, Jacob; Robinson, Jason W. A. (2 April 2015). "Superconducting spintronics". *Nature Physics* **11** (4): 307–315. doi:10.1038/nphys3242.

[51] Thomas et al, *Superconducting transmission lines – Sustainable electric energy transfer with higher public acceptance?* In: *Renewable and Sustainable Energy Reviews* 55, (2016), 59–72, doi:10.1016/j.rser.2015.10.041.

[52] "All Nobel Prizes in Physics". *Nobelprize.org*. Nobel Media AB 2014.

22.9   Further reading

- Hagen Kleinert (1989). "Superflow and Vortex Lines". *Gauge Fields in Condensed Matter* **1**. World Scientific. ISBN 9971-5-0210-0.

- Anatoly Larkin; Andrei Varlamov (2005). *Theory of Fluctuations in Superconductors*. Oxford University Press. ISBN 0-19-852815-9.

- A. G. Lebed (2008). *The Physics of Organic Superconductors and Conductors* **110** (1st ed.). Springer. ISBN 978-3-540-76667-4.

- Jean Matricon; Georges Waysand; Charles Glashausser (2003). *The Cold Wars: A History of Superconductivity*. Rutgers University Press. ISBN 0-8135-3295-7.

- "Physicist Discovers Exotic Superconductivity". ScienceDaily. 17 August 2006.

- Michael Tinkham (2004). *Introduction to Superconductivity* (2nd ed.). Dover Books. ISBN 0-486-43503-2.

- Terry Orlando; Kevin Delin (1991). *Foundations of Applied Superconductivity*. Prentice Hall. ISBN 978-0-201-18323-8.

- Paul Tipler; Ralph Llewellyn (2002). *Modern Physics* (4th ed.). W. H. Freeman. ISBN 0-7167-4345-0.

22.10   External links

- Everything about superconductivity: properties, research, applications with videos, animations, games

- Video about Type I Superconductors: R=0/transition temperatures/ B is a state variable/ Meissner effect/ Energy gap(Giaever)/ BCS model

- Superconductivity: Current in a Cape and Thermal Tights. An introduction to the topic for non-scientists National High Magnetic Field Laboratory

- Lectures on Superconductivity (series of videos, including interviews with leading experts)

- Superconductivity News Update

- Superconductor Week Newsletter – industry news, links, et cetera

- Superconducting Magnetic Levitation

- National Superconducting Cyclotron Laboratory at Michigan State University

- YouTube Video Levitating magnet

- International Workshop on superconductivity in Diamond and Related Materials (free download papers)

- New Diamond and Frontier Carbon Technology Volume 17, No.1 Special Issue on Superconductivity in CVD Diamond

- DoITPoMS Teaching and Learning Package – "Superconductivity"

- The Nobel Prize for Physics, 1901–2008

- folding hands-on activities about superconductivity

Chapter 23

Superfluidity

Superfluidity is a state of matter in which the matter behaves like a fluid with zero viscosity; where it appears to exhibit the ability to self-propel and travel in a way that defies the forces of gravity and surface tension. Superfluidity is found in astrophysics, high-energy physics, and theories of quantum gravity. The phenomenon is related to Bose–Einstein condensation, but neither is a specific type of the other: not all Bose-Einstein condensates can be regarded as superfluids, and not all superfluids are Bose–Einstein condensates.

23.1 Superfluidity of liquid helium

Main article: Superfluid helium-4

Superfluidity was originally discovered in liquid helium, by Pyotr Kapitsa and John F. Allen. It has since been described through phenomenology and microscopic theories. In liquid helium-4, the superfluidity occurs at far higher temperatures than it does in helium-3. Each atom of helium-4 is a boson particle, by virtue of its integer spin. A helium-3 atom is a fermion particle; it can form bosons only by pairing with itself at much lower temperatures. This process is similar to the electron pairing in superconductivity.

23.2 Ultracold atomic gases

Superfluidity in an ultracold fermionic gas was experimentally proven by Wolfgang Ketterle and his team who observed quantum vortices in ^{6}Li at a temperature of 50 nK at MIT in April 2005.[1][2] Such vortices had previously been observed in an ultracold bosonic gas using ^{87}Rb in 2000,[3] and more recently in two-dimensional gases.[4] As early as 1999 Lene Hau created such a condensate using sodium atoms[5] for the purpose of slowing light, and later stopping it completely.[6] Her team then subsequently used this system of compressed light[7] to generate the superfluid analogue of shock waves and tornadoes: "These dramatic excitations result in the formation of solitons that in turn decay into quantized vortices—created far out of equilibrium, in pairs of opposite circulation—revealing directly the process of superfluid breakdown in Bose-Einstein condensates. With a double light-roadblock setup, we can generate controlled collisions between shock waves resulting in completely unexpected, nonlinear excitations. We have observed hybrid structures consisting of vortex rings embedded in dark solitonic shells. The vortex rings act as 'phantom propellers' leading to very rich excitation dynamics."[8]

211

23.3   Superfluid in astrophysics

The idea that superfluidity exists inside neutron stars was first proposed by Arkady Migdal.[9][10] By analogy with electrons inside superconductors forming Cooper pairs due to electron-lattice interaction, it is expected that nucleons in a neutron star at sufficiently high density and low temperature can also form Cooper pairs due to the long-range attractive nuclear force and lead to superfluidity and superconductivity.[11]

23.4   Superfluidity in high-energy physics and quantum gravity

Main article: Superfluid vacuum theory

Superfluid vacuum theory (SVT) is an approach in theoretical physics and quantum mechanics where the physical vacuum is viewed as superfluid.

The ultimate goal of the approach is to develop scientific models that unify quantum mechanics (describing three of the four known fundamental interactions) with gravity. This makes SVT a candidate for the theory of quantum gravity and an extension of the Standard Model.

It is hoped that development of such theory would unify into a single consistent model of all fundamental interactions, and to describe all known interactions and elementary particles as different manifestations of the same entity, superfluid vacuum.

23.5   See also

- Boojum (superfluidity)

- Condensed matter physics

- Macroscopic quantum phenomena

- Quantum hydrodynamics

- Slow light

- Supersolid

23.6   References

[1] "MIT physicists create new form of matter". Retrieved November 22, 2010.

[2] Grimm, R. (2005). "Low-temperature physics: A quantum revolution". *Nature* **435** (7045): 1035–1036. doi:10.1038/4351035a. PMID 15973388.

[3] Madison, K.; Chevy, F.; Wohlleben, W.; Dalibard, J. (2000). "Vortex Formation in a Stirred Bose-Einstein Condensate". *Physical Review Letters* **84** (5): 806–809. arXiv:cond-mat/9912015. Bibcode:2000PhRvL..84..806M. doi:10.1103/PhysRevLett.84.806.PMID11017378.

[4] Burnett, K. (2007). "Atomic physics: Cold gases venture into Flatland". *Nature Physics* **3** (9): 589. Bibcode:2007NatPh...3..589B. doi:10.1038/nphys704.

[5] Hau, L. V.; Harris, S. E.; Dutton, Z.; Behroozi, C. H. (1999). "Light speed reduction to 17 metres per second in an ultracold atomic gas". *Nature* **397** (6720): 594–598. doi:10.1038/17561.

[6] "Lene Hau". Physicscentral.com. Retrieved 2013-02-10.

[7] Lene Vestergaard Hau (2003). "Frozen Light" (PDF). *Scientific American*: 44–51.

[8] Shocking Bose-Einstein Condensates with Slow Light

[9] A. B. Migdal (1959). "Superfluidity and the moments of inertia of nuclei". *Nucl. Phys.* **13**(5): 655–674. Bibcode:1959NucPh doi:10.1016/0029-5582(59)90264-0.

[10] A. B. Migdal (1960). *Soviet Phys. JETP* **10**: 176. Missing or empty |title= (help)

[11] U. Lombardo & H.-J. Schulze (2001). "Superfluidity in Neutron Star Matter". *Physics of Neutron Star Interiors*. Lecture Notes in Physics **578**. Springer. pp. 30–53. arXiv:astro-ph/0012209. doi:10.1007/3-540-44578-1_2.

23.7 Further reading

- Guénault, Antony M. (2003). *Basic superfluids*. London: Taylor & Francis. ISBN 0-7484-0891-6.

- Annett, James F. (2005). *Superconductivity, superfluids, and condensates*. Oxford: Oxford Univ. Press. ISBN 978-0-19-850756-7.

- Volovik, G. E. (2003). *The Universe in a helium droplet*. Int. Ser. Monogr. Phys. **117**. pp. 1–507. ISBN 978-0198507826.

Fig. 1. Helium II will "creep" along surfaces in order to find its own level—after a short while, the levels in the two containers will equalize. The Rollin film also covers the interior of the larger container; if it were not sealed, the helium II would creep out and escape.

Fig. 2. *The liquid helium is in the superfluid phase. As long as it remains superfluid, it creeps up the wall of the cup as a thin film. It comes down on the outside, forming a drop which will fall into the liquid below. Another drop will form—and so on—until the cup is empty.*

Chapter 24

Supersolid

A **supersolid** is a spatially ordered material with superfluid properties. Superfluidity is a special quantum state of matter in which a substance flows with zero viscosity.

24.1   Background

Liquid helium-4 was discovered by Pyotr Kapitza, John F. Allen, and Don Misener to exhibit property of superfluidity when it is cooled below a characteristic transition temperature called the lambda point. Superfluidity is also observed when superconductors are cooled below a critical temperature. However, before the recent observation of superfluid-like behavior in solid helium-4,[1] superfluidity was considered to be a property exclusive to the fluid state, e.g. superconducting electron and neutron fluids, gases with Bose–Einstein condensates, or unconventional liquids such as helium-4 or helium-3 at sufficiently low temperature.

Superfluidity in helium arises from the normal liquid by a second-order phase transition ("lambda transition"). In a dilute gas of Bose particles it comes about by a phase transition that belongs to the universality class of the spherical model. In thin liquid helium films, it arises from the normal liquid by a Kosterlitz-Thouless transition. In the case of helium-4, it has been conjectured since 1970 that it might be possible to create a supersolid.[2]

In most theories of this state, it is supposed that vacancies, empty sites normally occupied by particles in an ideal crystal, exist even at absolute zero. These vacancies are caused by zero-point energy, which also causes them to move from site to site as waves. Because vacancies are bosons, if such clouds of vacancies can exist at very low temperature, then a Bose–Einstein condensation of vacancies could occur at temperatures less than a few tenths of a kelvin. A coherent flow of vacancies is equivalent to a "superflow" (frictionless flow) of particles in the opposite direction. Despite the presence of the gas of vacancies, the ordered structure of a crystal is maintained, although with less than one particle on each lattice site on average.

24.2   Experiments

While several experiments yielded negative results, in the 1980s, John Goodkind from UCSD discovered the first 'anomaly' in a solid by using ultrasound.[3] Inspired by his observation, Eun-Seong Kim and Moses Chan at Pennsylvania State University saw phenomena which were interpreted as supersolid behavior.[4] Specifically, they observed what they later named Non-Classical Rotational Inertia, an unusual decoupling of the solid helium from a container's walls which could not be explained by classical models but which was consistent with a superfluid-like decoupling of a small percentage of the atoms from the rest of the atoms in the container. If such an interpretation is correct, it would signify the discovery of a new quantum phase of matter.

The experiment of Kim and Chan looked for superflow by means of a "torsional oscillator." To achieve this, a turntable is attached tightly to a spring-loaded spindle; then, instead of rotating at constant speed, the turntable is given an initial motion in one direction. The spring causes the table to oscillate similarly to a balance wheel. A toroid filled with solid helium-4 is attached to the table. The rate of oscillation of the turntable and toroid depend on the amount of solid moving with it. If there is frictionless superfluid inside, then the mass moving with the doughnut is less, and the oscillation will occur at a faster rate. In this way, one can measure the amount of superfluid existing at various temperatures. Kim and Chan found that up to about 2% of the material in the doughnut was superfluid. (Recent experiments have increased the percentage to over 20%). Similar experiments in other laboratories have confirmed these results.[3] A mysterious feature, not in agreement with the old theories, is that the transition continues to occur at high pressures.

High-precision measurements of the melting pressure of helium-4 have not resulted in any observation of a phase transition in the solid.[5]

Prior to 2007, many theorists performed calculations indicating that vacancies cannot exist at zero temperature in solid helium-4. While there is some debate, it seems more doubtful that what the experiments observed was the supersolid state.[3] Indeed, further experimentation, including that by Kim and Chan, has also cast some doubt on the existence of a true supersolid. One experiment found that repeated warming followed by slow cooling of the sample causes the effect to disappear. This annealing process removes flaws in the crystal structure. Furthermore, most samples of helium-4 contain a small amount of helium-3. When some of this helium-3 is removed, the superfluid transition occurs at a lower temperature, which suggests that the superflow is involved with actual fluid moving along imperfections in the crystal rather than a property of the perfect crystal.

In 2009, it was proposed to realize a supersolid in an optical lattice. Starting from a molecular quantum crystal, super-solidity is induced dynamically as an out-of-equilibrium state. While neighboring molecular wave functions overlap, two bosonic species simultaneously exhibit quasicondensation and long-range solid order, which is stabilized by their mass imbalance. This proposal can be realized[6] in present experiments with bosonic mixtures in an optical lattice that features simple on-site interactions.

Experimental and theoretical work continues in hopes of finally settling the question of the existence of a supersolid.

In 2012, Chan repeated his original experiments with a new apparatus that was designed to eliminate any contribution from elasticity of the helium. In this experiment, Chan and his coauthors found no evidence of supersolidity.[7] Researchers are continuing to search for conclusive evidence of supersolidity.

24.3 See also

- Superfluid film
- Superglass

24.4 References

[1] Ball, Philip (2004). "Glimpse of a new type of matter". *Nature*. doi:10.1038/news040112-7.

[2] Chester, G. V. (1970). "Speculations on Bose-Einstein Condensation and Quantum Crystals". *Physical Review A* **2**: 256. Bibcode:1970PhRvA...2..256C. doi:10.1103/PhysRevA.2.256.

[3] Chalmers, Matthew (2007-05-01). "The quantum solid that defies expectation". Physics World. Retrieved 2009-02-25.

[4] E. Kim & M. H. W. Chan (2004). "Probable Observation of a Supersolid Helium Phase". *Nature* **427** (6971): 225–227. Bibcode:2004Natur.427..225K. doi:10.1038/nature02220. PMID 14724632.

[5] Todoshchenko, I. A.; Alles, H.; Junes, H. J.; Parshin, A. Ya.; Tsepelin, V. (2007). "Absence of low-temperature anomaly on the melting curve of 4He" (PDF). *JETP Letters* **85** (9): 454. arXiv:cond-mat/0703743. Bibcode:2007JETPL..85..454T. doi:10.1134/S0021364007090093.

[6] Keilmann, Tassilo; Cirac, Ignacio; Roscilde, Tommaso (2009). "Dynamical Creation of a Supersolid in Asymmetric Mixtures of Bosons".*Physical Review Letters***102**(25).arXiv:0906.1110.Bibcode:2009PhRvL.102y5304K.doi:10.1103/PhysRevLett.102.

[7] News article describing new experiment that finds no evidence of supersolidity with link to PRL article

24.5   External links

- *Nature* story on a supersolid experiment (Premium Link)

- Penn State: What is a Supersolid?

- Phys. Rev. Lett. Vol.101, 8 August 2008

- (1969). "Destruction of Superflow in Unsaturated 4He Films and the Prediction of a New Crystalline Phase of 4He with Bose–Einstein Condensation", *Physics Letters*, Vol. 30, No. 5, November 3, 1969, pp. 300–301 Jack Sarfatti

Chapter 25

Quantum spin liquid

In condensed matter physics, **quantum spin liquid** is a state that can be achieved in a system of interacting quantum spins. The state is referred to as a "liquid" as it is a disordered state in comparison to a ferromagnetic spin state,[1] much in the way liquid water is in a disordered state compared to crystalline ice. However, unlike other disordered states, a quantum spin liquid state preserves its disorder to very low temperatures.[2]

The quantum spin liquid state was first proposed by physicist Phil Anderson in 1973 as the ground state for a system of spins on a triangular lattice that interact with their nearest neighbors via the so-called antiferromagnetic interaction. Quantum spin liquids generated further interest when in 1987 Anderson proposed a theory that described high temperature superconductivity in terms of a disordered spin-liquid state.[3] A quantum spin liquid state in κ-(BEDT-TTF)$_2$Cu$_2$(CN)$_3$ was first thoroughly mapped using muon spin spectroscopy by a team led by Dr Francis Pratt at ISIS neutron source, UK in March, 2011.[4]

25.1 Examples

Several physical models have a disordered ground state that can be described as a quantum spin liquid.

25.1.1 Frustrated magnetic moments

Localized spins are frustrated if there exist competing exchange interactions that can not all be satisfied at the same time, leading to a large degeneracy of the system's ground state. A triangle of Ising spins (meaning the only possible orientations of the spins are "up" and "down"), which interact antiferromagnetically, is a simple example for frustration. In the ground state, two of the spins can be antiparallel but the third one cannot. This leads to an increase of possible orientations (six in this case) of the spins in the ground state, enhancing fluctuations and thus suppressing magnetic ordering.
Some frustrated materials with different lattice structures and their Curie-Weiss temperature are listed in the table.[2] All of them are proposed spin liquid candidates.

25.1.2 Resonating valence bonds (RVB)

Main article: Resonating valence bond theory
 To build a ground state without magnetic moment, valence bond states can be used, where two electron spins form a spin 0 singlet due to the antiferromagnetic interaction. If every spin in the system is bound like this, the state of the system as a whole has spin 0 too and is non-magnetic. The two spins forming the bond are maximally entangled, while not being entangled with the other spins. If all spins are distributed to certain localized static bonds, this is called a **valence bond solid** (VBS).

There are two things that still distinguish a VBS from a spin liquid: First, by ordering the bonds in a certain way, the

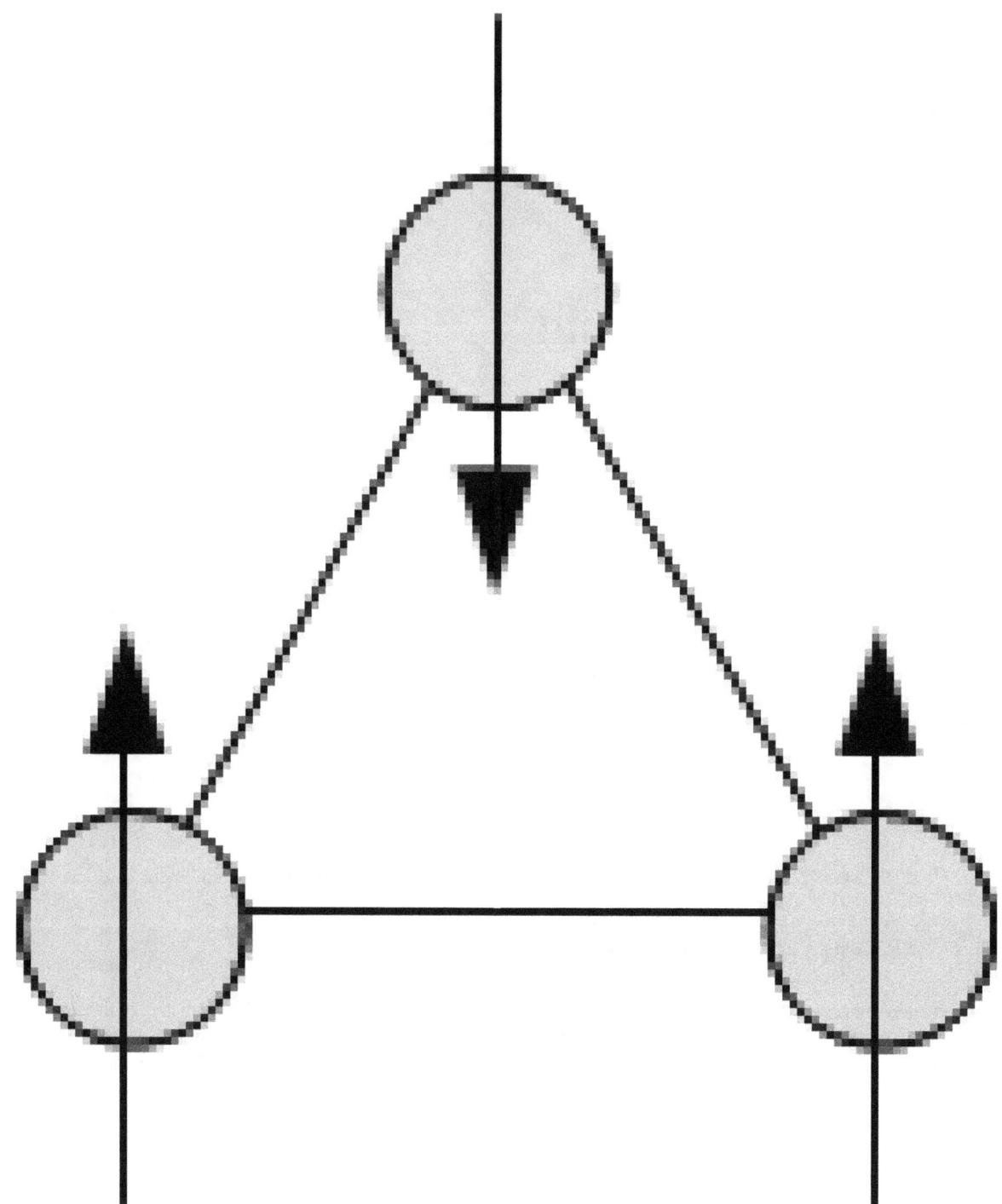

Frustrated Ising spins on a triangle.

lattice symmetry is usually broken, which is not the case for a spin liquid. Second, this ground state lacks long-range entanglement. To achieve this, quantum mechanical fluctuations of the valence bonds must be allowed, leading to a ground state consisting of a superposition of many different partitionings of spins into valence bonds. If the partitionings are equally distributed (with the same quantum amplitude), there is no preference for any specific partitioning ("valence bond liquid"). This kind of ground state wavefunction was proposed by P. W. Anderson in 1973 as the ground state of spin liquids[7] and is called a **resonating valence bond** (RVB) state. These states are of great theoretical interest as they

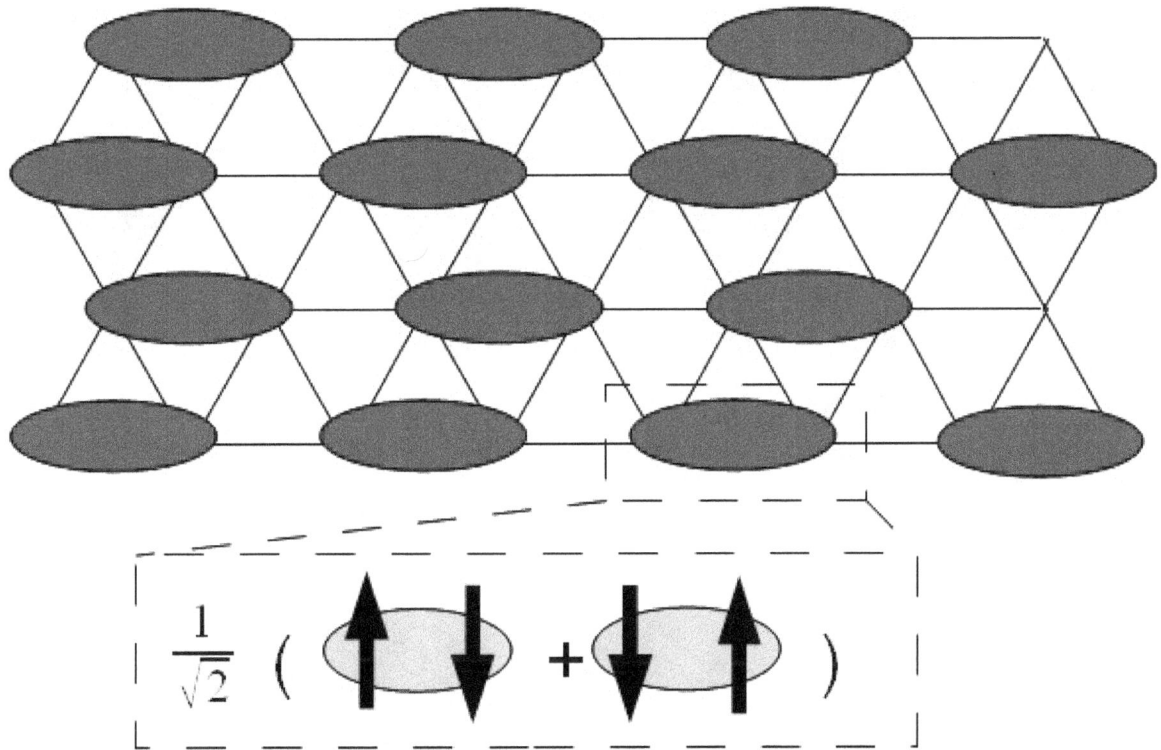

Valence bond solid. The bonds form a specific pattern and consist of pairs of entangled spins.

are proposed to play a key role in high-temperature superconductor physics.[8]

- One possible short-range pairing of spins in a RVB state.

- Long-range pairing of spins.

Excitations

The valence bonds do not have to be formed by nearest neighbors only and their distributions may vary in different materials. Ground states with large contributions of long range valence bonds have more low-energy spin excitations, as those valence bonds are easier to break up. On breaking, they form two free spins. Other excitations rearrange the valence bonds, leading to low-energy excitations even for short-range bonds. Very special about spin liquids is, that they support **exotic excitations**, meaning excitations with fractional quantum numbers. A prominent example is the excitation of spinons which are neutral in charge and carry spin $S = 1/2$. In spin liquids, a spinon is created if one spin is not paired in a valence bond. It can move by rearranging nearby valence bonds at low energy cost.

Realizations of (stable) RVB states

The first discussion of the RVB state on square lattice using the RVB picture[9] only consider nearest neighbour bonds that connect different sub-lattices. The constructed RVB state is an equal amplitude superposition of all the nearest-neighbour bond configurations. Such a RVB state is believed to contain emergent gapless $U(1)$ gauge field which may confine the spinons etc. So the equal-amplitude nearest-neighbour RVB state on square lattice is unstable and may describe a critical phase transition point between two stable phases. A version of RVB state which is stable and contains deconfined spinons is the chiral spin state.[10][11] Later, another version of stable RVB state with deconfined spinons, the Z2 spin liquid, is proposed,[12][13] which realizes the simplest topological order -- Z2 topological order. Both chiral spin state and Z2 spin liquid state have RVB bonds that connect the same sub-lattice. In chiral spin state, different bond configurations can have

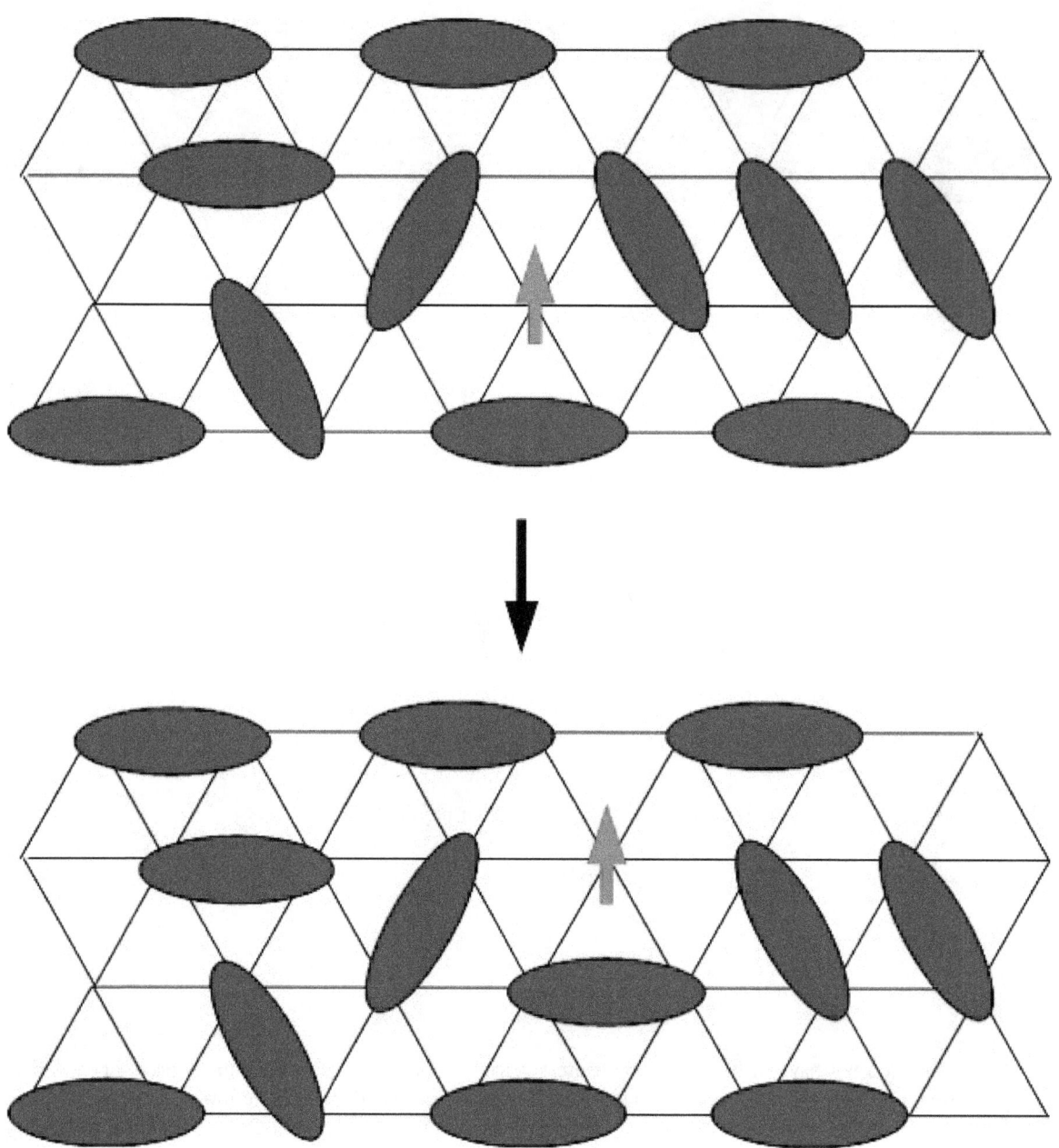

Spinon moving in spin liquids.

complex amplitudes, while in Z2 spin liquid state, different bond configurations only have real amplitudes. The RVB state on triangle lattice also realizes the Z2 spin liquid,[14] where different bond configurations only have real amplitudes. The toric code model is yet another realization of Z2 spin liquid (and Z2 topological order) that explicitly breaks the spin rotation symmetry and is exactly soluble.[15]

25.2 Identification in Experiments

Since there is no single experimental feature which identifies a material as a spin liquid, several experiments have to be conducted to gain information on different properties which characterize a spin liquid. An indication is given by a large value of the **frustration parameter** $f > 100$, which is defined as

$$f = \frac{|\Theta_{cw}|}{T_c}$$

where Θ_{cw} is the Curie-Weiss temperature and T_c is the temperature below which magnetic order begins to develop.

One of the most direct evidence for absence of magnetic ordering give NMR or μSR experiments. If there is a local magnetic field present, the nuclear or muon spin would be affected which can be measured. ^{1}H-NMR measurements [16] on κ-(BEDT-TTF)$_2$Cu$_2$(CN)$_3$ have shown no sign of magnetic ordering down to 32 mK, which is four orders of magnitude smaller than the coupling constant J≈250 K [17] between neighboring spins in this compound. Further investigations include:

- **Specific heat measurements** give information about the low-energy density of states, which can be compared to theoretical models.

- **Thermal transport measurements** can determine if excitations are localized or itinerant.

- **Neutron scattering** gives information about the nature of excitations and correlations (e.g. spinons).

- **Reflectance measurements** can uncover spinons, which couple via emergent gauge fields to the electromagnetic field, giving rise to a power-law optical conductivity. [18]

Herbertsmithite, the mineral whose ground state was shown to have QSL behaviour

25.2.1 Observation of fractionalization

In 2012, Young Lee and his collaborators at MIT and the National Institute of Standards and Technology artificially developed a crystal of herbertsmithite, a crystal with kagome lattice ordering, on which they were able to perform neutron scattering experiments. [19] The experiments revealed evidence for spin-state fractionalization, a predicted property of quantum spin-liquid type states.[20] The observation has been described as a hallmark for the quantum spin liquid state in herbertsmithite.[21] Data indicate that the strongly correlated quantum spin liquid, a specific form of quantum spin liquid, is realized in Herbertsmithite.[22]

25.3   Applications

Materials supporting quantum spin liquid states may have applications in data storage and memory.[23] In particular, it is possible to realize topological quantum computation by means of spin-liquid states.[24] Developments in quantum spin liquids may also help in the understanding of high temperature superconductivity.[25]

25.4   References

[1] Wilkins, Alasdair (August 15, 2011). "A Strange New Quantum State of Matter: Spin Liquids". *io9*. Retrieved 23 December 2012.

[2] Leon Balents (2010). "Spin liquids in frustrated magnets". *Nature* **464** (7286): 199–208. Bibcode:2010Natur.464..199B. doi:10.1038/nature08917. PMID 20220838.

[3] Trafton, Anne (March 28, 2011). "A new spin on superconductivity?". *MIT News*. Retrieved 24 December 2012.

[4] "Quantum mapmakers complete first voyage through spin liquid". *ISIS neutron and muon source*. Retrieved 23 August 2013.

[5] Emily A. Nytko, Joel S. Helton, Peter Müller, and Daniel G. Nocera *A Structurally Perfect S = 1/2 Metal−Organic Hybrid Kagome Antiferromagnet J. Am. Chem. Soc.* 2008; 130(10), pp 2922–2923 doi:10.1021/ja709991u

[6] Matan, K.; Ono, T.; Fukumoto, Y.; Sato, T. J.; et al. (2010). "Pinwheel valence-bond solid and triplet excitations in the two-dimensional deformed kagome lattice". *Nature Physics* **6**: 865–869. arXiv:1007.3625. Bibcode:2010NatPh...6..865M. doi:10.1038/nphys1761.

[7] P. W. Anderson (1973). "Resonating valence bonds: A new kind of insulator?". *Mater. Res. Bull.* **8** (2). doi:10.1016/0025-5408(73)90167-0.

[8] P. W. Anderson (1987). "The resonating valence bond state in La_2CuO_4 and superconductivity". *Science* **235** (4793): 1196–1198. Bibcode:1987Sci...235.1196A. doi:10.1126/science.235.4793.1196. PMID 17818979.

[9] Kivelson, Steven A.; Rokhsar, Daniel S.; Sethna, James P. (1987). "Topology of the resonating valence-bond state: Solitons and high-Tc superconductivity". *Phys. Rev. B* **35**: 8865. Bibcode:1987PhRvB..35.8865K. doi:10.1103/physrevb.35.8865.

[10] Kalmeyer, V.; Laughlin, R. B. (1987). "Equivalence of the resonating-valence-bond and fractional quantum Hall states". , *Phys. Rev. Lett.* **59**: 2095. Bibcode:1987PhRvL..59.2095K. doi:10.1103/physrevlett.59.2095.

[11] Wen, Xiao-Gang; Wilczek, F.; Zee, A. (1989). "Chiral Spin States and Superconductivity". , *Phys. Rev. B* **39**: 11413. Bibcode:1989PhRvB..3911413W. doi:10.1103/physrevb.39.11413.

[12] Read, N.; Sachdev, Subir (1991). "Large-N expansion for frustrated quantum antiferromagnets". , *Phys. Rev. Lett.* **66**: 1773. Bibcode:1991PhRvL..66.1773R. doi:10.1103/physrevlett.66.1773.

[13] Wen, Xiao-Gang (1991). "Mean Field Theory of Spin Liquid States with Finite Energy Gaps". , *Phys. Rev. B* **44**: 2664. Bibcode:1991PhRvB..44.2664W. doi:10.1103/physrevb.44.2664.

[14] R. Moessner, S. L. Sondhi, Phys. Rev. Lett 86, 1881 (2001); arXiv:cond-mat/0205029

[15] A. Yu. Kitaev, Annals Phys., 303, 2 (2003); arXiv:quant-ph/9707021

[16] Y. Shimizu; K. Miyagawa; K. Kanoda; M. Maesato; et al. (2003). "Spin Liquid State in an Organic Mott Insulator with a Triangular Lattice".*Phys. Rev. Lett.***91**(10).arXiv:cond-mat/0307483.Bibcode:2003PhRvL..91j7001S.doi:10.1103/PhysRevLet

[17] In literature, the value of J is commonly given in units of temperature (J/k_B) instead of energy.

[18] T. Ng & P. A. Lee (2007). "Power-Law Conductivity inside the Mott Gap: Application to κ-(BEDT-TTF)$_2$Cu$_2$(CN)$_3$". *Phys. Rev. Lett.* **99** (15). arXiv:0706.0050. Bibcode:2007PhRvL..99o6402N. doi:10.1103/PhysRevLett.99.156402.

[19] Anthony, Sebastian. "MIT discovers a new state of matter, a new kind of magnetism". *Extremetech*. Retrieved 23 August 2013.

[20] "Third State Of Magnetism Discovered By MIT Researchers". *Red Orbit*. December 21, 2012. Retrieved 24 December 2012.

[21] Han, Tiang-Heng; Young S. Lee; et al. (2012). "Fractionalized excitations in the spin-liquid state of a kagome-lattice antiferromagnet". *Nature* **492** (7429). arXiv:1307.5047. Bibcode:2012Natur.492..406H. doi:10.1038/nature11659. Retrieved 24 December 2012.

[22] Amusia, M., Popov, K., Shaginyan, V., Stephanovich, V. (2014). "Theory of Heavy-Fermion Compounds - Theory of Strongly Correlated Fermi-Systems". Springer. ISBN 978-3-319-10825-4.

[23] Aguilar, Mario (December 20, 2012). "This Weird Crystal Demonstrates a New Magnetic Behavior That Works Like Magic". *Gizmodo*. Retrieved 24 December 2012.

[24] Fendley, Paul. "Topological Quantum Computation from non-abelian anyons" (PDF). University of Virginia. Retrieved 24 December 2012.

[25] Chandler, David (December 20, 2012). "New kind of magnetism discovered: Experiments demonstrate 'quantum spin liquid'". *Phys.org*. Retrieved 24 December 2012.

Chapter 26

String-net liquid

In condensed matter physics, a **string-net** is an extended object whose collective behavior has been proposed as a physical mechanism for topological order by Michael A. Levin and Xiao-Gang Wen. A particular string-net model may involve only closed loops; or networks of oriented, labeled strings obeying branching rules given by some gauge group; or still more general networks.[1]

Their model purports to show the derivation of photons, electrons, and U(1) gauge charge, small (relative to the planck mass) but nonzero masses, and suggestions that the leptons, quarks, and gluons, can be modeled in the same way. In other words, string-net condensation provides an unification of photon and electron (or gauge bosons and fermions). It can be viewed as an origin of light and electron (or gauge interactions and Fermi statistics). However, their model does not account for the chiral coupling between the fermions and the SU(2) gauge bosons in the standard model.

For strings labeled by the positive integers, string-nets are the spin networks studied in loop quantum gravity. This has led to the proposal by Levin and Wen,[2] and Smolin, Markopoulou and Konopka[3] that loop quantum gravity's spin networks can give rise to the standard model of particle physics through this mechanism, along with fermi statistics and gauge interactions. To date, a rigorous derivation from LQG's spin networks to Levin and Wen's spin lattice has yet to be done, but the project to do so is called "quantum graphity", and in a more recent paper, Tomasz Konopka, Fotini Markopoulou, Simone Severini argued that there are some similarities to spin networks (but not necessarily an exact equivalence) that gives rise to U(1) gauge charge and electrons in the string net mechanism.[4]

Herbertsmithite may be an example of string-net matter.[5][6]

26.1 Examples

26.1.1 Z2 spin liquid

Z2 spin liquid obtained using slave-particle approach may be the first theoretical example of string-net liquid.[7][8]

26.1.2 The toric code

The toric code is a two dimensional spin lattice, that acts as a quantum error-correcting code. It is defined on a two dimensional lattice with toric boundary conditions with a spin-1/2 on each link. It can be shown that the ground state of the standard toric code Hamiltonian is an *equal weight superposition* of closed string states.[9] Such a ground state is an example of a string-net condensate,[10] which has the same topological order as the Z2 spin liquid above.

26.2 References

[1] Levin, Michael A. & Xiao-Gang Wen (12 January 2005). "String-net condensation: A physical mechanism for topological phases".*Physical Review B***71**(045110): 21.arXiv:cond-mat/0404617.Bibcode:2005PhRvB..71d5110L.doi:10.1103/Phys

[2] Photons and electrons as emergent phenomena Michael Levin, Xiao-Gang Wen http://arxiv.org/abs/cond-mat/0407140 page 8 "loop quantum gravity appears to be a string net condensation..."

[3] Tomasz Konopka; Fotini Markopoulou; Lee Smolin (2006). "Quantum Graphity". arXiv:hep-th/0611197 [hep-th]. we argue, but do not prove, that loop quantum gravity's spin networks can reproduce Levin and Wen's string net condensation in quantum gravity...

[4] Quantum Graphity: a model of emergent locality http://arxiv.org/abs/0801.0861 page #19 "the excitation of the ground state...is expected to give rise to U(1) gauge charge, ...main difference between this model and the original model of Levin and Wen is that in the present case the background is dynamical, and has hexagonal rather than square plaquettes

[5] Bowles, Claire. "Have researchers found a new state of matter?". Eureka Alert. Retrieved 29 January 2012.

[6] Merali, Zeeya (2007-03-17). "The universe is a string-net liquid". *New Scientist* **193** (2595): 8–9. doi:10.1016/s0262-4079(07)60640-x. Retrieved 29 January 2012.

[7] N. Read and Subir Sachdev, Large-N expansion for frustrated quantum antiferromagnets, Phys. Rev. Lett. 66 1773 (1991)

[8] Xiao-Gang Wen, Mean Field Theory of Spin Liquid States with Finite Energy Gaps and Topological Orders, Phys. Rev. B44, 2664 (1991).

[9] Kitaev, Alexei, Y.; Chris Laumann (2009). "Topological phases and quantum computation". arXiv:0904.2771.

[10] Morimae, Tomoyuki (2012). "Quantum computational tensor network on string-net condensate". *Physical Review A* **85**: 062328. arXiv:1012.1000. Bibcode:2012PhRvA..85f2328M. doi:10.1103/PhysRevA.85.062328.

Chapter 27

Supercritical fluid

A **supercritical fluid** (SCF[1]) is any substance at a temperature and pressure above its critical point, where distinct liquid and gas phases do not exist. It can effuse through solids like a gas, and dissolve materials like a liquid. In addition, close to the critical point, small changes in pressure or temperature result in large changes in density, allowing many properties of a supercritical fluid to be "fine-tuned". Supercritical fluids are suitable as a substitute for organic solvents in a range of industrial and laboratory processes. Carbon dioxide and water are the most commonly used supercritical fluids, being used for decaffeination and power generation, respectively.

27.1 Properties

In general terms, supercritical fluids have properties between those of a gas and a liquid. In Table 1, the critical properties are shown for some components, which are commonly used as supercritical fluids.

Table 2 shows density, diffusivity and viscosity for typical liquids, gases and supercritical fluids.

In addition, there is no surface tension in a supercritical fluid, as there is no liquid/gas phase boundary. By changing the pressure and temperature of the fluid, the properties can be "tuned" to be more liquid- or more gas-like. One of the most important properties is the solubility of material in the fluid. Solubility in a supercritical fluid tends to increase with density of the fluid (at constant temperature). Since density increases with pressure, solubility tends to increase with pressure. The relationship with temperature is a little more complicated. At constant density, solubility will increase with temperature. However, close to the critical point, the density can drop sharply with a slight increase in temperature. Therefore, close to the critical temperature, solubility often drops with increasing temperature, then rises again.[3]

All supercritical fluids are completely miscible with each other so for a mixture a single phase can be guaranteed if the critical point of the mixture is exceeded. The critical point of a binary mixture can be estimated as the arithmetic mean of the critical temperatures and pressures of the two components,

Tc(mix) = (mole fraction *A*) x $T_c A$ + (mole fraction *B*) x $T_c B$.

For greater accuracy, the critical point can be calculated using equations of state, such as the Peng Robinson, or group contribution methods. Other properties, such as density, can also be calculated using equations of state.[4]

27.2 Phase diagram

Figures 1 and 2 show projections of a phase diagram. In the pressure-temperature phase diagram (Fig. 1) the boiling separates the gas and liquid region and ends in the critical point, where the liquid and gas phases disappear to become a single supercritical phase. This can be observed in the density-pressure phase diagram for carbon dioxide, as shown in

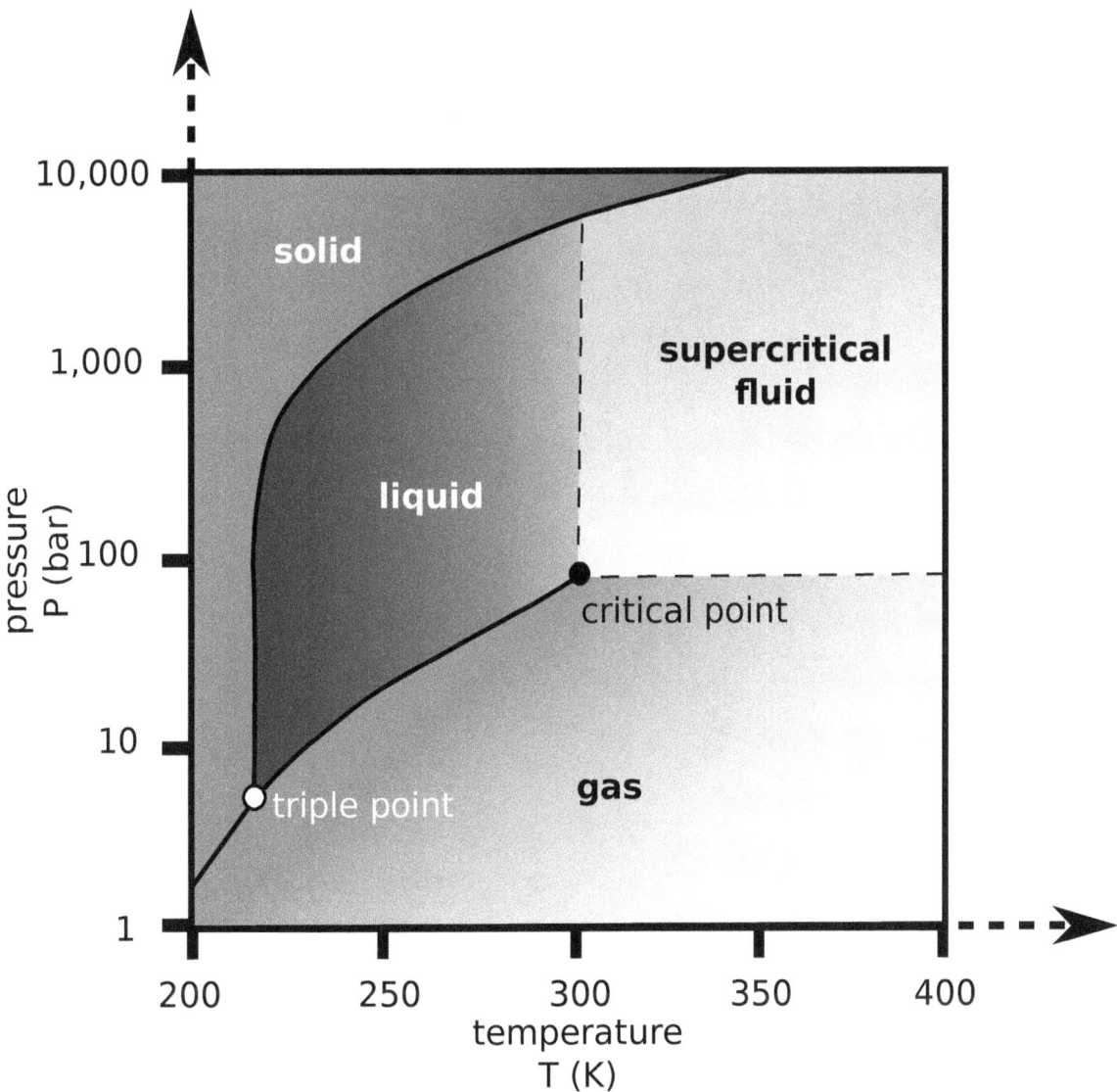

Figure 1. *Carbon dioxide pressure-temperature phase diagram*

Figure 2. At well below the critical temperature, e.g., 280K, as the pressure increases, the gas compresses and eventually (at just over 40 bar) condenses into a much denser liquid, resulting in the discontinuity in the line (vertical dotted line). The system consists of 2 phases in equilibrium, a dense liquid and a low density gas. As the critical temperature is approached (300K), the density of the gas at equilibrium becomes higher, and that of the liquid lower. At the critical point, (304.1 K and 7.38 MPa (73.8 bar)), there is no difference in density, and the 2 phases become one fluid phase. Thus, above the critical temperature a gas cannot be liquefied by pressure. At slightly above the critical temperature (310K), in the vicinity of the critical pressure, the line is almost vertical. A small increase in pressure causes a large increase in the density of the supercritical phase. Many other physical properties also show large gradients with pressure near the critical point, e.g. viscosity, the relative permittivity and the solvent strength, which are all closely related to the density. At higher temperatures, the fluid starts to behave like a gas, as can be seen in Figure 2. For carbon dioxide at 400 K, the density increases almost linearly with pressure.

Many pressurised gases are actually supercritical fluids. For example, nitrogen has a critical point of 126.2K (- 147 °C) and 3.4 MPa (34 bar). Therefore, nitrogen (or compressed air) in a gas cylinder above this pressure is actually a supercritical fluid. These are more often known as permanent gases. At room temperature, they are well above their critical temperature, and therefore behave as a gas, similar to CO_2 at 400K above. However, they cannot be liquified by

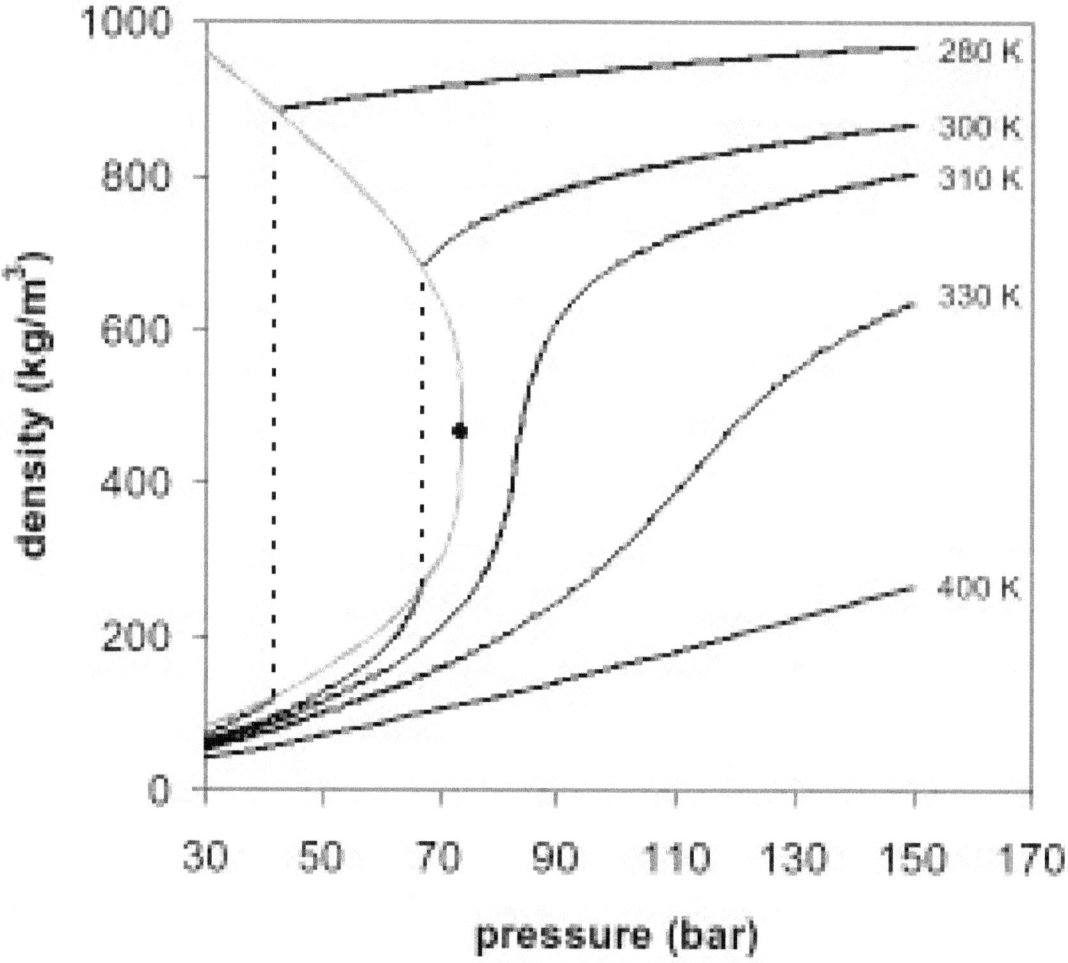

Figure 2. *Carbon dioxide density-pressure phase diagram*

pressure unless cooled below their critical temperature.

27.3   Thermodynamics

In recent years, a significant effort has been devoted to investigation of various properties of supercritical fluids. This has been an exciting field with a long history since 1822 when Baron Charles Cagniard de la Tour discovered supercritical fluids while conducting experiments involving the discontinuities of the sound in a sealed cannon barrel filled with various fluids at high temperature. More recently, supercritical fluids have started to be deployed in several important applications, ranging from the extraction of floral fragrance from flowers to applications in food science such as creating decaffeinated coffee, functional food ingredients, pharmaceuticals, cosmetics, polymers, powders, bio- and functional materials, nano-systems, natural products, biotechnology, fossil and bio-fuels, microelectronics, energy and environment. Much of the excitement and interest of the past decade is due to the enormous progress made in increasing the power of relevant experimental tools. The development of new experimental methods and improvement of existing ones continues to play an important role in this field, with recent research focusing on dynamic properties of fluids.

Dima Bolmatov, V. V. Brazhkin and K. Trachenko discovered that specific heat shows a crossover between two different dynamic regimes of the low-temperature rigid-liquid and high temperature non-rigid gas-like fluid.[5] Rigid liquids are rigid like a solid on short time scales, but flow like a liquid on long time scales;[6] while a supercritical gas-like fluid has the dynamic motions of a gas but is able to dissolve materials, like a liquid. The crossover challenges the currently held belief that no difference can be made between a gas and a liquid above the critical point and that the supercritical state is homogeneous in terms of physical properties. Bolmatov with colleagues formulated a theory of system thermodynamics and heat capacity above the crossover. In that theory, energy and heat capacity are governed by the minimal length of the longitudinal mode in the system only, and do not depend on system-specific structure and interactions. Dima Bolmatov with colleagues predicted relationship between supercritical exponents of heat capacity and viscosity and derived a power law for the supercritical state.[5]

The Fisher-Widom line allows to distinguish liquid-like and gas-like states within the supercritical fluid.

27.4 Natural occurrence

27.4.1 Hydrothermal circulation

See also: Hydrothermal circulation

Hydrothermal circulation occurs within the Earth's crust wherever fluid becomes heated and begins to convect. These fluids are thought to reach supercritical conditions under a number of different settings, such as in the formation of porphyry copper deposits or high temperature circulation of seawater in the sea floor. At mid-ocean ridges, this circulation is most evident by the appearance of hydrothermal vents known as "black smokers". These are large (metres high) chimneys of sulfide and sulfate minerals which vent fluids up to 400 °C. The fluids appear like great black billowing clouds of smoke due to the precipitation of dissolved metals in the fluid. It is likely that at depth many of these vent sites reach supercritical conditions, but most cool sufficiently by the time they reach the sea floor to be subcritical. One particular vent site, Turtle Pits, has displayed a brief period of supercriticality at the vent site. A further site, Beebe, in the Cayman Trough, is thought to display sustained supercriticality at the vent orifice.[7]

27.4.2 Planetary atmospheres

The atmosphere of Venus is 96.5% carbon dioxide and 3.5% nitrogen. The surface pressure is 9.3 MPa (93 bar) and the surface temperature is 735 K, above the critical points of both major constituents and making the surface atmosphere a supercritical fluid.

The interior atmospheres of the solar system's gas giant planets are composed mainly of hydrogen and helium at temperatures well above their critical points. The gaseous outer atmospheres of Jupiter and Saturn transition smoothly into the fluid interior, while the nature of the transition zones of Neptune and Uranus are unknown. Theoretical models of extrasolar planets 55 Cancri e and Gliese 876 d have posited an ocean of pressurized, supercritical fluid water with a sheet of solid high pressure water ice at the bottom.

27.5 Applications

27.5.1 Supercritical fluid extraction

The advantages of supercritical fluid extraction (compared with liquid extraction) are that it is relatively rapid because of the low viscosities and high diffusivities associated with supercritical fluids. The extraction can be selective to some extent by controlling the density of the medium, and the extracted material is easily recovered by simply depressurizing, allowing the supercritical fluid to return to gas phase and evaporate leaving little or no solvent residues. Carbon dioxide is the most common supercritical solvent. It is used on a large scale for the decaffeination of green coffee beans, the extraction of hops for beer production,[8] and the production of essential oils and pharmaceutical products from plants.

A few laboratory test methods include the use of supercritical fluid extraction as an extraction method instead of using traditional solvents.[9][10][11]

27.5.2   Supercritical fluid decomposition

Supercritical water can be used to decompose biomass via supercritical water gasification of biomass[12] This type of biomass gasification can be used to produce hydrocarbon fuels for use in an efficient combustion device or to produce hydrogen for use in a fuel cell. In the latter case, hydrogen yield can be much higher than the hydrogen content of the biomass due to steam reforming where water is a hydrogen-providing participant in the overall reaction.

27.5.3   Dry-cleaning

Supercritical carbon dioxide (SCD) can be used instead of PERC (perchloroethylene) or other undesirable solvents for dry-cleaning. Supercritical carbon dioxide sometimes intercalates into buttons, and, when the SCD is depressurized, the buttons pop, or break apart. Detergents that are soluble in carbon dioxide improve the solvating power of the solvent.[13]

27.5.4   Supercritical fluid chromatography

Supercritical fluid chromatography (SFC) can be used on an analytical scale, where it combines many of the advantages of high performance liquid chromatography (HPLC) and gas chromatography (GC). It can be used with non-volatile and thermally labile analytes (unlike GC) and can be used with the universal flame ionization detector (unlike HPLC), as well as producing narrower peaks due to rapid diffusion. In practice, the advantages offered by SFC have not been sufficient to displace the widely used HPLC and GC, except in a few cases such as chiral separations and analysis of high-molecular-weight hydrocarbons.[14] For manufacturing, efficient preparative simulated moving bed units are available.[15] The purity of the final products is very high, but the cost makes it suitable only for very high-value materials such as pharmaceuticals.

27.5.5   Chemical reactions

Changing the conditions of the reaction solvent can allow separation of phases for product removal, or single phase for reaction. Rapid diffusion accelerates diffusion controlled reactions. Temperature and pressure can tune the reaction down preferred pathways, e.g., to improve yield of a particular chiral isomer.[16] There are also significant environmental benefits over conventional organic solvents.

An electrochemical carboxylation of a para-isobutylbenzyl chloride to Ibuprofen is promoted under supercritical carbon dioxide.[17]

27.5.6   Impregnation and dyeing

Impregnation is, in essence, the converse of extraction. A substance is dissolved in the supercritical fluid, the solution flowed past a solid substrate, and is deposited on or dissolves in the substrate. Dyeing, which is readily carried out on polymer fibres such as polyester using disperse (non-ionic) dyes, is a special case of this. Carbon dioxide also dissolves in many polymers, considerably swelling and plasticising them and further accelerating the diffusion process.

27.5.7   Nano and micro particle formation

See also: micronization

The formation of small particles of a substance with a narrow size distribution is an important process in the pharmaceutical and other industries. Supercritical fluids provide a number of ways of achieving this by rapidly exceeding the saturation point of a solute by dilution, depressurization or a combination of these. These processes occur faster in su-percritical fluids than in liquids, promoting nucleation or spinodal decomposition over crystal growth and yielding very small and regularly sized particles. Recent supercritical fluids have shown the capability to reduce particles up to a range of 5-2000 nm.[18]

27.5.8 Generation of pharmaceutical cocrystals

Supercritical fluids act as a new media for the generation of novel crystalline forms of APIs (Active Pharmaceutical Ingredients) named as pharmaceutical cocrystals. Supercritical fluid technology offers a new platform that allows a single-step generation of particles that are difficult or even impossible to obtain by traditional techniques. The generation of pure and dried new cocrystals (crystalline molecular complexes comprising the API and one or more conformers in the crystal lattice) can be achieved due to unique properties of SCFs by using different supercritical fluid properties: supercritical CO_2 solvent power, anti-solvent effect and its atomization enhancement.[1][19]

27.5.9 Supercritical drying

See also: Critical point drying

Supercritical drying is a method of removing solvent without surface tension effects. As a liquid dries, the surface tension drags on small structures within a solid, causing distortion and shrinkage. Under supercritical conditions there is no surface tension, and the supercritical fluid can be removed without distortion. Supercritical drying is used for manufacture of aerogels and drying of delicate materials such as archeological samples and biological samples for electron microscopy.

27.5.10 Supercritical water oxidation

Supercritical water oxidation uses supercritical water as a medium in which to oxidize hazardous waste, eliminating production of toxic combustion products that burning can produce.

The waste product to be oxidised is dissolved in the supercritical water along with molecular oxygen (or an oxidising agent that gives up oxygen upon decomposition, e.g. Hydrogen Peroxide) at which point the oxidation reaction occurs.

27.5.11 Supercritical water hydrolysis

Supercritical hydrolysis is a method of converting all biomass polysaccharides as well the associated lignin into low molec-ular compounds by contacting with water alone under supercritical conditions. The supercritical water, acts as a solvent, a supplier of bond-breaking thermal energy, a heat transfer agent and as a source of hydrogen atoms. All polysaccharides are converted into simple sugars in near-quantitative yield in a second or less. The aliphatic inter-ring linkages of lignin are also readily cleaved into free radicals that are stabilized by hydrogen originating from the water. The aromatic rings of the lignin are unaffected under short reaction times so that the lignin-derived products are low molecular weight mixed phenols. To take advantage of the very short reaction times needed for cleavage a continuous reaction system must be devised. The amount of water heated to a supercritical state is thereby minimized.

27.5.12 Supercritical water gasification

Supercritical water gasification is a process of exploiting the beneficial effect of supercritical water to convert aqueous biomass streams into clean water and gases like H_2, CH_4, CO_2, CO etc.[20]

27.5.13 Supercritical fluid in power generation

The efficiency of a heat engine is ultimately dependent on the temperature difference between heat source and sink (Carnot cycle). To improve efficiency of power stations the operating temperature must be raised. Using water as the working fluid, this takes it into supercritical conditions.[21] Efficiencies can be raised from about 39% for subcritical operation to about 45% using current technology.[22] Supercritical water reactors (SCWRs) are promising advanced nuclear systems that offer similar thermal efficiency gains. Carbon dioxide can also be used in supercritical cycle nuclear power plants, with similar efficiency gains.[23] Many coal-fired supercritical steam generators are operational all over the world, and have enhanced the efficiency of traditional steam-power plants.

27.5.14 Biodiesel production

Conversion of vegetable oil to biodiesel is via a transesterification reaction, where the triglyceride is converted to the methyl ester plus glycerol. This is usually done using methanol and caustic or acid catalysts, but can be achieved using supercritical methanol without a catalyst. The method of using supercritical methanol for biodiesel production was first studied by Saka and his coworkers. This has the advantage of allowing a greater range and water content of feedstocks (in particular, used cooking oil), the product does not need to be washed to remove catalyst, and is easier to design as a continuous process.[24]

27.5.15 Enhanced oil recovery and carbon capture and storage

Supercritical carbon dioxide is used to enhance oil recovery in mature oil fields. At the same time, there is the possibility of using "clean coal technology" to combine enhanced recovery methods with carbon sequestration. The CO_2 is separated from other flue gases, compressed to the supercritical state, and injected into geological storage, possibly into existing oil fields to improve yields.

At present, only schemes isolating fossil CO_2 from natural gas actually use carbon storage, (e.g., Sleipner gas field),[25] but there are many plans for future CCS schemes involving pre- or post- combustion CO_2.[26][27][28][29] There is also the possibility to reduce the amount of CO_2 in the atmosphere by using biomass to generate power and sequestering the CO_2 produced.

27.5.16 Enhanced geothermal system

Main article: Enhanced geothermal system § CO2 EGS

The use of supercritical carbon dioxide, instead of water, has been examined as a geothermal working fluid.

27.5.17 Refrigeration

Supercritical carbon dioxide is also an important emerging refrigerant, being used in new, low-carbon solutions for domestic heat pumps.[30] These systems are undergoing continuous development with supercritical carbon dioxide heat pumps already being successfully marketed in Asia. The EcoCute systems from Japan, developed by consortium of companies including Mitsubishi, develop high-temperature domestic water with small inputs of electric power by moving heat into the system from their surroundings. Their success makes a future use in other world regions possible.[31]

27.5.18 Supercritical fluid deposition

Supercritical fluids can be used to deposit functional nanostructured films and nanometer-size particles of metals onto surfaces. The high diffusivities and concentrations of precursor in the fluid as compared to the vacuum systems used in chemical vapour deposition allow deposition to occur in a surface reaction rate limited regime, providing stable and uniform interfacial growth.[32] This is crucial in developing more powerful electronic components, and metal particles deposited in this way are also powerful catalysts for chemical synthesis and electrochemical reactions. Additionally, due to the high rates of precursor transport in solution, it is possible to coat high surface area particles which under chemical vapour deposition would exhibit depletion near the outlet of the system and also be likely to result in unstable interfacial growth features such as dendrites. The result is very thin and uniform films deposited at rates much faster than atomic layer deposition, the best other tool for particle coating at this size scale.[33]

27.5.19 Antimicrobial properties

CO_2 at high pressures has antimicrobial properties.[34] While its effectiveness has been shown for various applications, the mechanisms of inactivation have not been fully understood although they have been investigated for more than 60 years.[35]

27.6 History

In 1822, Baron Charles Cagniard de la Tour discovered the critical point of a substance in his famous cannon barrel experiments. Listening to discontinuities in the sound of a rolling flint ball in a sealed cannon filled with fluids at various temperatures, he observed the critical temperature. Above this temperature, the densities of the liquid and gas phases become equal and the distinction between them disappears, resulting in a single supercritical fluid phase.[36]

27.7 See also

- Transcritical cycle
- Critical point (thermodynamics)
- Iceland Deep Drilling Project

27.8 References

[1] L. Padrela, M.A. Rodrigues, S.P. Velaga, H.A. Matos and E.G. Azevedo (2009). "Formation of indomethacin–saccharin cocrystals using supercritical fluid technology".*European Journal of Pharmaceutical Sciences*.**38**, pp. 9–17.doi:10.1016/j.ejps.2009.

[2] Edit Székely. "What is a supercritical fluid?". Budapest University of Technology and Economics. Retrieved 2014-06-26.

[3] "Supercritical Fluid Extraction, Density Considerations". Retrieved 2007-11-20.

[4] A.A. Clifford (2007-12-04). "Calculation of Thermodynamic Properties of CO_2 using Peng–Robinson equation of state.". Critical Processes Ltd. Retrieved 2007-11-20.

[5] Bolmatov, D.; Brazhkin, V. V.; Trachenko, K. (2013). "Thermodynamic behaviour of supercritical matter". *Nature Communications* **4**. arXiv:1303.3153v3. Bibcode:2013NatCo...4E2331B. doi:10.1038/ncomms3331.

[6] Bolmatov, D.; Brazhkin, V. V.; Trachenko, K. (2012). "The phonon theory of liquid thermodynamics". *Scientific Reports* **2**. arXiv:1202.0459. Bibcode:2012NatSR...2E.421B. doi:10.1038/srep00421. Lay summary – *Physics World*.

[7] Webber, A.P.; Murton, B.; Roberts, S.; Hodgkinson, M. "Supercritical Venting and VMS Formation at the Beebe Hydrothermal Field, Cayman Spreading Centre". *Goldschmidt Conference Abstracts 2014*. Geochemical Society. Retrieved 29 July 2014.

[8] "The Naked Scientist Interviews". Retrieved 2007-11-20.

[9] U.S.EPA Method 3560 Supercritical Fluid Extraction of Total Recoverable Hydrocarbons. http://www.epa.gov/SW-846/pdfs/3560.pdf

[10] U.S.EPA Method 3561 Supercritical Fluid Extraction of Polycyclic Aromatic Hydrocarbons. http://www.epa.gov/SW-846/pdfs/3561.pdf

[11] Use of Ozone Depleting Substances in Laboratories. TemaNord 2003:516. http://www.norden.org/pub/ebook/2003-516.pdf

[12] "Supercritical water gasification of biomas". Retrieved 201-11-17. Check date values in: |access-date= (help)

[13] "Science News Online". Retrieved 2007-11-20.

[14] Bart, C. J. (2005). "Chapter 4: Separation Techniques". *Additives in Polymers: industrial analysis and applications.* John Wiley and Sons. p. 212. doi:10.1002/0470012064.ch4. ISBN 978-0-470-01206-2.

[15] "Simulated Moving Bed Theory" (PDF). Retrieved 2007-11-20.

[16] R. Scott Oakes; Anthony A. Clifford; Keith D. Bartle; Mark Thornton Pett & Christopher M. Rayner (1999). "Sulfur oxidation in supercritical carbon dioxide: dramatic pressure dependent enhancement of diastereoselectivity for sulphoxidation of cysteine derivatives". *Chemical Communications* (3): 247–248. doi:10.1039/a809434i.

[17] Sakakura, Toshiyasu; Choi, Jun-Chul; Yasuda, Hiroyuki (13 June 2007). "Transformation of Carbon dioxide". *Chemical Reviews* (American Chemical Society) **107** (6): 2365–2387. doi:10.1021/cr068357u. PMID 17564481.

[18] Sang-Do Yeo & Erdogan Kiran (2005). "Formation of polymer particles with supercritical fluids: A review". *The Journal of Supercritical Fluids* **34** (3): 287–308. doi:10.1016/j.supflu.2004.10.006.

[19] L. Padrela, M.A. Rodrigues, S.P. Velaga, H.A. Matos and E.G. Azevedo (2009). "Screening for pharmaceutical cocrystals using the supercritical fluid enhanced atomization process". *Journal of Supercritical Fluids.* article in press, corrected proof. doi:10.1016/j.supflu.2010.01.010

[20] http://www.btgworld.com/en/rtd/technologies/supercritical-water-reforming. Missing or empty |title= (help)

[21] Malhotra, Ashok and Satyakam,R, 2000,Influence of climatic parameters on optimal design of supercritical power plants,IECEC, Energy Conversion Engineering Conference, pp. 1053–1058,

[22] "Supercritical steam cycles for power generation applications" (PDF). Retrieved 2007-11-20.

[23] V. Dostal; M.J. Driscoll; P. Hejzlar. "A Supercritical Carbon Dioxide Cycle for Next Generation Nuclear Reactors" (PDF). *MIT-ANP-TR-100.* MIT-ANP-Series. Retrieved 2007-11-20.

[24] Kunchana Bunyakiat; Sukunya Makmee; Ruengwit Sawangkeaw & Somkiat Ngamprasertsith (2006). "Continuous Production of Biodiesel via Transesterification from Vegetable Oils in Supercritical Methanol". *Energy and Fuels* **20** (2): 812–817. doi:10.1021/ef050329b.

[25] "Saline Aquifer CO_2 Storage". Retrieved 2007-12-10.

[26] "The Hydrogen Economy: Opportunities, Costs, Barriers, and R&D Needs", p. 84 (2004)

[27] FutureGen Technology

[28] Øyvind Vessia: "Fischer- Tropsch reactor fed by syngas"

[29] Intergovernmental Panel on Climate Change IPCC Special Report on Carbon Dioxide Capture and Storage.

[30] FAQs – Supercritical CO_2 in heat pumps and other applications

[31] Eco Cute hot water heat pumps in Japan

[32] Ye, Xiang-Rong; Lin, YH & Wai, CM (2003). "Supercritical fluid fabrication of metal nanowires and nanorods templated by multiwalled carbon nanotubes". *Advanced Materials* **15** (4): 316–319. doi:10.1002/adma.200390077.

[33] "SFD compared to CVD". *navolta.com.* Navolta. Retrieved 3 October 2014.

[34] Cinquemani, Boyle, et al. (2007). "Inactivation of microbes using compressed carbon dioxide – An environmentally sound disinfection process of medical fabrics." Journal of Supercritical fluids 42(3): 1–6.

[35] Fraser, D. (1951). "Bursting bacteria by release of gas pressure." *Nature* 167: 33–34.

[36] Berche, Bertrand; Henkel, Malte; Kenna, Ralph (2009). "Critical phenomena: 150 years since Cagniard de la Tour". *Journal of Physical Studies* **13** (3): 3001-1–3001-4. arXiv:0905.1886. Bibcode:2009arXiv0905.1886B.

27.9 Further reading

- Brunner, G. (2010). "Applications of Supercritical Fluids". *Annual Review of Chemical and Biomolecular Engineering* **1**: 321–342. doi:10.1146/annurev-chembioeng-073009-101311. PMID 22432584.

27.10 External links

- Handy calculator for density, enthalpy, entropy and other thermodynamic data of supercritical CO_2

- animated presentation describing what a supercritical fluid is (broken link)

- NewScientist Environment FOUND:The hottest water on Earth

- Poliakoff, Martyn (28 April 2008). "Supercritical fluids". *Test Tube*. Brady Haran for the University of Nottingham.

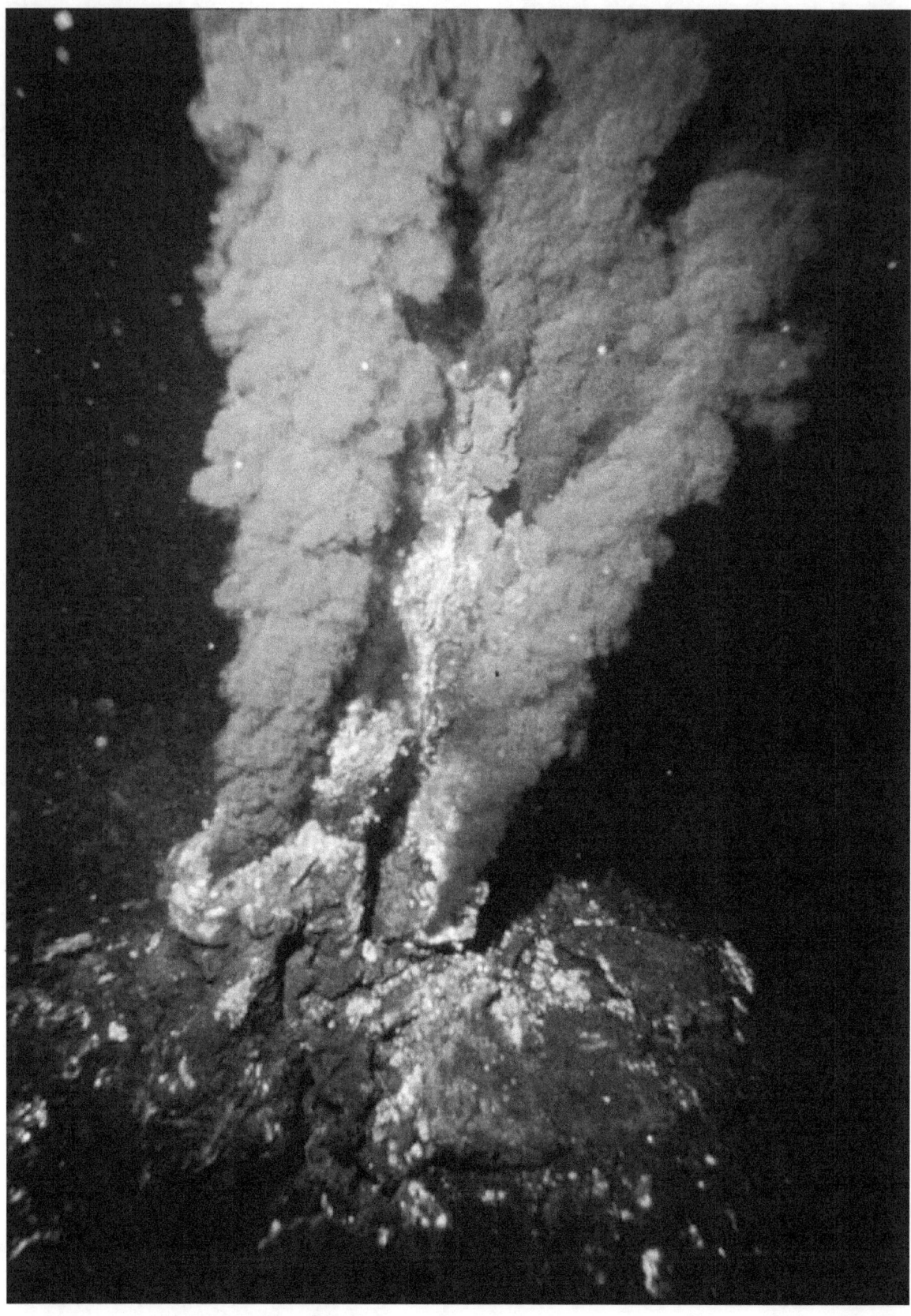

A black smoker, a type of hydrothermal vent

Chapter 28

Dropleton

A **dropleton** or **quantum droplet** is a quasiparticle comprising a collection of electrons and places without them inside a semiconductor. Dropletons give the first known quasiparticle characterization where the quasiparticle behaves like a liquid.[1] The creation of dropletons was announced on 26 February 2014 in a *Nature* article, which presented evidence for the creation of dropletons in an electron–hole plasma inside a gallium arsenide quantum well by ultrashort laser pulses.[2] Their existence was not predicted before the experiment.

Despite the relatively short lifetime of about 25 picoseconds, the dropletons are stable enough to be studied[1] and possess favorable properties for certain investigations of quantum mechanics. They are approximately 200 nanometers wide, the size of the smallest bacteria, for which reason the discoverers have expressed hope that they might one day actually see quantum droplets.[1]

28.1 References

[1] Clara Moskowitz (26 February 2014). "Meet the Dropleton—a "Quantum Droplet" That Acts Like a Liquid". *Scientific American*. Retrieved 26 February 2014.

[2] A. E. Almand-Hunter, H. Li, S. T. Cundiff, M. Mootz, M. Kira & S. W. Koch (26 February 2014). "Quantum droplets of electrons and holes". *Nature*. Bibcode:2014Natur.506..471A. doi:10.1038/nature12994. Retrieved 26 February 2014.

Chapter 29

Jahn–Teller effect

The **Jahn–Teller effect**, sometimes also known as **Jahn–Teller distortion**, describes the geometrical distortion of molecules and ions that is associated with certain electron configurations. This electronic effect is named after Hermann Arthur Jahn and Edward Teller, who proved, using group theory, that orbital nonlinear spatially degenerate molecules *cannot* be stable.[1] The **Jahn–Teller theorem** essentially states that any nonlinear molecule with a spatially degenerate electronic ground state will undergo a geometrical distortion that removes that degeneracy, because the distortion lowers the overall energy of the species. For a description of another type of geometrical distortion that occurs in crystals with substitutional impurities see article off-center ions.

29.1   Transition metal chemistry

The Jahn–Teller effect is most often encountered in octahedral complexes of the transition metals.[3] The phenomenon is very common in six-coordinate copper(II) complexes.[4] The d^9 electronic configuration of this ion gives three electrons in the two degenerate *eg* orbitals, leading to a doubly degenerate electronic ground state. Such complexes distort along one of the molecular fourfold axes (always labelled the *z* axis), which has the effect of removing the orbital and electronic degeneracies and lowering the overall energy. The distortion normally takes the form of elongating the bonds to the ligands lying along the *z* axis, but occasionally occurs as a shortening of these bonds instead (the Jahn–Teller theorem does not predict the direction of the distortion, only the presence of an unstable geometry). When such an elongation occurs, the effect is to lower the electrostatic repulsion between the electron-pair on the Lewis basic ligand and any electrons in orbitals with a *z* component, thus lowering the energy of the complex. If the undistorted complex would be expected to have an inversion centre, this is preserved after the distortion.

In octahedral complexes, the Jahn–Teller effect is most pronounced when an odd number of electrons occupy the *eg* orbitals. This situation arises in complexes with the configurations d^9, low-spin d^7 or high-spin d^4 complexes, all of which have doubly degenerate ground states. In such compounds the *eg* orbitals involved in the degeneracy point directly at the ligands, so distortion can result in a large energetic stabilisation. Strictly speaking, the effect also occurs when there is a degeneracy due to the electrons in the *t2g* orbitals (*i.e.* configurations such as d^1 or d^2, both of which are triply degenerate). In such cases, however, the effect is much less noticeable, because there is a much smaller lowering of repulsion on taking ligands further away from the *t2g* orbitals, which do not point *directly* at the ligands (see the table below). The same is true in tetrahedral complexes (e.g. manganate: distortion is very subtle because there is less stabilisation to be gained because the ligands are not pointing directly at the orbitals.

The expected effects for octahedral coordination are given in the following table:

w: weak Jahn–Teller effect (*t2g* orbitals unevenly occupied)

s: strong Jahn–Teller effect expected (*eg* orbitals unevenly occupied)

blank: no Jahn–Teller effect expected.

The Jahn–Teller effect is manifested in the UV-VIS absorbance spectra of some compounds, where it often causes splitting

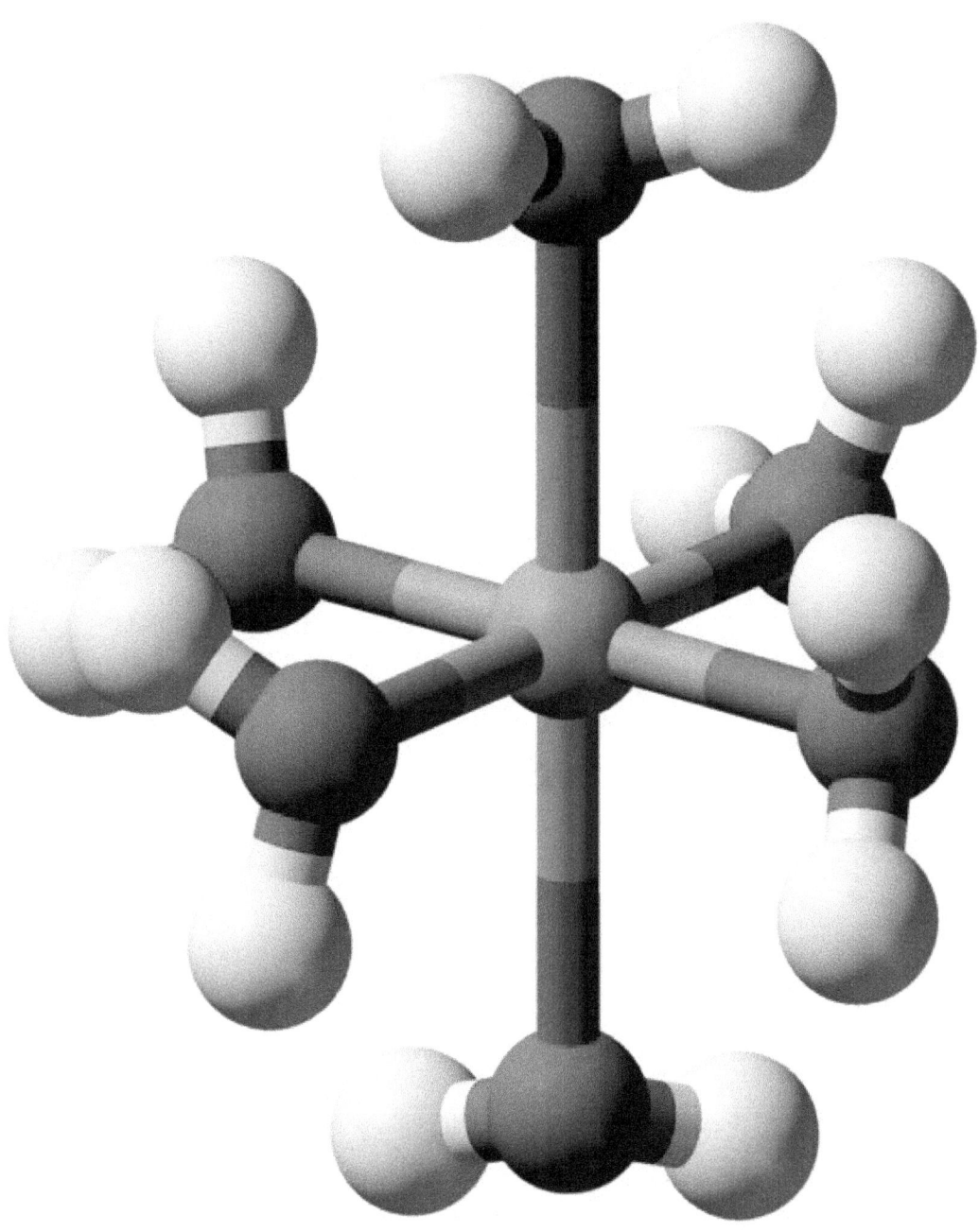

The Jahn–Teller effect is responsible for the tetragonal distortion of the hexaaquacopper(II) complex ion, $[Cu(OH_2)_6]^{2+}$, which might otherwise possess octahedral geometry. The two axial Cu–O distances are 238 pm, whereas the four equatorial Cu–O distances are ~195 pm.

of bands. It is readily apparent in the structures of many copper(II) complexes.[2] Additional, detailed information about the anisotropy of such complexes and the nature of the ligand binding can be however obtained from the fine structure of the low-temperature electron spin resonance spectra.

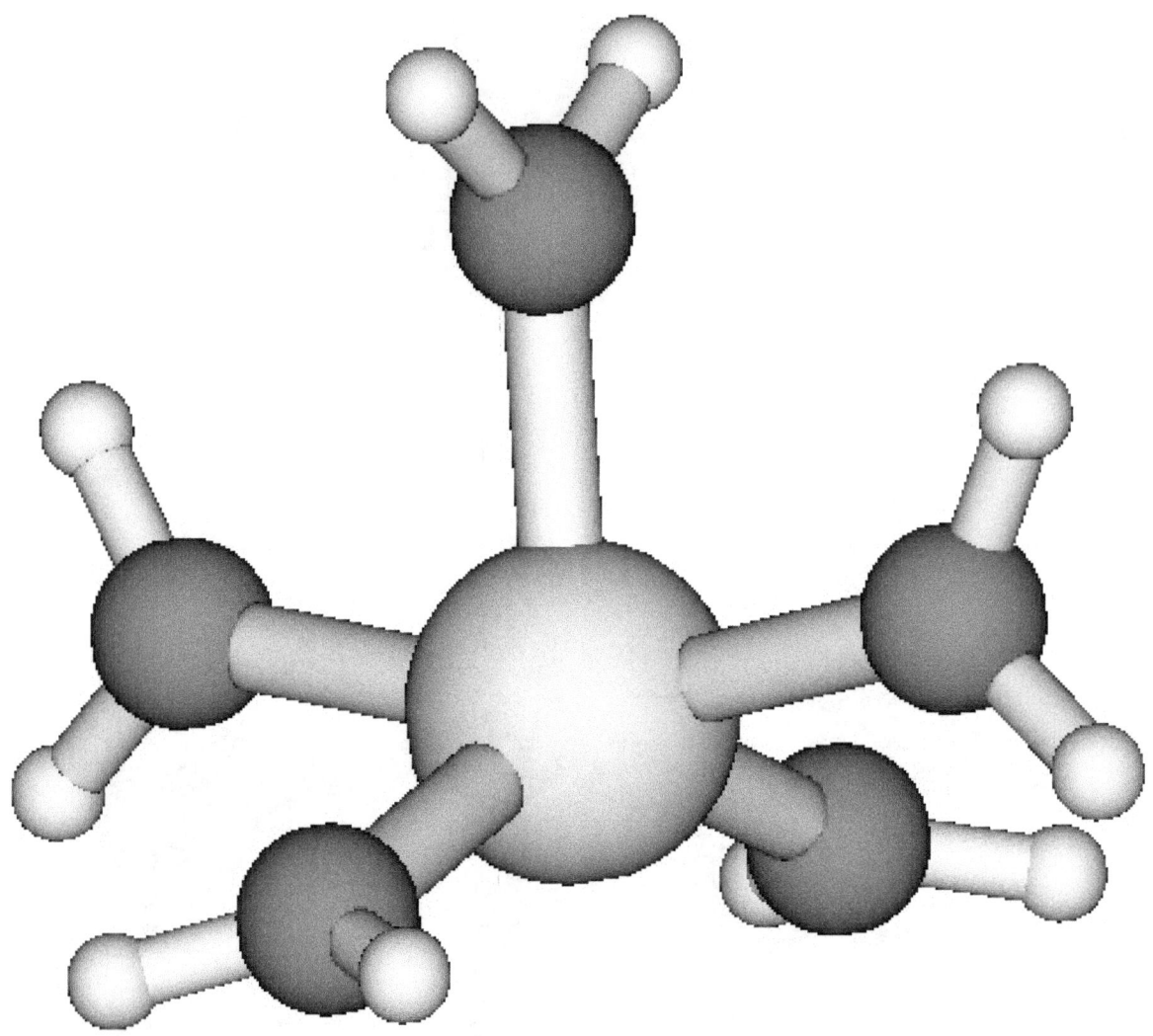

A recent study shows that the copper(II) ion coordinates five water molecules in an elongated square pyramid with four Cu-Oeq bonds (2x1.98 Å and 2x1.95 Å) and a long Cu-Oax bond (2.35 Å). The four equatorial ligands were distorted from the mean equatorial plane by ± 17 degrees.[2]

29.2   Related effects

The underlying cause of the Jahn–Teller effect is the presence of molecular orbitals which are both degenerate and open shell. This situation is not unique to coordination complexes and can be encountered in other areas of chemistry. In organic chemistry the phenomenon of antiaromaticity has the same cause and also often sees molecules distorting; as in the case of cyclobutadiene[5] and cyclooctatetraene (COT).[6]

29.3   See also

- Crystal field theory

- Potential energy surface

- Diabatic

- Conical intersection

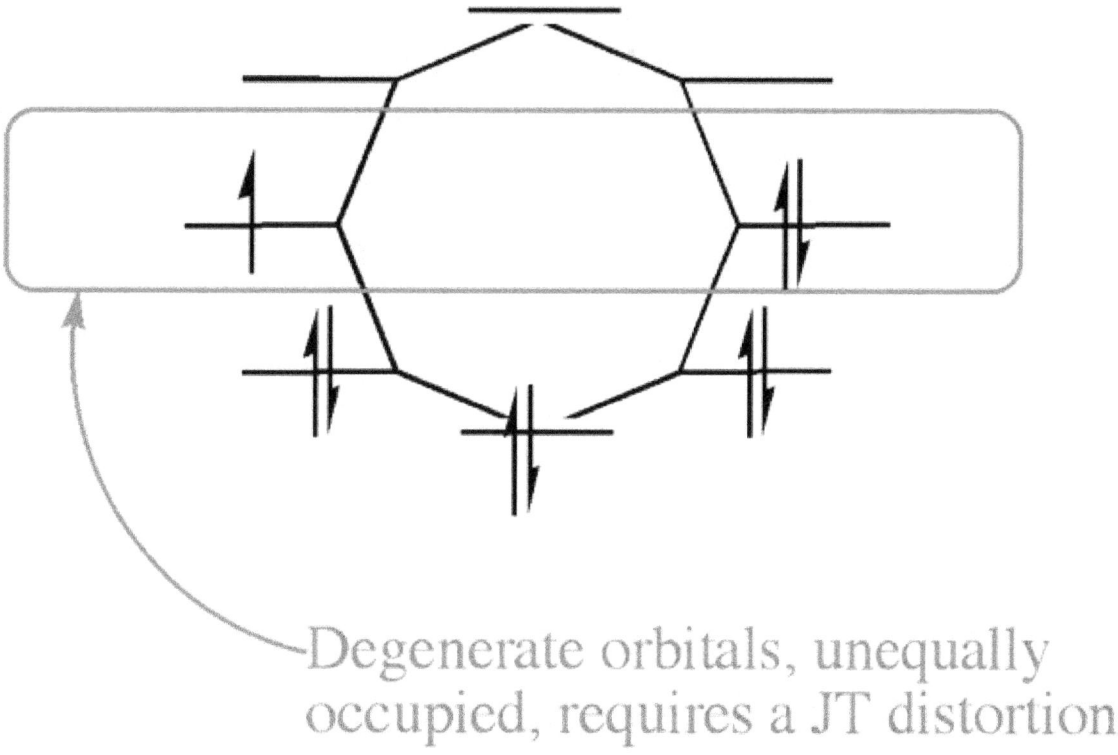

Degenerate orbitals, unequally occupied, requires a JT distortion

The Jahn–Teller effect forces the radical anion of cyclooctatetraene (−1) to be non-symmetric (see text)

- Avoided crossing

- Vibronic coupling

- Renner-Teller effect

29.4   References

[1] Jahn, H.; Teller, E. (1937). "Stability of Polyatomic Molecules in Degenerate Electronic States. I. Orbital Degeneracy". *Proceedings of the Royal Society A* **161** (905): 220–235. Bibcode:1937RSPSA.161..220J. doi:10.1098/rspa.1937.0142.

[2] Frank, Patrick; Benfatto, Maurizio; Szilagyi, Robert K.; D'Angelo, Paola; Della Longa, Stefano; Hodgson, Keith O. (2005). "The Solution Structure of $[Cu(aq)]^{2+}$ and Its Implications for Rack-Induced Bonding in Blue Copper Protein Active Sites". *Inorganic Chemistry* **44**: 1922–1933. doi:10.1021/ic0400639.

[3] Shriver, D. F.; Atkins, P. W. (1999). *Inorganic Chemistry* (3rd ed.). Oxford University Press. pp. 235–236. ISBN 0-19-850330-X.

[4] Janes, Rob; Moore, Elaine A. (2004). *Metal-ligand bonding*. Royal Society of Chemistry. ISBN 0-85404-979-7.

[5] Senn, Peter (October 1992). "A simple quantum mechanical model that illustrates the Jahn-Teller effect". *Journal of Chemical Education* **69** (10): 819. Bibcode:1992JChEd..69..819S. doi:10.1021/ed069p819.

[6] Klärner, Frank-Gerrit (2001). "About the Antiaromaticity of Planar Cyclooctatetraene". *Angewandte Chemie, Int. Ed. Eng.* **40** (21): 3977–3981. doi:10.1002/1521-3773(20011105)40:21<3977::AID-ANIE3977>3.0.CO;2-N.

29.5   External links

- Pavarini, Eva (2012). "Crystal-field Theory, Tight-binding Method, and Jahn-Teller Effect" (PDF). In Pavarini, E.; Koch, E.; Anders, F.; Jarrell, M. *Correlated Electrons: From Models to Materials*. Jülich: Forschungszentrum Jülich. ISBN 978-3-89336-796-2.

- "Jahn-Teller effect". *Citizendium*. (under development)

Chapter 30

Quark–gluon plasma

A **quark–gluon plasma** (**QGP**) or **quark soup**[1] is a state of matter in quantum chromodynamics (QCD) which is hypothesized to exist at extremely high temperature, density, or both temperature and density. This state is thought to consist of asymptotically free quarks and gluons, which are several of the basic building blocks of matter. It is believed that up to a few milliseconds after the Big Bang, known as the Quark epoch, the Universe was in a quark–gluon plasma state. In June 2015, an international team of physicists have produced quark-gluon plasma at the Large Hadron Collider by colliding protons with lead nuclei at high energy inside the supercollider's Compact Muon Solenoid detector. They also discovered that this new state of matter behaves like a fluid.[2]

The strength of the color force means that unlike the gas-like plasma, quark–gluon plasma behaves as a near-ideal Fermi liquid, although research on flow characteristics is ongoing.[3] In the quark matter phase diagram, QGP is placed in the high-temperature, high-density regime; whereas, ordinary matter is a cold and rarefied mixture of nuclei and vacuum, and the hypothetical quark stars would consist of relatively cold, but dense quark matter.

Experiments at CERN's Super Proton Synchrotron (SPS) first tried to create the QGP in the 1980s and 1990s: the results led CERN to announce indirect evidence for a "new state of matter"[4] in 2000. Current experiments (2011) at the Brookhaven National Laboratory's Relativistic Heavy Ion Collider (RHIC) on Long Island (NY, USA) and at CERN's recent Large Hadron Collider near Geneva (Switzerland) are continuing this effort,[5][6] by colliding relativistically accelerated gold (at RHIC) or lead (at LHC) with each other or with protons. Although the results have yet to be independently verified as of February 2010, scientists at Brookhaven RHIC have tentatively claimed to have created a quark–gluon plasma with an approximate temperature of 4 trillion (4×10^{12}) degrees Kelvin.[6]

As already mentioned, three new experiments running on CERN's Large Hadron Collider (LHC), on the spectrometers ALICE,[7] ATLAS and CMS, will continue studying properties of QGP. Starting in November 2010, CERN temporarily ceased colliding protons, and began colliding lead Ions for the ALICE experiment. They were looking to create a QGP and were expected to stop December 6, colliding protons again in January.[8] A new record breaking temperature was set by ALICE: A Large Ion Collider Experiment at CERN on August, 2012 in the ranges of 5.5 trillion (5.5×10^{12}) degrees Kelvin as claimed in their Nature PR.[9]

30.1 General introduction

Quark–gluon plasma is a state of matter in which the elementary particles that make up the hadrons of baryonic matter are freed of their strong attraction for one another under extremely high energy densities. These particles are the quarks and gluons that compose baryonic matter.[10] In normal matter quarks are *confined*; in the QGP quarks are *deconfined*. In classical QCD quarks are the Fermionic components of mesons and baryons while the gluons are considered the Bosonic components of such particles. The gluons are the force carriers, or bosons, of the QCD color force, while the quarks by themselves are their Fermionic matter counterparts.

Although the experimental high temperatures and densities predicted as producing a quark–gluon plasma have been realized in the laboratory, the resulting matter does *not* behave as a quasi-ideal state of free quarks and gluons, but, rather, as an almost perfect dense fluid.[11] Actually, the fact that the quark–gluon plasma will not yet be "free" at temperatures realized at present accelerators was predicted in 1984 as a consequence of the remnant effects of confinement.[12][13]

30.1.1   Relation to normal plasma

A plasma is matter in which charges are screened due to the presence of other mobile charges; for example: Coulomb's Law is suppressed by the screening to yield a distance-dependent charge (Q -> Q × exp(-r/α), i.e, the charge Q is reduced exponentially with the distance divided by a screening length α). In a QGP, the color charge of the quarks and gluons is screened. The QGP has other analogies with a normal plasma. There are also dissimilarities because the color charge is non-abelian, whereas the electric charge is abelian. Outside a finite volume of QGP the color-electric field is not screened, so that a volume of QGP must still be color-neutral. It will therefore, like a nucleus, have integer electric charge.

30.1.2   Theory

One consequence of this difference is that the color charge is too large for perturbative computations which are the mainstay of QED. As a result, the main theoretical tools to explore the theory of the QGP is lattice gauge theory.[14] The transition temperature (approximately 175 MeV) was first predicted by lattice gauge theory. Since then lattice gauge theory has been used to predict many other properties of this kind of matter. The AdS/CFT correspondence conjecture may provide insights in QGP, moreover the ultimate goal of the fluid/gravity correspondence is to understand QGP. The QGP is believed to be a phase of QCD which is completely locally thermalized and thus suitable for an effective fluid dynamic description.

30.1.3   Production

The QGP can be created by heating matter up to a temperature of 2×10^{12} K, which amounts to 175 MeV per particle. This can be accomplished by colliding two large nuclei at high energy (note that 175 MeV is not the energy of the colliding beam). Lead and gold nuclei have been used for such collisions at CERN SPS and BNL RHIC, respectively. The nuclei are accelerated to ultrarelativistic speeds (contracting their length) and directed towards each other, creating a "fireball", in the rare event of a collision. Hydrodynamic simulation predicts this fireball will expand under its own pressure, and cool while expanding. By carefully studying the spherical and elliptic flow, experimentalists put the theory to test.

30.1.4   How the QGP fits into the general scheme of physics

QCD is one part of the modern theory of particle physics called the Standard Model. Other parts of this theory deal with electroweak interactions and neutrinos. The theory of electrodynamics has been tested and found correct to a few parts in a billion. The theory of weak interactions has been tested and found correct to a few parts in a thousand. Perturbative forms of QCD have been tested to a few percent. In contrast, non-perturbative forms of QCD have barely been tested. The study of the QGP is part of this effort to consolidate the grand theory of particle physics.

The study of the QGP is also a testing ground for finite temperature field theory, a branch of theoretical physics which seeks to understand particle physics under conditions of high temperature. Such studies are important to understand the early evolution of our universe: the first hundred microseconds or so. It is crucial to the physics goals of a new generation of observations of the universe (WMAP and its successors). It is also of relevance to Grand Unification Theories which seek to unify the three fundamental forces of nature (excluding gravity).

30.2   Expected properties

30.2.1 Thermodynamics

The cross-over temperature from the normal hadronic to the QGP phase is about 175 MeV. This "crossover" may actually *not* be only a qualitative feature, but instead one may have to do with a true (second order) phase transition, e.g. of the universality class of the three-dimensional Ising model, as some theorists say, e.g. Frithjof Karsch and coworkers from the university of Bielefeld. The phenomena involved correspond to an energy density of a little less than 1 GeV/fm^3. For relativistic matter, pressure and temperature are not independent variables, so the equation of state is a relation between the energy density and the pressure. This has been found through lattice computations, and compared to both perturbation theory and string theory. This is still a matter of active research. Response functions such as the specific heat and various quark number susceptibilities are currently being computed.

30.2.2 Flow

The equation of state is an important input into the flow equations. The speed of sound is currently under investigation in lattice computations. The mean free path of quarks and gluons has been computed using perturbation theory as well as string theory. Lattice computations have been slower here, although the first computations of transport coefficients have recently been concluded. These indicate that the mean free time of quarks and gluons in the QGP may be comparable to the average interparticle spacing: hence the QGP is a liquid as far as its flow properties go. This is very much an active field of research, and these conclusions may evolve rapidly. The incorporation of dissipative phenomena into hydrodynamics is another recent development that is still in an active stage.

30.2.3 Excitation spectrum

Does the QGP really contain (almost) free quarks and gluons? The study of thermodynamic and flow properties would indicate that this is an over-simplification. Many ideas are currently being evolved and will be put to test in the near future. It has been hypothesized recently that some mesons built from heavy quarks do not dissolve until the temperature reaches about 350 MeV. This has led to speculation that many other kinds of bound states may exist in the plasma. Some static properties of the plasma (similar to the Debye screening length) constrain the excitation spectrum.

30.2.4 Glasma hypothesis

Since 2008, there is a discussion about a hypothetical precursor state of the Quark–gluon plasma, the so-called "Glasma", where the dressed particles are condensed into some kind of glassy (or amorphous) state, below the genuine transition between the confined state and the plasma liquid. This would be analogous to the formation of metallic glasses, or amorphous alloys of them, below the genuine onset of the liquid metallic state.

30.3 Experimental situation

Those forms of the QGP that are easiest to compute are not those that are easiest to verify experimentally. While the balance of evidence points towards the QGP being the origin of the detailed properties of the fireball produced at SPS (CERN), in the RHIC and at LHC, this is the main barrier which prevents experimentalists from declaring a sighting of the QGP. For a summary see 2005 RHIC Assessment.

The important classes of experimental observations are

- Single particle spectra (photons and dileptons)

- Strangeness production

- Photon and muon rates (and J/ψ melting)

- Elliptic flow

- Jet quenching

- Fluctuations

- Hanbury Brown and Twiss effect and Bose–Einstein correlations

In short, a quark–gluon plasma flows like a splat of liquid, and because it's not "transparent" with respect to quarks, it can attenuate jets emitted by collisions. Furthermore, once formed, a ball of quark–gluon plasma, like any hot object, transfers heat internally by radiation. However, unlike in everyday objects, there is enough energy available that gluons (particles mediating the strong force) collide and produce an excess of the heavy (i.e. high-energy) strange quarks. Whereas, if the QGP didn't exist and there was a pure collision, the same energy would be converted into even heavier quarks such as charm quarks or bottom quarks.

30.4   Formation of quark matter

In April 2005, formation of quark matter was tentatively confirmed by results obtained at Brookhaven National Laboratory's Relativistic Heavy Ion Collider (RHIC). The consensus of the four RHIC research groups was that they had created a quark–gluon liquid of very low viscosity. However, contrary to what was at that time still the widespread assumption, it is yet unknown from theoretical predictions whether the QCD "plasma", especially close to the transition temperature, should behave like a gas or liquid. Authors favoring the weakly interacting interpretation derive their assumptions from the lattice QCD calculation, where the entropy density of quark–gluon plasma approaches the weakly interacting limit. However, since both energy density and correlation shows significant deviation from the weakly interacting limit, it has been pointed out by many authors that there is in fact no reason to assume a QCD "plasma" close to the transition point should be weakly interacting, like electromagnetic plasma (see, e.g.,[15]). That being said, systematically improvable perturbative QCD quasiparticle models do a very good job of reproducing the lattice data for thermodynamical observables (pressure, entropy, quark susceptibility), including the aforementioned "significant deviation from the weakly interacting limit", down to temperatures on the order of 2 to 3 times the critical temperature for the transition.[16][17][18]

30.5   See also

- Hadrons (that is mesons and baryons) and confinement

- Hadronization

- List of plasma (physics) articles

- Neutron stars

- Plasma physics

- QCD matter

- Quantum electrodynamics

- Quantum chromodynamics

- Quantum hydrodynamics

- Relativistic plasma

- Relativistic nuclear collision

- Strangeness production

- Strange matter

- Color-glass condensate

30.6 References

[1] Bohr, Henrik; Nielsen, H. B. (1977). "Hadron production from a boiling quark soup: quark model predicting particle ratios in hadronic collisions". *Nuclear Physics B* **128** (2): 275. Bibcode:1977NuPhB.128..275B. doi:10.1016/0550-3213(77)90032-3.

[2] LHC creates liquid from Big Bang

[3] Quark-gluon plasma goes liquid - physicsworld.com

[4] A New State of Matter - Experiments

[5] Relativistic Heavy Ion Collider, RHIC

[6] http://www.bnl.gov/rhic/news2/news.asp?a=1074&t=pr 'Perfect' Liquid Hot Enough to be Quark Soup

[7] Alice Experiment: Welcome to ALICE Portal

[8] CERN Press Release November 4th 2010

[9] Hot stuff: CERN physicists create record-breaking subatomic soup : Nature News Blog

[10] The Indian Lattice Gauge Theory Initiative

[11] WA Zajc (2008). "The fluid nature of quark-gluon plasma". *Nuclear Physics A* **805**: 283c–294c. arXiv:0802.3552. Bibcode:200 doi:10.1016/j.nuclphysa.2008.02.285.

[12] Plümer, M.; Raha, S. & Weiner, R. M. (1984). "How free is the quark-gluon plasma". *Nucl. Phys. A* **418**: 549–557. Bibcode:1984NuPhA.418..549P. doi:10.1016/0375-9474(84)90575-X..

[13] Plümer, M.; Raha, S. & Weiner, R. M. (1984). "Effect of confinement on the sound velocity in a quark-gluon plasma". *Phys. Lett. B* **139** (3): 198–202. Bibcode:1984PhLB..139..198P. doi:10.1016/0370-2693(84)91244-9..

[14] Lattice-QCD calculations of the Quark-Gluon Plasma have been reviewed in and in

[15] Miklos Gyulassy (2004). "The QGP Discovered at RHIC". arXiv:nucl-th/0403032 [nucl-th].

[16] Andersen; Leganger; Strickland; Su (2011). "NNLO hard-thermal-loop thermodynamics for QCD". *Physics Letters B* **696** (5): 468. arXiv:1009.4644. Bibcode:2011PhLB..696..468A. doi:10.1016/j.physletb.2010.12.070.

[17] Andersen; Michael Strickland; Nan Su (2010). "Gluon Thermodynamics at Intermediate Coupling". *Physical Review Letters* **104** (12). arXiv:0911.0676. Bibcode:2010PhRvL.104l2003A. doi:10.1103/PhysRevLett.104.122003.

[18] Blaizot; Iancu; Rebhan (2003). "Thermodynamics of the high-temperature quark-gluon plasma". arXiv:hep-ph/0303185 [hep-ph].

30.7 External links

- The Relativistic Heavy Ion Collider at Brookhaven National Laboratory

- The Alice Experiment at CERN

- The Indian Lattice Gauge Theory Initiative

- Quark matter reviews: 2004 theory, 2004 experiment

- Quark-Gluon Plasma reviews: 2011 theory

- Lattice reviews: 2003, 2005

- BBC article mentioning Brookhaven results (2005)

- Physics News Update article on the quark-gluon liquid, with links to preprints

- Read for free : "Hadrons and Quark-Gluon Plasma" by Jean Letessier and Johann Rafelski Cambridge University Press (2002) ISBN 0-521-38536-9, Cambridge, UK;

Chapter 31

Strongly symmetric matter

Strongly symmetric matter: If the predictions of supersymmetry and more so, string theory are correct then during the time of the Planck Epoch (10^{-43} seconds after the Big Bang) all four fundamental forces were of equal strength and united into a single fundamental force. However, shortly thereafter, during the grand unification epoch (10^{-33} seconds after the Big Bang) gravity began to separate from the other three forces of electromagnetism, the strong, and weak nuclear force. After gravity had separated, the strong nuclear force was quick to follow and separated itself from the electroweak force, ending the epoch. The phrase strongly symmetric matter refers to the united forces present during this time.

31.1   See also

- List of phases of matter

- Timeline of the Big Bang

Chapter 32

Glass

This article is about the material. For other uses, see Glass (disambiguation).

Glass is a non-crystalline amorphous solid that is often transparent and has widespread practical, technological, and

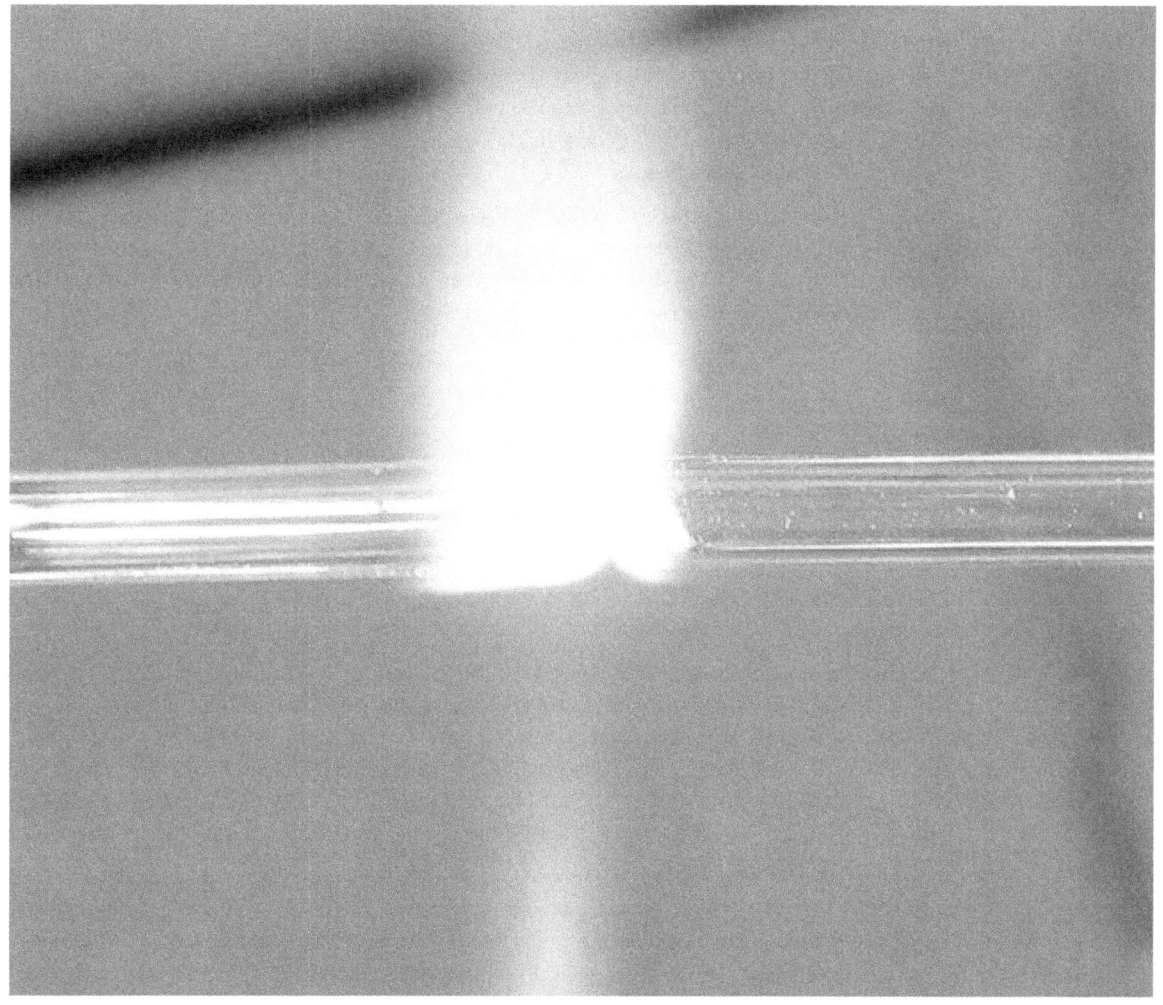

The joining of two tubes made of lead glass during glass welding.

decorative usage in things like window panes, tableware, and optoelectronics. Scientifically, the term "glass" is often defined in a broader sense, encompassing every solid that possesses a non-crystalline (that is, amorphous) structure at the atomic scale and that exhibits a glass transition when heated towards the liquid state.

The most familiar, and historically the oldest, types of glass are based on the chemical compound silica (silicon dioxide), the primary constituent of sand. The term *glass*, in popular usage, is often used to refer only to this type of material, which is familiar from use as window glass and in glass bottles. Of the many silica-based glasses that exist, ordinary glazing and container glass is formed from a specific type called soda-lime glass, composed of approximately 75% silicon dioxide (SiO_2), sodium oxide (Na_2O) from sodium carbonate (Na_2CO_3), calcium oxide, also called lime (CaO), and several minor additives. A very clear and durable quartz glass can be made from pure silica which is very tough and resistant to thermal shock, being able to survive immersion in water while red hot. However, quartz must be heated to well over 3,000 °F (1,650 °C) (white hot) before it begins to melt, and it has a very narrow glass transition, making glassblowing and hot working difficult. In glasses like soda lime, the other compounds are used to lower the melting temperature and improve the temperature workability of the product at a cost in the toughness, thermal stability, and optical transmittance.

Many applications of silicate glasses derive from their optical transparency, which gives rise to one of silicate glasses' primary uses as window panes. Glass will transmit, reflect and refract light; these qualities can be enhanced by cutting and polishing to make optical lenses, prisms, fine glassware, and optical fibers for high speed data transmission by light. Glass can be colored by adding metallic salts, and can also be painted. These qualities have led to the extensive use of glass in the manufacture of art objects and in particular, stained glass windows. Although brittle, silicate glass is extremely durable, and many examples of glass fragments exist from early glass-making cultures. Because glass can be formed or molded into any shape, and also because it is a sterile product, it has been traditionally used for vessels: bowls, vases, bottles, jars and drinking glasses. In its most solid forms it has also been used for paperweights, marbles, and beads. When extruded as glass fiber and matted as glass wool in a way to trap air, it becomes a thermal insulating material, and when these glass fibers are embedded into an organic polymer plastic, they are a key structural reinforcement part of the composite material fiberglass. Some objects are so commonly made of glass that they are simply called by the name of the material, such as drinking glasses and reading glasses.

In science, porcelains and many polymer thermoplastics familiar from everyday use are glasses too. These sorts of glasses can be made of quite different kinds of materials than silica: metallic alloys, ionic melts, aqueous solutions, molecular liquids, and polymers. For many applications, like glass bottles or eyewear, polymer glasses (acrylic glass, polycarbonate or polyethylene terephthalate) are a lighter alternative than traditional glass.

32.1   Silicate glass

32.1.1   Ingredients

Silica (the chemical compound SiO_2) is a common fundamental constituent of glass. In nature, vitrification of quartz occurs when lightning strikes sand, forming hollow, branching rootlike structures called fulgurite.

Fused quartz is a glass made from chemically-pure silica. However, its high melting-temperature (1723 °C) and viscosity make it difficult to work with.[1] Normally, other substances are added to simplify processing. One is sodium carbonate (Na_2CO_3, "soda"), which lowers the glass transition temperature. The soda makes the glass water-soluble, which is usually undesirable, so lime (calcium oxide [CaO], generally obtained from limestone), some magnesium oxide (MgO) and aluminium oxide (Al_2O_3) are added to provide for a better chemical durability. The resulting glass contains about 70 to 74% silica by weight and is called a soda-lime glass.[2] Soda-lime glasses account for about 90% of manufactured glass.

Most common glass contains other ingredients to change its properties. Lead glass or flint glass is more 'brilliant' because the increased refractive index causes noticeably more specular reflection and increased optical dispersion. Adding barium also increases the refractive index. Thorium oxide gives glass a high refractive index and low dispersion and was formerly used in producing high-quality lenses, but due to its radioactivity has been replaced by lanthanum oxide in modern eyeglasses. Iron can be incorporated into glass to absorb infrared energy, for example in heat absorbing filters for movie projectors, while cerium(IV) oxide can be used for glass that absorbs UV wavelengths.[3]

The following is a list of the more common types of silicate glasses, and their ingredients, properties, and applications:

1. Fused quartz, also called **fused silica glass, vitreous silica glass**, is silica (SiO_2) in vitreous or glass form (i.e., its molecules are disordered and random, without crystalline structure). It has very low thermal expansion, is very hard, and resists high temperatures (1000–1500 °C). It is also the most resistant against weathering (caused in other glasses by alkali ions leaching out of the glass, while staining it). Fused quartz is used for high temperature applications such as furnace tubes, lighting tubes, melting crucibles, etc.

2. **Soda-lime-silica glass, window glass**: silica 72% + sodium oxide (Na_2O) 14.2% + lime (CaO) 10.0% + magnesia (MgO) 2.5% + alumina (Al_2O_3) 0.6%. Is transparent, easily formed and most suitable for window glass (see flat glass). It has a high thermal expansion and poor resistance to heat (500–600 °C). It is used for windows, some low temperature incandescent light bulbs, and tableware. Container glass is a soda-lime glass that is a slight variation on flat glass, which uses more alumina and calcium, and less sodium and magnesium which are more water-soluble. This makes it less susceptible to water erosion.

3. **Sodium borosilicate glass, Pyrex**: silica 81% + boric oxide (B_2O_3) 12% + soda (Na_2O) 4.5% + alumina (Al_2O_3) 2.0%. Stands heat expansion much better than window glass. Used for chemical glassware, cooking glass, car head lamps, etc. Borosilicate glasses (e.g. Pyrex) have as main constituents silica and boron oxide. They have fairly low coefficients of thermal expansion (7740 Pyrex CTE is $3.25{\times}10^{-6}/°C$[4] as compared to about $9{\times}10^{-6}/°C$ for a typical soda-lime glass[5]), making them more dimensionally stable. The lower coefficient of thermal expansion (CTE) also makes them less subject to stress caused by thermal expansion, thus less vulnerable to cracking from thermal shock. They are commonly used for reagent bottles, optical components and household cookware.

4. **Lead-oxide glass, crystal glass**: silica 59% + lead oxide (PbO) 25% + potassium oxide (K_2O) 12% + soda (Na_2O) 2.0% + zinc oxide (ZnO) 1.5% + alumina 0.4%. Because of its high density (resulting in a high electron density) it has a high refractive index, making the look of glassware more brilliant (called "crystal", though of course it is a glass and not a crystal). It also has a high elasticity, making glassware 'ring'. It is also more workable in the factory, but cannot stand heating very well.

5. **Aluminosilicate glass**: silica 57% + alumina 16% + lime 10% + magnesia 7.0% + barium oxide (BaO) 6.0% + boric oxide (B_2O_3) 4.0%. Extensively used for fiberglass, used for making glass-reinforced plastics (boats, fishing rods, etc.) and for halogen bulb glass.

6. **Oxide glass**: alumina 90% + germanium oxide (GeO_2) 10%. Extremely clear glass, used for fiber-optic waveguides in communication networks. Light loses only 5% of its intensity through 1 km of glass fiber.[6] Most optical fiber is based on silica, as are all the glasses above.

Another common glass ingredient is crushed alkali glass or "cullet" ready for recycled glass. The recycled glass saves on raw materials and energy. Impurities in the cullet can lead to product and equipment failure. Fining agents such as sodium sulfate, sodium chloride, or antimony oxide may be added to reduce the number of air bubbles in the glass mixture.[2] Glass batch calculation is the method by which the correct raw material mixture is determined to achieve the desired glass composition.

- Moldavite, a natural glass formed by meteorite impact, from Besednice, Bohemia

- Tube fulgurites

- Quartz sand (silica) is the main raw material in commercial glass production

- Trinitite, a glass made by the Trinity nuclear weapon test.

32.2 Physical properties

See also: List of physical properties of glass

32.2.1 Optical properties

Glass is in widespread use largely due to the production of glass compositions that are transparent to visible light. In contrast, polycrystalline materials do not generally transmit visible light.[7] The individual crystallites may be transparent, but their facets (grain boundaries) reflect or scatter light resulting in diffuse reflection. Glass does not contain the internal subdivisions associated with grain boundaries in polycrystals and hence does not scatter light in the same manner as a polycrystalline material. The surface of a glass is often smooth since during glass formation the molecules of the supercooled liquid are not forced to dispose in rigid crystal geometries and can follow surface tension, which imposes a microscopically smooth surface. These properties, which give glass its clearness, can be retained even if glass is partially light-absorbing—i.e., colored.[8]

Glass has the ability to refract, reflect, and transmit light following geometrical optics, without scattering it. It is used in the manufacture of lenses and windows. Common glass has a refraction index around 1.5. This may be modified by adding low-density materials such as boron, which lowers the index of refraction (see crown glass), or increased (to as much as 1.8) with high-density materials such as (classically) lead oxide (see flint glass and lead glass), or in modern uses, less toxic oxides of zirconium, titanium, or barium. These high-index glasses (inaccurately known as "crystal" when used in glass vessels) causes more chromatic dispersion of light, and are prized for their diamond-like optical properties.

According to Fresnel equations, the reflectivity of a sheet of glass is about 4% per surface (at normal incidence in air), and the transmissivity of one element (two surfaces) is about 90%. Glass with high germanium oxide content also finds application in optoelectronics—e.g., for light-transmitting optical fibers.

- A wine glass

- Simple optical device: the magnifying glass

- Glass petri dish

32.2.2 Other properties

In the process of manufacture, silicate glass can be poured, formed, extruded and molded into forms ranging from flat sheets to highly intricate shapes. The finished product is brittle and will fracture, unless laminated or specially treated, but is extremely durable under most conditions. It erodes very slowly and can withstand the action of water. It is resilient to chemical attack and is an ideal material for the manufacture of containers for foodstuffs and most chemicals.

32.3 Contemporary production

Main articles: Glass production, Float glass, Flat glass and Glazier

Following the glass batch preparation and mixing, the raw materials are transported to the furnace. Soda-lime glass for mass production is melted in gas fired units. Smaller scale furnaces for specialty glasses include electric melters, pot furnaces, and day tanks.[2] After melting, homogenization and refining (removal of bubbles), the glass is formed. Flat glass for windows and similar applications is formed by the float glass process, developed between 1953 and 1957 by Sir Alastair Pilkington and Kenneth Bickerstaff of the UK's Pilkington Brothers, who created a continuous ribbon of glass using a molten tin bath on which the molten glass flows unhindered under the influence of gravity. The top surface of the glass is subjected to nitrogen under pressure to obtain a polished finish.[9] Container glass for common bottles and jars is formed by blowing and pressing methods. This glass is often slightly modified chemically (with more alumina and calcium oxide) for greater water resistance. Further glass forming techniques are summarized in the table Glass forming techniques.

Once the desired form is obtained, glass is usually annealed for the removal of stresses. Surface treatments, coatings or lamination may follow to improve the chemical durability (glass container coatings, glass container internal treatment), strength (toughened glass, bulletproof glass, windshields), or optical properties (insulated glazing, anti-reflective coating).

- Impurities give the glass its color

- Some of the many color possibilities of glass

- Transparent and opaque examples

- Glass can be blown into an infinite number of shapes

32.3.1 Color

Main article: Glass coloring and color marking

Color in glass may be obtained by addition of electrically charged ions (or color centers) that are homogeneously distributed, and by precipitation of finely dispersed particles (such as in photochromic glasses).[10] Ordinary soda-lime glass appears colorless to the naked eye when it is thin, although iron(II) oxide (FeO) impurities of up to 0.1 wt%[11] produce a green tint, which can be viewed in thick pieces or with the aid of scientific instruments. Further FeO and Cr_2O_3 additions may be used for the production of green bottles. Sulfur, together with carbon and iron salts, is used to form iron polysulfides and produce amber glass ranging from yellowish to almost black.[12] A glass melt can also acquire an amber color from a reducing combustion atmosphere. Manganese dioxide can be added in small amounts to remove the green tint given by iron(II) oxide. When used in art glass or studio glass is colored using closely guarded recipes that involve specific combinations of metal oxides, melting temperatures and 'cook' times. Most colored glass used in the art market is manufactured in volume by vendors who serve this market although there are some glassmakers with the ability to make their own color from raw materials.

32.4 History of silicate glass

Main article: History of glass
See also: Architectural glass, Stained glass, Glass art, Art glass and Studio glass

Naturally occurring glass, especially the volcanic glass obsidian, has been used by many Stone Age societies across the globe for the production of sharp cutting tools and, due to its limited source areas, was extensively traded. But in general, archaeological evidence suggests that the first true glass was made in coastal north Syria, Mesopotamia or Ancient Egypt.[13] The earliest known glass objects, of the mid third millennium BCE, were beads, perhaps initially created as accidental by-products of metal-working (slags) or during the production of faience, a pre-glass vitreous material made by a process similar to glazing.[14]

Glass remained a luxury material, and the disasters that overtook Late Bronze Age civilizations seem to have brought glass-making to a halt. Indigenous development of glass technology in South Asia may have begun in 1730 BCE.[15] In ancient China, though, glassmaking seems to have a late start, compared to ceramics and metal work. The term *glass* developed in the late Roman Empire. It was in the Roman glassmaking center at Trier, now in modern Germany, that the late-Latin term *glesum* originated, probably from a Germanic word for a transparent, lustrous substance.[16] Glass objects have been recovered across the Roman empire in domestic, industrial and funerary contexts.

Glass was used extensively during the Middle Ages. Anglo-Saxon glass has been found across England during archaeological excavations of both settlement and cemetery sites. Glass in the Anglo-Saxon period was used in the manufacture of a range of objects including vessels, beads, windows and was also used in jewelry. From the 10th-century onwards, glass was employed in stained glass windows of churches and cathedrals, with famous examples at Chartres Cathedral and the Basilica of Saint Denis. By the 14th-century, architects were designing buildings with walls of stained glass such as Sainte-Chapelle, Paris, (1203–1248)[17] and the East end of Gloucester Cathedral.[18] Stained glass had a major revival with Gothic Revival architecture in the 19th-century. With the Renaissance, and a change in architectural style, the use of large stained glass windows became less prevalent. The use of domestic stained glass increased until most substantial houses had glass windows. These were initially small panes leaded together, but with the changes in technology, glass could be manufactured relatively cheaply in increasingly larger sheets. This led to larger window panes, and, in the 20th-century, to much larger windows in ordinary domestic and commercial buildings.

In the 20th century, new types of glass such as laminated glass, reinforced glass and glass bricks have increased the use of glass as a building material and resulted in new applications of glass. Multi-storey buildings are frequently constructed with curtain walls made almost entirely of glass. Similarly, laminated glass has been widely applied to vehicles for windscreens. While glass containers have always been used for storage and are valued for their hygienic properties, glass has been utilized increasingly in industry. Optical glass for spectacles has been used since the late Middle Ages. The production of lenses has become increasingly proficient, aiding astronomers as well as having other application in medicine and science. Glass is also employed as the aperture cover in many solar energy systems.

From the 19th century, there was a revival in many ancient glass-making techniques including Cameo glass, achieved for the first time since the Roman Empire and initially mostly used for pieces in a neo-classical style. The Art Nouveau movement made great use of glass, with René Lalique, Émile Gallé, and Daum of Nancy producing colored vases and similar pieces, often in cameo glass, and also using luster techniques. Louis Comfort Tiffany in America specialized in stained glass, both secular and religious, and his famous lamps. The early 20th-century saw the large-scale factory production of glass art by firms such as Waterford and Lalique. From about 1960 onwards there have been an increasing number of small studios hand-producing glass artworks, and glass artists began to class themselves as in effect sculptors working in glass, and their works as part fine arts.

In the 21st century, scientists observing the properties of ancient stained glass windows, in which suspended nanoparticles prevent UV light from causing chemical reactions that change image colors, are developing photographic techniques that use similar stained glass to capture true color images of Mars for the 2019 ESA Mars Rover mission.[19]

32.4.1   Chronology of advances in architectural glass

- 1226 – "Broad Sheet" first produced in Sussex.

- 1330 – "Crown glass" for art work and vessels first produced in Rouen, France. "Broad Sheet" also produced. Both were also supplied for export.

- 1500s – The method of making mirrors out of plate glass was invented by Venetian glassmakers on the island of Murano, who covered the back of the glass with mercury, obtaining near-perfect and undistorted reflection.

- 1620 – "Blown Plate" first produced in London. Used for mirrors and coach plates.

- 1678 – "Crown Glass" first produced in London. This process dominated until the 19th century.

- 1843 – An early form of "Float Glass" invented by Henry Bessemer, pouring glass onto liquid tin. Expensive and not a commercial success.

- 1874 – Tempered glass is developed by Francois Barthelemy Alfred Royer de la Bastie (1830-1901) of Paris, France by quenching almost molten glass in a heated bath of oil or grease.

- 1888 – "Machine Rolled" glass introduced allowing patterns to be introduced.

- 1898 – "Wired Cast" glass invented by Pilkington for use where safety or security was an issue.

- 1959 – "Float Glass" launched in UK. Invented by Sir Alastair Pilkington.[20]

- Mouth-blown window-glass in Sweden Kosta Glasbruk, (1742) with a pontil mark from the glassblower's pipe

- A building in Canterbury, England, which displays its long history in different building styles and glazing of every century from the 16th to the 20th included.

- Windows in the choir of the Basilica of Saint Denis, one of the earliest uses of extensive areas of glass. (early 13th-century architecture with restored glass of the 19th-century)

- *"Hardwick Hall, more glass than wall"*. (late 16th-century)

- Windows at Österreichische Postsparkasse, Vienna, (early 20th-century)

- Westin Bonaventure Hotel, USA, show the extensive use of glass as a building material in the 20th-21st centuries

32.5 Other types of glass

New chemical glass compositions or new treatment techniques can be initially investigated in small-scale laboratory experiments. The raw materials for laboratory-scale glass melts are often different from those used in mass production because the cost factor has a low priority. In the laboratory mostly pure chemicals are used. Care must be taken that the raw materials have not reacted with moisture or other chemicals in the environment (such as alkali or alkaline earth metal oxides and hydroxides, or boron oxide), or that the impurities are quantified (loss on ignition).[21] Evaporation losses during glass melting should be considered during the selection of the raw materials, e.g., sodium selenite may be preferred over easily evaporating SeO_2. Also, more readily reacting raw materials may be preferred over relatively inert ones, such as $Al(OH)_3$ over Al_2O_3. Usually, the melts are carried out in platinum crucibles to reduce contamination from the crucible material. Glass homogeneity is achieved by homogenizing the raw materials mixture (glass batch), by stirring the melt, and by crushing and re-melting the first melt. The obtained glass is usually annealed to prevent breakage during processing.[21][22]

To make glass from materials with poor glass forming tendencies, novel techniques are used to increase cooling rate, or reduce crystal nucleation triggers. Examples of these techniques include aerodynamic levitation (cooling the melt whilst it floats on a gas stream), splat quenching (pressing the melt between two metal anvils) and roller quenching (pouring the melt through rollers).

32.5.1 Network glasses

Some glasses that do not include silica as a major constituent may have physico-chemical properties useful for their application in fiber optics and other specialized technical applications. These include fluoride glasses, aluminosilicates, phosphate glasses, borate glasses, and chalcogenide glasses.

There are three classes of components for oxide glasses: network formers, intermediates, and modifiers. The network formers (silicon, boron, germanium) form a highly cross-linked network of chemical bonds. The intermediates (titanium, aluminium, zirconium, beryllium, magnesium, zinc) can act as both network formers and modifiers, according to the glass composition. The modifiers (calcium, lead, lithium, sodium, potassium) alter the network structure; they are usually present as ions, compensated by nearby non-bridging oxygen atoms, bound by one covalent bond to the glass network and holding one negative charge to compensate for the positive ion nearby. Some elements can play multiple roles; e.g. lead can act both as a network former (Pb^{4+} replacing Si^{4+}), or as a modifier.

The presence of non-bridging oxygens lowers the relative number of strong bonds in the material and disrupts the network, decreasing the viscosity of the melt and lowering the melting temperature.

The alkali metal ions are small and mobile; their presence in glass allows a degree of electrical conductivity, especially in molten state or at high temperature. Their mobility decreases the chemical resistance of the glass, allowing leaching by water and facilitating corrosion. Alkaline earth ions, with their two positive charges and requirement for two non-bridging oxygen ions to compensate for their charge, are much less mobile themselves and also hinder diffusion of other ions, especially the alkalis. The most common commercial glasses contain both alkali and alkaline earth ions (usually sodium and calcium), for easier processing and satisfying corrosion resistance.[24] Corrosion resistance of glass can be achieved by dealkalization, removal of the alkali ions from the glass surface by reaction with e.g. sulfur or fluorine compounds. Presence of alkaline metal ions has also detrimental effect to the loss tangent of the glass, and to its electrical resistance; glasses for electronics (sealing, vacuum tubes, lamps...) have to take this in account.

Addition of lead(II) oxide lowers melting point, lowers viscosity of the melt, and increases refractive index. Lead oxide also facilitates solubility of other metal oxides and is used in colored glasses. The viscosity decrease of lead glass melt is very significant (roughly 100 times in comparison with soda glasses); this allows easier removal of bubbles and working at lower temperatures, hence its frequent use as an additive in vitreous enamels and glass solders. The high ionic radius of the Pb^{2+} ion renders it highly immobile in the matrix and hinders the movement of other ions; lead glasses therefore have high electrical resistance, about two orders of magnitude higher than soda-lime glass ($10^{8.5}$ vs $10^{6.5}$ Ohm·cm, DC at 250 °C). For more details, see lead glass.[25]

Addition of fluorine lowers the dielectric constant of glass. Fluorine is highly electronegative and attracts the electrons in the lattice, lowering the polarizability of the material. Such silicon dioxide-fluoride is used in manufacture of integrated circuits as an insulator. High levels of fluorine doping lead to formation of volatile SiF_2O and such glass is then thermally unstable. Stable layers were achieved with dielectric constant down to about 3.5–3.7.[26]

32.5.2   Amorphous metals

In the past, small batches of amorphous metals with high surface area configurations (ribbons, wires, films, etc.) have been produced through the implementation of extremely rapid rates of cooling. This was initially termed "splat cooling" by doctoral student W. Klement at Caltech, who showed that cooling rates on the order of millions of degrees per second is sufficient to impede the formation of crystals, and the metallic atoms become "locked into" a glassy state. Amorphous metal wires have been produced by sputtering molten metal onto a spinning metal disk. More recently a number of alloys have been produced in layers with thickness exceeding 1 millimeter. These are known as bulk metallic glasses (BMG). Liquidmetal Technologies sell a number of zirconium-based BMGs. Batches of amorphous steel have also been produced that demonstrate mechanical properties far exceeding those found in conventional steel alloys.[27][28][29]

In 2004, NIST researchers presented evidence that an isotropic non-crystalline metallic phase (dubbed "q-glass") could be grown from the melt. This phase is the first phase, or "primary phase," to form in the Al-Fe-Si system during rapid cooling. Interestingly, experimental evidence indicates that this phase forms by a *first-order transition*. Transmission electron microscopy (TEM) images show that the q-glass nucleates from the melt as discrete particles, which grow spherically with a uniform growth rate in all directions. The diffraction pattern shows it to be an isotropic glassy phase. Yet there is a nucleation barrier, which implies an interfacial discontinuity (or internal surface) between the glass and the melt.[30][31]

32.5.3   Electrolytes

Electrolytes or molten salts are mixtures of different ions. In a mixture of three or more ionic species of dissimilar size and shape, crystallization can be so difficult that the liquid can easily be supercooled into a glass. The best-studied example is $Ca_{0.4}K_{0.6}(NO_3)_{1.4}$.

32.5.4   Aqueous solutions

Some aqueous solutions can be supercooled into a glassy state, for instance $LiCl:RH_2O$ in the composition range $4<R<8$.

32.5.5   Molecular liquids

A *molecular liquid* is composed of molecules that do not form a covalent network but interact only through weak van der Waals forces or through transient hydrogen bonds. Many molecular liquids can be supercooled into a glass; some are excellent glass formers that normally do not crystallize.

A widely known example is sugar glass.

Under extremes of pressure and temperature solids may exhibit large structural and physical changes that can lead to polyamorphic phase transitions.[32] In 2006 Italian scientists created an amorphous phase of carbon dioxide using extreme pressure. The substance was named amorphous carbonia(a-CO_2) and exhibits an atomic structure resembling that of silica.[33]

32.5.6   Polymers

Important polymer glasses include amorphous and glassy pharmaceutical compounds. These are useful because the solubility of the compound is greatly increased when it is amorphous compared to the same crystalline composition. Many emerging pharmaceuticals are practically insoluble in their crystalline forms.

32.5.7 Colloidal glasses

Concentrated colloidal suspensions may exhibit a distinct glass transition as function of particle concentration or density. In cell biology there is recent evidence suggesting that the cytoplasm behaves like a colloidal glass approaching the liquid-glass transition.[37] During periods of low metabolic activity, as in dormancy, the cytoplasm vitrifies and prohibits the movement to larger cytoplasmic particles while allowing the diffusion of smaller ones throughout the cell.[37]

32.5.8 Glass-ceramics

Glass-ceramic materials share many properties with both non-crystalline glass and crystalline ceramics. They are formed as a glass, and then partially crystallized by heat treatment. For example, the microstructure of whiteware ceramics frequently contains both amorphous and crystalline phases. Crystalline grains are often embedded within a non-crystalline intergranular phase of grain boundaries. When applied to whiteware ceramics, vitreous means the material has an extremely low permeability to liquids, often but not always water, when determined by a specified test regime.[38][39]

The term mainly refers to a mix of lithium and aluminosilicates that yields an array of materials with interesting thermomechanical properties. The most commercially important of these have the distinction of being impervious to thermal shock. Thus, glass-ceramics have become extremely useful for countertop cooking. The negative thermal expansion coefficient (CTE) of the crystalline ceramic phase can be balanced with the positive CTE of the glassy phase. At a certain point (~70% crystalline) the glass-ceramic has a net CTE near zero. This type of glass-ceramic exhibits excellent mechanical properties and can sustain repeated and quick temperature changes up to 1000 °C.[38][39]

32.6 Structure

Main article: Structure of liquids and glasses

As in other amorphous solids, the atomic structure of a glass lacks any long-range translational periodicity. Due to chemical bonding characteristics glasses do possess a high degree of short-range order with respect to local atomic polyhedra.[40]

32.6.1 Formation from a supercooled liquid

Main article: Glass transition

In physics, the standard definition of a glass (or vitreous solid) is a solid formed by rapid melt quenching.[41][42][43][44][45] The term glass is often used to describe any amorphous solid that exhibits a glass transition temperature T_g. If the cooling is sufficiently rapid (relative to the characteristic crystallization time) then crystallization is prevented and instead the disordered atomic configuration of the supercooled liquid is frozen into the solid state at T_g. The tendency for a material to form a glass while quenched is called glass-forming ability. This ability can be predicted by the rigidity theory.[46] Generally, the structure of a glass exists in a metastable state with respect to its crystalline form, although in certain circumstances, for example in atactic polymers, there is no crystalline analogue of the amorphous phase.[47]

Some people consider glass to be a liquid due to its lack of a first-order phase transition[48][49] where certain thermodynamic variables such as volume, entropy and enthalpy are discontinuous through the glass transition range. The glass transition may be described as analogous to a second-order phase transition where the intensive thermodynamic variables such as the thermal expansivity and heat capacity are discontinuous.[42] Nonetheless, the equilibrium theory of phase transformations does not entirely hold for glass, and hence the glass transition cannot be classed as one of the classical equilibrium phase transformations in solids.[44][45]

Glass is an amorphous solid. It exhibits an atomic structure close to that observed in the supercooled liquid phase but displays all the mechanical properties of a solid.[48][50] The notion that glass flows to an appreciable extent over extended periods of time is not supported by empirical research or theoretical analysis (see viscosity of amorphous materials). Laboratory measurements of room temperature glass flow do show a motion consistent with a material viscosity on the order of 10^{17}–10^{18} Pa s.[51]

Although the atomic structure of glass shares characteristics of the structure in a supercooled liquid, glass tends to behave as a solid below its glass transition temperature.[52] A supercooled liquid behaves as a liquid, but it is below the freezing point of the material, and in some cases will crystallize almost instantly if a crystal is added as a core. The change in heat capacity at a glass transition and a melting transition of comparable materials are typically of the same order of magnitude, indicating that the change in active degrees of freedom is comparable as well. Both in a glass and in a crystal it is mostly only the vibrational degrees of freedom that remain active, whereas rotational and translational motion is arrested. This helps to explain why both crystalline and non-crystalline solids exhibit rigidity on most experimental time scales.

32.6.2   Behavior of antique glass

The observation that old windows are sometimes found to be thicker at the bottom than at the top is often offered as supporting evidence for the view that glass flows over a timescale of centuries, the assumption being that the glass has exhibited the liquid property of flowing from one shape to another.[54] This assumption is incorrect, as once solidified, glass stops flowing. The reason for the observation is that in the past, when panes of glass were commonly made by glassblowers, the technique used was to spin molten glass so as to create a round, mostly flat and even plate (the crown glass process, described above). This plate was then cut to fit a window. The pieces were not absolutely flat; the edges of the disk became a different thickness as the glass spun. When installed in a window frame, the glass would be placed with the thicker side down both for the sake of stability and to prevent water accumulating in the lead cames at the bottom of the window.[55] Occasionally such glass has been found installed with the thicker side at the top, left or right.[56]

Mass production of glass window panes in the early twentieth century caused a similar effect. In glass factories, molten glass was poured onto a large cooling table and allowed to spread. The resulting glass is thicker at the location of the pour, located at the center of the large sheet. These sheets were cut into smaller window panes with nonuniform thickness, typically with the location of the pour centered in one of the panes (known as "bull's-eyes") for decorative effect. Modern glass intended for windows is produced as float glass and is very uniform in thickness.

Several other points can be considered that contradict the "cathedral glass flow" theory:

- Writing in the *American Journal of Physics*, the materials engineer Edgar D. Zanotto states "... the predicted relaxation time for GeO_2 at room temperature is 10^{32} years. Hence, the relaxation period (characteristic flow time) of cathedral glasses would be even longer."[57] (10^{32} years is many times longer than the estimated age of the universe.)

- If medieval glass has flowed perceptibly, then ancient Roman and Egyptian objects should have flowed proportionately more—but this is not observed. Similarly, prehistoric obsidian blades should have lost their edge; this is not observed either (although obsidian may have a different viscosity from window glass).[48]

- If glass flows at a rate that allows changes to be seen with the naked eye after centuries, then the effect should be noticeable in antique telescopes. Any slight deformation in the antique telescopic lenses would lead to a dramatic decrease in optical performance, a phenomenon that is not observed.[48]

- There are many examples of centuries-old glass shelving that has not bent, even though it is under much higher stress from gravitational loads than vertical window glass.

32.7   Gallery

- Ear Stud, ca. 1390-1353 B.C.E., 48.66.30, Brooklyn Museum. The shafts of these brightly colored studs were inserted through a hole in the earlobe to display the studs' circular heads.

- Phoenician glass necklace 5-6th century BC

- Roman glass amphoriskoi 1st century AD–2nd century AD

- Blue Head Flask (Roman, AD 300-500, cast glass)

- Lombardic glass drinking horn 6th century AD–7th century AD

- Two cups Cobalt Blue Glass with gilt floral decoration from India, Mughal, circa 1700-1775

- Base for a Water Pipe India, Mughal, circa 1700-1775

- Venetian Goblet made in Italy in the early 19th century.

- Bracelets with Peacocks Delhi, Enameled silver inlaid with gemstones and glass 19th century

- Jug, 1876, James Powell & Sons

- Siphon seltzer water 1922

- New Martinsville Glass Hostmaster Tea Cup, cobalt blue 1930

- Perfume set from Soviet union ca. 1965

- Murano Millefiori Glass Vase provided

- Murano Millefiori Glass Vase

- Window glass

- Detail from a glass chandelier

- Glass chandelier

- The Rotunda, or main entrance, of the Victoria and Albert Museum now sports a 30 ft high, blown glass chandelier by Dale Chihuly

32.8 See also

- Caneworking

- Fabrication and testing of optical components

- Fire glass

- Knitted glass

- Glass recycling

- Kimberley points

- Murrine

- Optical lens design

- Prince Rupert's Drop

- Superglass

- Tektite

- Vitrified sand

- Toughened glass

- Mirror

- Stained glass

32.9   References

[1] "Glass - Chemistry Encyclopedia". Retrieved 1 April 2015.

[2] B. H. W. S. de Jong, "Glass"; in "Ullmann's Encyclopedia of Industrial Chemistry"; 5th edition, vol. A12, VCH Publishers, Weinheim, Germany, 1989, ISBN 978-3-527-20112-9, pp. 365–432.

[3] Heinz G. Pfaender (1996). *Schott guide to glass*. Springer. pp. 135, 186. ISBN 978-0-412-62060-7. Retrieved 8 February 2011.

[4] Corning, Inc. Pyrex data sheet. (PDF). Retrieved 2012-05-15.

[5] AR-GLAS Schott, N.A., Inc data sheet

[6] Mining the sea sand. Seafriends.org.nz (1994-02-08). Retrieved 2012-05-15.

[7] Michel W. Barsoum (2003). *Fundamentals of ceramics* (2 ed.). Bristol: IOP. ISBN 0-7503-0902-4.

[8] Donald R. Uhlmann, Norbert J. Kreidl, ed. (1991). *Optical properties of glass*. Westerville, OH: American Ceramic Society. ISBN 0-944904-35-1.

[9] "PFG Glass". Pfg.co.za. Retrieved 24 October 2009.

[10] Werner Vogel (1994). *Glass Chemistry* (2 ed.). Springer-Verlag Berlin and Heidelberg GmbH & Co. K. ISBN 3-540-57572-3.

[11] Thomas P. Seward, ed. (2005). *High temperature glass melt property database for process modeling*. Westerville, Ohio: American Ceramic Society. ISBN 1-57498-225-7.

[12] David M Issitt. Substances Used in the Making of Coloured Glass 1st.glassman.com.

[13] "Glass Online: The History of Glass". Retrieved 29 October 2007.

[14] True glazing over a ceramic body was not used until many centuries after the production of the first glass.

[15] J.A.J. Gowlett (1997). *High Definition Archaeology: Threads Through the Past*. Routledge. ISBN 0-415-18429-0.

[16] Douglas, R. W. (1972). *A history of glassmaking*. Henley-on-Thames: G T Foulis & Co Ltd. ISBN 0-85429-117-2.

[17] Rene Hughe, *Byzantine and Medieval Art*, Paul Hamlyn, (1963)

[18] John Harvey, *English Cathedrals*, Batsford, (1961)

[19] Ellie Zolfagharifard (15 October 2013). "How medieval stained-glass is creating the ultimate SPACE camera: Nanoparticles used in church windows will help scientists see Mars' true colours under extreme UV light". *Daily Mail* (London).

[20] History of Glass Manufacture: London Crown Glass co.

[21] "Glass melting, Pacific Northwest National Laboratory". Depts.washington.edu. Retrieved 24 October 2009.

[22] Alexander Fluegel. "Glass melting in the laboratory". Glassproperties.com. Retrieved 24 October 2009.

[23] Greer, A. Lindsay; Mathur, N (2005). "Materials science: Changing Face of the Chameleon". *Nature* **437** (7063): 1246–1247. Bibcode:2005Natur.437.1246G. doi:10.1038/4371246a. PMID 16251941.

[24] Eric Le Bourhis (2007). *Glass: Mechanics and Technology*. Wiley-VCH. p. 74. ISBN 3-527-31549-7.

[25] James F. Shackelford, Robert H. Doremus (2008). *Ceramic and Glass Materials: Structure, Properties and Processing*. Springer. p. 158. ISBN 0-387-73361-2.

[26] Robert Doering, Yoshio Nishi (2007). *Handbook of semiconductor manufacturing technology*. CRC Press. pp. 12–3. ISBN 1-57444-675-4.

[27] W. Klement, Jr., R. H. Willens and Pol Duwez (1960). "Non-crystalline Structure in Solidified Gold-Silicon Alloys". *Nature* **187** (4740): 869. Bibcode:1960Natur.187..869K. doi:10.1038/187869b0.

[28] H. Liebermann and C. Graham (1976). "Production of Amorphous Alloy Ribbons and Effects of Apparatus Parameters on Ribbon Dimensions". *IEEE Transactions on Magnetics* **12**(6): 921. Bibcode:1976ITM....12..921L. doi:10.1109/TMAG.1976.1059

[29] V. Ponnambalam, S. Joseph Poon and Gary J. Shiflet (2004). "Fe-based bulk metallic glasses with diameter thickness larger than one centimeter". *Journal of Materials Research* **19** (5): 1320. Bibcode:2004JMatR..19.1320P. doi:10.1557/JMR.2004.0176.

[30] "Metallurgy Division Publications". *NIST Interagency Report 7127.*

[31] M. I. Mendelev, J. Schmalian, C. Z. Wang, J. R. Morris, and K. M. Ho (2006). "Interface Mobility and the Liquid-Glass Transition in a One-Component System".*Physical Review B***74**(10).Bibcode:2006PhRvB..74j4206M.doi:10.1103/PhysRevB.74.10

[32] P. F. McMillan (2004). "Polyamorphic Transformations in Liquids and Glasses". *Journal of Materials Chemistry* **14** (10): 1506–1512. doi:10.1039/b401308p.

[33] Carbon dioxide glass created in the lab. *NewScientist.* 15 June 2006.

[34] P. N. Pusey and W. van Megen (1987). "Observation of a glass transition in suspensions of spherical colloidal particles". *Physical Review Letters* **59** (18): 2083–2086. Bibcode:1987PhRvL..59.2083P. doi:10.1103/PhysRevLett.59.2083. PMID 10035413.

[35] W. Van Megen and S. Underwood (1993). "Dynamic-light-scattering study of glasses of hard colloidal spheres". *Physical Review E* **47**: 248. Bibcode:1993PhRvE..47..248V. doi:10.1103/PhysRevE.47.248.

[36] H. Löwen (1996). A. K. Arora, B. V. R. Tata, ed. "Dynamics of charged colloidal suspensions across the freezing and glass transition" (PDF). *Ordering and Phase Transitions in Charged Colloids.* VCH Series of Textbooks on "Complex Fluids and Fluid Microstructures" (New York): 207–234.

[37] http://www.cell.com/abstract/S0092-8674%2813%2901479-7

[38] W.D. Kingery; H.K. Bowen, D.R. Uhlmann (1976). *Introduction to ceramics* (2 ed.). New York: Wiley. ISBN 978-0471478607.

[39] David W. Richerson (1992). *Modern ceramic engineering : properties, processing and use in design* (2 ed.). New York, NY: Dekker. ISBN 0-8247-8634-3.

[40] P. S. Salmon (2002). "Order within disorder". *Nature Materials* **1** (2): 87–8. doi:10.1038/nmat737. PMID 12618817.

[41] ASTM definition of glass from 1945; also: DIN 1259, Glas – Begriffe für Glasarten und Glasgruppen, September 1986

[42] Zallen, R. (1983). *The Physics of Amorphous Solids.* New York: John Wiley. ISBN 0-471-01968-2.

[43] Cusack, N. E. (1987). *The physics of structurally disordered matter: an introduction.* Adam Hilger in association with the University of Sussex press. ISBN 0-85274-829-9.

[44] Elliot, S. R. (1984). *Physics of Amorphous Materials.* Longman group ltd.

[45] Horst Scholze (1991). *Glass – Nature, Structure, and Properties.* Springer. ISBN 0-387-97396-6.

[46] J.C. Phillips (1979). "Topology of covalent non-crystalline solids I: Short-range order in chalcogenide alloys". *Journal of Non-Crystalline Solids* **34** (2): 153. Bibcode:1979JNCS...34..153P. doi:10.1016/0022-3093(79)90033-4.

[47] J. C. W. Folmer and Stefan Franzen (2003). "Study of polymer glasses by modulated differential scanning calorimetry in the undergraduate physical chemistry laboratory". *Journal of Chemical Education* **80** (7): 813. Bibcode:2003JChEd..80..813F. doi:10.1021/ed080p813.

[48] Philip Gibbs. "Is glass liquid or solid?". Retrieved 21 March 2007.

[49] Jim Loy. "Glass Is A Liquid?". Retrieved 21 March 2007.

[50] "Philip Gibbs" *Glass Worldwide,* (May/June 2007), pp. 14–18

[51] M. Vannoni, A. Sordini, G. Molesini (2011). "Relaxation time and viscosity of fused silica glass at room temperature". *Eur. Phys. J. E* **34**: 9–14. doi:10.1140/epje/i2011-11092-9.

[52] Florin Neumann. "Glass: Liquid or Solid – Science vs. an Urban Legend". Retrieved 8 April 2007.

[53] P. W. Anderson (1995). "Through the Glass Lightly". *Science* **267** (5204): 1615. doi:10.1126/science.267.5204.1615-e.

[54] Kenneth Chang (29 July 2008). "The Nature of Glass Remains Anything but Clear". *The New York Times.* Retrieved 29 July 2008.

[55] "Dr Karl's Homework: Glass Flows". Australia: ABC. 26 January 2000. Retrieved 24 October 2009.

[56] H. Halem. "Does Glass Flow". Retrieved 2 September 2010.

[57]Zanotto, Edgar Dutra (1998). "Do Cathedral Glasses Flow?".*American Journal of Physics***66**(5): 392–396.Bibcode:1998AmJ doi:10.1119/1.19026.

32.10   Further reading

- Carboni, Stefano & Whitehouse, David (2001). *Glass of the Sultans*. New York: The Metropolitan Museum of Art. ISBN 0-87099-986-9.

- Ghosh, Amalananda (1990). *An Encyclopaedia of Indian Archaeology*. BRILL. ISBN 90-04-09262-5.

- Gowlett, J. A. J. (1997). *High Definition Archaeology: Threads Through the Past*. Routledge. ISBN 0-415-18429-0.

- Noel C. Stokes; *The Glass and Glazing Handbook*; Standards Australia; SAA HB125–1998

- staff (2 November 2011). "Robot speeds up glass development" (press release). Fraunhofer Institute. Retrieved 10 December 2011. (reprinted by *R&D Magazine*)

- Stookey, S. Donald. *Explorations in Glass: An Autobiography*. Wiley, 2000. ISBN 978-1-57498-124-7

- Vogel, Werner. *Chemistry of Glass*. Wiley, 1985. ISBN 978-0-916094-73-7

32.11   External links

- The Story of Glass Making in Canada from The Canadian Museum of Civilization.

- *"How Your Glass Ware Is Made"* by George W. Waltz, February 1951, Popular Science.

- All About Glass from the Corning Museum of Glass: a collection of articles, multimedia, and virtual books all about glass, including the Glass Dictionary.

- Glass Encyclopedia from 20th Century Glass: a comprehensive guide to all types of antique and collectible glass, with information, pictures and references.

- National Glass Association the largest trade association representing the flat (architectural), auto glass, and window & door industries

Bohemian flashed and engraved ruby glass (19th-century)

Wine Goblet, mid-19th century. Qajar dynasty. Brooklyn Museum.

Roman Cage cup from the 4th century CE

Studio glass. Multiple colors within a single object increase the difficulty of production, as glasses of different colors have different chemical and physical properties when molten.

A CD-RW (CD). Chalcogenide glasses form the basis of rewritable CD and DVD solid-state memory technology.[23]

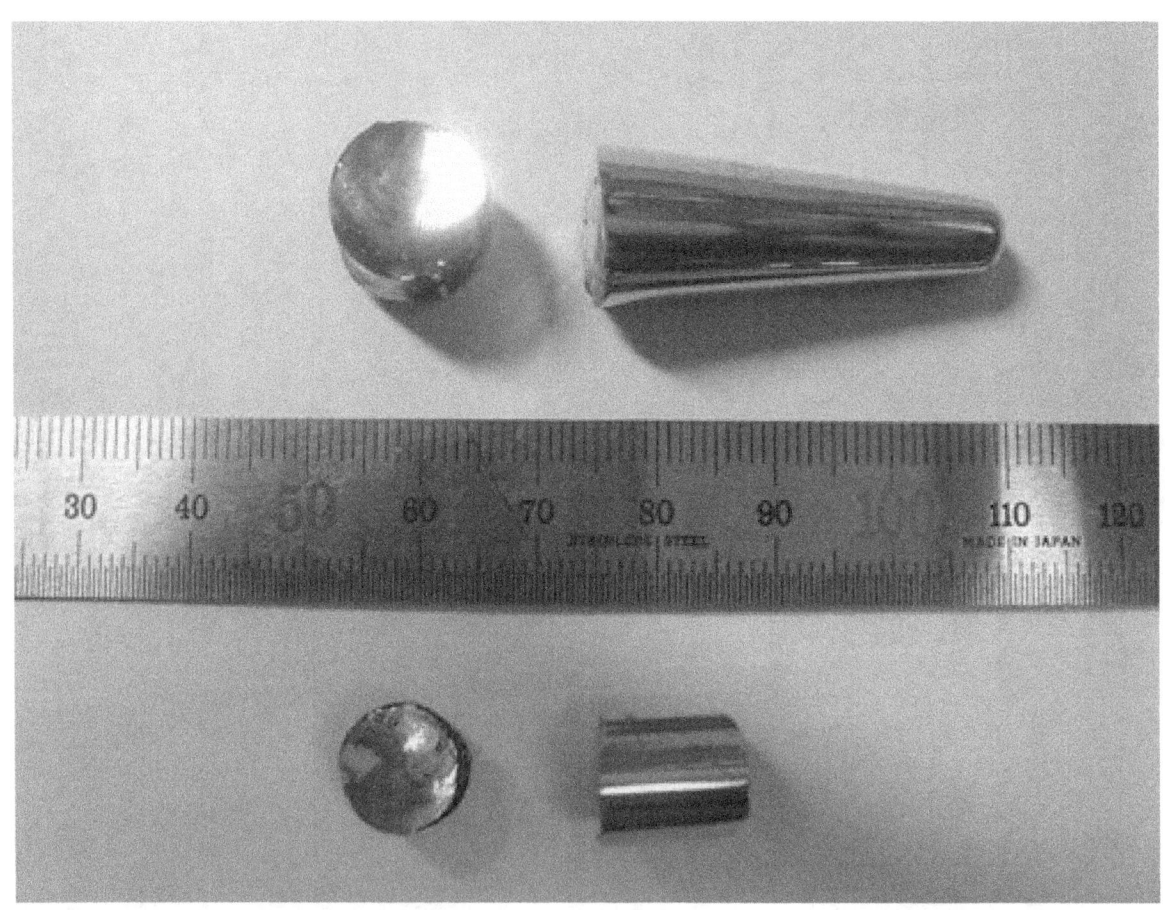

Samples of amorphous metal, with millimeter scale

A high-strength glass-ceramic cooktop with negligible thermal expansion.

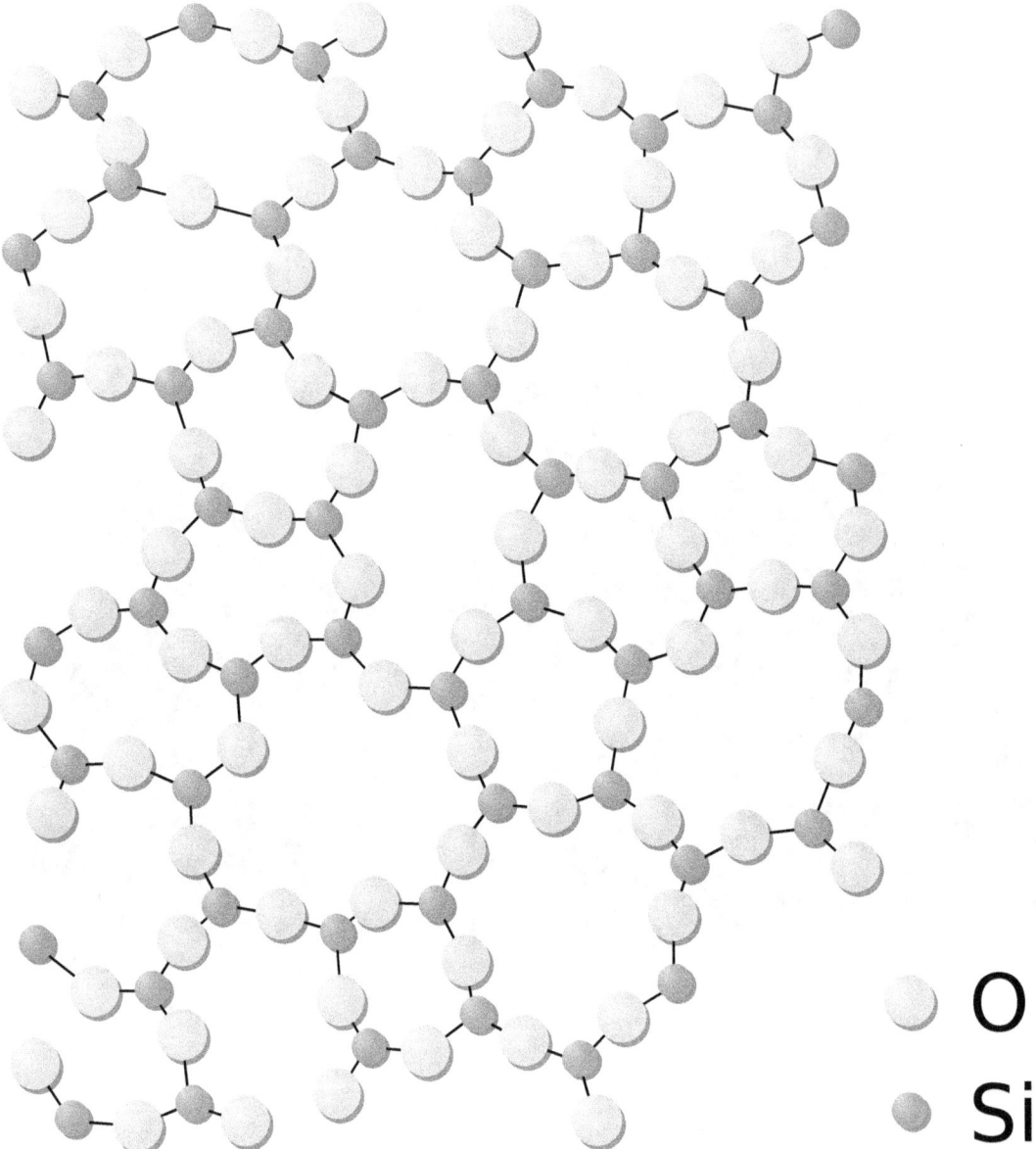

The amorphous structure of glassy silica (SiO$_2$) in two dimensions. No long-range order is present, although there is local ordering with respect to the tetrahedral arrangement of oxygen (O) atoms around the silicon (Si) atoms.

Chapter 33

Superglass

A **superglass** is a phase of matter which is characterized (at the same time) by superfluidity and a frozen amorphous structure.[1]

33.1 See also

- Superfluid

33.2 References

[1] Giulio Biroli; Claudio Chamon; Francesco Zamponi (2008). "Theory of the superglass phase". *Physical Review B* **78** (22): 19. arXiv:0807.2458. Bibcode:2008PhRvB..78v4306B. doi:10.1103/PhysRevB.78.224306.

33.3 External links

- Superglass could be new state of matter (subscription required)

- A new quantum glass phase: the superglass

- Phys. Rev. Lett. Vol.101, 8th Aug 2008

33.4 Text and image sources, contributors, and licenses

33.4.1 Text

- **State of matter** *Source:* https://en.wikipedia.org/wiki/State_of_matter?oldid=698119776 *Contributors:* CYD, Zardoz, Olof, Zundark, William Avery, Heron, Ubiquity, Ixfd64, Docu, Julesd, Seani, Reddi, Alexfiles, Robbot, Mayooranathan, Academic Challenger, Pepijn Schmitz, Rursus, Giftlite, Smjg, Art Carlson, Jackol, SoWhy, Rdsmith4, Oneiros, JimWae, DragonflySixtyseven, Glogger, ArthurDenture, Mike Rosoft, Discospinster, Rich Farmbrough, Vsmith, Walden, Rgdboer, Shanes, Smalljim, Zetawoof, Nsaa, Alansohn, Hydriotaphia, Wtmitchell, Mikeo, Vuo, Kazvorpal, Oleg Alexandrov, LOL, Pol098, Ruud Koot, MONGO, Yuriybrisk, Rjwilmsi, DoubleBlue, ElfQrin, Srleffler, Kri, The-Sun, Imnotminkus, King of Hearts, Moocha, DVdm, Mercury McKinnon, Wavelength, Phantomsteve, Jeffhoy, Arado, Madkayaker, Gaius Cornelius, CambridgeBayWeather, Pseudomonas, NawlinWiki, Dhollm, BOT-Superzerocool, Wknight94, 2over0, Closedmouth, Pb30, Willtron, Yonir, Moomoomoo, JDspeeder1, Cmglee, Knowledgeum, SmackBot, McGeddon, Edgar181, Skizzik, Quinsareth, Miquonranger03, Colonies Chris, Darth Panda, Addshore, RedHillian, Ghiraddje, Valenciano, Bombshell, Keramos, Kipala, A. Parrot, Beetstra, Cratylus3, Mark999, BranStark, Iridescent, Joseph Solis in Australia, Newone, Igoldste, RekishiEJ, Courcelles, Tltltetd, Dycedarg, MarsRover, Cydebot, Mato, Xxanthippe, Bazzargh, Christian75, DumbBOT, Mtpaley, Epbr123, Urdna, N5iln, Headbomb, Kathovo, James086, Yettie0711, PJtP, Escarbot, Mentifisto, Daniels220, Jayron32, Jj137, Gökhan, JAnDbot, D99figge, YK Times, Bongwarrior, VoABot II, JNW, Father Goose, Rich257, Animum, Dirac66, Philg88, Khalid Mahmood, FisherQueen, Anaxial, Sm8900, Jonathan Hall, R'n'B, CommonsDelinker, Wiki Raja, Mausy5043, J.delanoy, Pharaoh of the Wizards, MITBeaverRocks, Uncle Dick, BobEnyart, Victuallers, Myrin1, Gombang, Chemicalrubber, NewEnglandYankee, Kraftlos, Bonadea, Ja 62, CrazyRob926, Useight, Martial75, Funandtrvl, CWii, ABF, Flyingidiot, Jeff G., Indubitably, Barneca, Philip Trueman, TXiKiBoT, Monkey Bounce, Dendodge, Corvus cornix, Cremepuff222, RiverStyx23, Venny85, Madhero88, Blurpeace, Falcon8765, Pageman~enwiki, Brianga, Universaladdress, HPeugeot~enwiki, NHRHS2010, Cryonic07, SieBot, Alessgrimal, Nubiatech, Euryalus, Paradoctor, Winchelsea, Yintan, Happysailor, Flyer22 Reborn, Socal gal at heart, TrufflesTheLamb, Dillard421, Svick, Mygerardromance, Pinkadelica, Denisarona, Escape Orbit, Kanonkas, ShajiA, Troy 07, Ainlina, Faithlessthewonderboy, Elassint, Clue-Bot, Wookie501, Rumping, GorillaWarfare, Snigbrook, The Thing That Should Not Be, Ariadacapo, Tizeff, Polyamorph, GoEThe, Dylan620, Mad.martian999, Prognitor~enwiki, Quantumspinhall, Skihatboatbike, Djr32, Excirial, Alexbot, Coralmizu, Ykhwong, Pot, Singhalawap, Dekisugi, The Red, Oswald07, JasonAQuest, Mikhailov Kusserow, Thingg, Horselover Frost, PCHS-NJROTC, Burner0718, SoxBot III, Jmanigold, XLinkBot, Spitfire, PseudoOne, Leftspk, Andypandy2020, Saeed.Veradi, Libcub, Skarebo, SilvonenBot, Noctibus, JinJian, ZooFari, Thatguyflint, HexaChord, IBeAiMpErS0naT0r, Xvijayx, Addbot, Xp54321, Some jerk on the Internet, Jojhutton, PaterMcFly, Ronhjones, TutterMouse, Fieldday-sunday, Bertrc, Icelazer3, Shirtwaist, Vishnava, CanadianLinuxUser, NjardarBot, Download, PranksterTurtle, Glane23, Chzz, FCSundae, Kyle1278, CuteHappyBrute, Eh kia, Tide rolls, Gail, Micki, Cesaar, Angrysockhop, Legobot, Luckas-bot, Yobot, Dr. Footfoe, 2D, Andreasmperu, II MusLiM HyBRiD II, THEN WHO WAS PHONE?, Empireheart, IW.HG, Tempodivalse, Nintendo06, Quangbao, Hairhorn, Daniele Pugliesi, Jim1138, Lewismith3, Piano non troppo, Kingpin13, Dinesh smita, Westonpark, Materialscientist, Spirit469, ImperatorExercitus, 90 Auto, Citation bot, OllieFury, ArthurBot, Xqbot, TinucherianBot II, Beaserbebbeb, Pvkeller, Grim23, GrouchoBot, Charliekarst, Hamhat, Backpackadam, Amaury, Logger9, JediMaster362, Wasteman1066, Lagooncopperorange, FrescoBot, LucienBOT, Lothar von Richthofen, Hielor, Cannolis, Citation bot 1, Athenanoenvoy, Pinethicket, Boulaur, Overthinkingly, Tom.Reding, ContinueWithCaution, Johann137, Jauhienij, Gamewizard71, FoxBot, Yunshui, Zvn, January, Allen4names, Theo10011, Reaper Eternal, Jeffrd10, Tbhotch, Hornlitz, Marie Poise, DARTH SIDIOUS 2, NerdyScienceDude, Chuanenlin123, Jack Schlederer, Kenvancleve, WikitanvirBot, Ajraddatz, DillonLMcCabe, Mordgier, Katherine, Gcastellanos, The Sharminator, Tommy2010, Wikipelli, John Cline, Zane45177, Alpha Quadrant (alt), DidgeGuy, AndrewN, Yamumyadad, Jay-Sebastos, Jesanj, Brandmeister, Donner60, Ashunigam, Orange Suede Sofa, ChuispastonBot, Llightex, ShatteredSpiral, Isocliff, ClueBot NG, Gareth Griffith-Jones, MelbourneStar, UAK32, Rtucker913, Yourmomblah, Theimmaculatechemist, 123Hedgehog456, Whazie, Lfmm11, Mtheodric, Kumamah, ساجد امجد ساجد, Friend20gan, MerlIwBot, Helpful Pixie Bot, Candleabracadabra, Bibcode Bot, Mingmingla, Mark Fole, BZTMPS, Lowercase sigmabot, Yetisyny, Hallows AG, MusikAnimal, Frze, AvocatoBot, Planetary Chaos Redux, Tatchell, Nsda, ChE Fundamentalist, Ugncreative Usergname, Joydeep, Averysamantha, Snow Blizzard, ImhotepBallZ, Mantike, Kydon Shadow, Th4n3r, The Illusive Man, ChrisGualtieri, EuroCarGT, Sidsandyy, Lugia2453, CaSJer, Frosty, SFK2, Jamesx12345, Perfecttwoegan, Reatlas, Passengerpigeon, Epicgenius, I am One of Many, Lsmll, Surfer43, Tentinator, Cebr1979, Amdz96, Backendgaming, Lince celeste, DavidLeighEllis, Ginsuloft, WikiMannWikiMann, Annieminnie21, Rickdutta, Meteor sandwich yum, Muneeb abdelhadi, ShahryarAhmad27900, Nijuthomasgeorge, Bilorv, Saaaaaaaaaa, Ookami-to-Koneko, J73364, S r u fejvd u ub creating u of, Amortias, Natdeso, Adenine2k, Lalalalalala is too great, ElmoLover88545, Darkmatter1435, Chetan082, Alango1998, XxX$pace5Xxx, Cambles13, Johno2000, Jsbdjejnwn, Weegeerunner, Slayeredwarrior, Account50, Blissy2005, Minnie431, KasparBot and Anonymous: 823

- **List of states of matter** *Source:* https://en.wikipedia.org/wiki/List_of_states_of_matter?oldid=697307735 *Contributors:* CYD, Olof, Ike9898, Korath, Lowellian, Gkochanowsky, Cyrius, BrendanRyan, Jason Quinn, Sigfpe, Gene s, Sam Hocevar, Deglr6328, Randwicked, FT2, Max Terry, Bobo192, Zyqqh, Knowledge Seeker, RJFJR, Kazvorpal, Umofomia, Ems57fcva, Watcharakorn, Pathoschild, Consumed Crustacean, RussBot, Tryforceful, Ocicatmuseum, Dhollm, Ethan lowter, StuRat, Serendipodous, Shadowin, SmackBot, Jrockley, Yamaguchi⊡⊡, Gilliam, Colonies Chris, Moreovaltine, Pirkid, EdC~enwiki, Laplace's Demon, Kwg06516, Mtpaley, Headbomb, Mentifisto, WinBot, Cgingold, Dirac66, Quanticle, Chiswick Chap, Largoplazo, Funandtrvl, Hy Brasil, PlanetStar, Paradoctor, Wing gundam, Tones22, Erebus Morgaine, Doprendek, Nekiko, Addbot, WikDefender, Wenlongtian, Cowgoesmoo2, Xqbot, I dream of horses, Hessamnia, Merehap, Wikipelli, Hhhippo, ClueBot NG, Widr, Wormulon, OsCeZrCd, FracturedRetina, Kind Tennis Fan, Dough34, Captaindoodletime, Mousenight, TheRaknok, Bhavya velani, RCDGamer, Nigerfuk and Anonymous: 96

- **Solid** *Source:* https://en.wikipedia.org/wiki/Solid?oldid=693643869 *Contributors:* Mav, Olof, Css, Heron, FlorianMarquardt, Luckymama58, Lir, Tim Starling, SebastianHelm, Ahoerstemeier, Suisui, Andrewa, Cgs, Glenn, Andres, Smack, Charles Matthews, Tantalate, Dysprosia, Taxman, Shizhao, Jusjih, Digital-h~enwiki, Donarreiskoffer, Gentgeen, Mayooranathan, Tosha, Giftlite, Bensaccount, Yekrats, Youngoat, Joeblakesley, Icairns, Discospinster, Vsmith, MisterSheik, El C, Joanjoc~enwiki, Bobo192, Smalljim, Eddideigel, Jumbuck, Zachlipton, Danski14, Alansohn, Keenan Pepper, Lectonar, Walkerma, Gene Nygaard, Woohookitty, Mindmatrix, WadeSimMiser, The Wordsmith, Kmg90, Graham87, BD2412, FreplySpang, Rjwilmsi, Vary, Bruce1ee, ElKevbo, FlaBot, Mathbot, Nihiltres, Lmatt, TheSun, Chobot, Sharkface217, DVdm, Stoive, Gwernol, Roboto de Ajvol, YurikBot, Wavelength, RussBot, Salsb, NawlinWiki, Grafen, Anetode, Zwobot, Kooky, Bota47, Wknight94, 2over0, StuRat, Dspradau, Willtron, RunOrDie, Ybbor, Bdve, SmackBot, FocalPoint, C.Fred, Bomac, Jrockley, Delldot, Vincent de Ruijter, Edgar181, Yamaguchi⊡⊡, Gilliam, Chaojoker, Andy M. Wang, Chris the speller, Bluebot, MalafayaBot, SchfiftyThree, Akanemoto,

PureRED, DHN-bot~enwiki, Gracenotes, Jahiegel, PeteShanosky, DMacks, SashatoBot, ArglebargleIV, Kuru, John, Kipala, Gobonobo, Yms, Optakeover, Jose77, Levineps, G1076, NativeForeigner, Amalas, Dycedarg, Nunquam Dormio, Black and White, MarsRover, WeggeBot, Neelix, Equendil, HawkShark, Fl, Xxanthippe, ST47, Pinky sl, Epbr123, MarkBuckles, Headbomb, A3RO, Rhrad, Huigh4444, Mentifisto, AntiVandalBot, Seaphoto, Fatkat61, JAnDbot, Leuko, Wizardboy777, VoABot II, Esmhead, Animum, Cgingold, Dirac66, User A1, Shijualex, JaGa, MartinBot, Anaxial, R'n'B, CommonsDelinker, Tgeairn, J.delanoy, Trusilver, NightFalcon90909, Dbiel, AgainErick, Afluegel, Belovedfreak, Darrendeng, Kaleau, Alan012, Fatkid72, Squids and Chips, CardinalDan, RAult, Youbejoking, VolkovBot, Lear's Fool, Ryan032, Philip Trueman, TXiKiBoT, Rei-bot, Warlord dehacker, Monkey Bounce, Dilos100, JhsBot, Brandonrush, Synthebot, Falcon8765, AlleborgoBot, Sparkylad, Sfztang, SieBot, Nubiatech, Caltas, Yintan, Toddst1, Lightmouse, Fratrep, Mojoworker, ClueBot, Venske, Snigbrook, The Thing That Should Not Be, Newsjunker, Polyamorph, Blanchardb, Mvp08, Djr32, Excirial, Jusdafax, Anon lynx, PixelBot, California847, Cenarium, SchreiberBike, Unmerklich, Vanished user uih38riiw4hjlsd, Vanished user 1004, DumZiBoT, Steven McTowelie the second, BodhisattvaBot, Rror, Addbot, Narayansg, Eric Drexler, Fleetsbridge, Nath1991, X VeNTe, Ballervision, Ilovelakai, EconoPhysicist, AtheWeatherman, Numbo3-bot, Tide rolls, Lrrasd, Quantumobserver, Luckas-bot, Zaereth, Thatgains, A Stop at Willoughby, Kingpin13, Bonusballs, Materialscientist, Spirit469, The High Fin Sperm Whale, Citation bot, OllieFury, Xqbot, Sketchmoose, Capricorn42, Blindgrapefruit2, J04n, GrouchoBot, Amaury, Logger9, Shadowjams, Methcub, FrescoBot, LucienBOT, Fahed ahmed, Fast kartwheels, Tetraedycal, Citation bot 1, Krish Dulal, Pinethicket, I dream of horses, Boulaur, Tom.Reding, BigDwiki, Gruntler, Attack 2400, Jauhienij, ActivExpression, Gamewizard71, FoxBot, TobeBot, Yunshui, Lotje, Javierito92, Vrenator, Theo10011, Ghjthgh, Marie Poise, DARTH SIDIOUS 2, Ripchip Bot, VernoWhitney, Solidboy123, DASHBot, EmausBot, WikitanvirBot, Immunize, Super48paul, Hplovecraftpwns, Dewritech, RenamedUser01302013, Solarra, Tommy2010, Wikipelli, K6ka, Lou1986, Xman0444, Hhhippo, JSquish, Gr8xoz, Tolly4bolly, Asiaglass, Ventilate, Donner60, Jan 1922, Lovetinkle, ClueBot NG, Bidpald253, MelbourneStar, This lousy T-shirt, Baseball Watcher, O.Koslowski, Widr, Theopolisme, Lugia77, Pbuffat, Helpful Pixie Bot, Bibcode Bot, Vagobot, MusikAnimal, Writ Keeper, Djhfsdhf, Klilidiplomus, BattyBot, WhiteNebula, Ducknish, Frosty, AgaRed, Reatlas, Faizan, Hareesa, Harlem Baker Hughes, JamesMoose, PhantomTech, EvergreenFir, TCMemoire, Aravinduser, Hisashiyarouin, Ukulelemollyjade, Cyberalchemyst, Poopoohead54321, Poopyhead321, Lofle, JHeid2014, JP216, PotatoNinja, Aryaman SInghi, EduacationDino, Stixky, GeneralizationsAreBad, KasparBot, Ethan's biggest fan, Chris1238, Kuantingl, KSFT, Rainkiss666 and Anonymous: 374

- **Amorphous solid** *Source:* https://en.wikipedia.org/wiki/Amorphous_solid?oldid=686078349 *Contributors:* AxelBoldt, Olof, MarXidad, Andre Engels, Ben-Zin~enwiki, Stevertigo, Ellywa, Glenn, Andres, Smack, Ike9898, Robbot, Jondel, Acegikmo1, Wikibot, Tgiles, Giftlite, Epton, Hagedis, Orsanyuksek, AHM, JHCC, Deglr6328, Vsmith, Kjoonlee, Edwinstearns, Pinzo, ~K, Maurreen, Jag123, Hooperbloob, DV8 2XL, Gene Nygaard, Redvers, Ceyockey, Joriki, Woohookitty, Benbest, Polyparadigm, SDC, Evilmoo, Graham87, Rjwilmsi, Syndicate, SMC, FlaBot, LAk loho, Srleffler, Chobot, YurikBot, TexasAndroid, Jamesmorrison, BOT-Superzerocool, Jhchang, Eyal0, Schizobullet, TravisTX, FocalPoint, Kmarinas86, DHN-bot~enwiki, Para, Stevenmitchell, Smokefoot, Clicketyclack, John, JorisvS, Zarniwoot, EXeption, Dr Smith, Courcelles, Stifynsemons, Porterjoh, Neelix, Reywas92, ST47, Odie5533, VPliousnine, Thijs!bot, Vertium, Mumby, Aadal, AntiVandalBot, Seaphoto, Leftynm, The Transhumanist, MSBOT, Acroterion, Magioladitis, VoABot II, Soulbot, EagleFan, User A1, Lawrie.skinner, MartinBot, R'n'B, J.delanoy, Carre, Whirling Sands, Afluegel, EdC2, Ojovan, VolkovBot, Skaraoke, TXiKiBoT, Alex rosenberg35, Someguy1221, Flopster2, MackSalmon, Don4of4, LeaveSleaves, Venny85, Enigmaman, MCTales, Serprex, SieBot, Nubiatech, BotMultichill, Matthew Yeager, Keilana, Qst, Faradayplank, WingkeeLEE, Soyseñorsnibbles, ClueBot, Uncle Milty, Polyamorph, Zulu Inuoe, StigBot, Puppy8800, California847, DhananSekhar, HowStellar, DanielPharos, Thingg, BodhisattvaBot, Skarebo, WikiDao, Addbot, Tanhabot, Glane23, SpBot, SPat, WikiDreamer Bot, Peko, Luckas-bot, Yobot, Ptbotgourou, AnomieBOT, ArthurBot, Xqbot, Capricorn42, 661kts, Anna Frodesiak, SassoBot, Logger9, Mnmngb, FrescoBot, Citation bot 1, MBirkholz, BigDwiki, 124Nick, Giles247, TheGrimReaper NS, Uchihatobisaga, Marie Poise, Andrea105, RjwilmsiBot, Knuffels, Slon02, MrDenberg, John of Reading, Manojhande 2196, Optiguy54, Wikipelli, John Cline, Josve05a, Traxs7, Fizicist, Erianna, Just granpa, Tekstovi, Tieo8, ClueBot NG, Ontist, Very trivial, Curb Chain, Calabe1992, Bibcode Bot, BG19bot, MusikAnimal, Zedshort, Shephd, Egdevoie, Bharu12, SreejanAlapati, دوعسمیمرا, Anrnusna, Martinmagnuson, KasparBot and Anonymous: 198

- **Crystal** *Source:* https://en.wikipedia.org/wiki/Crystal?oldid=690919068 *Contributors:* Vicki Rosenzweig, Mav, Olof, Mirwin, Eclecticology, William Avery, Ben-Zin~enwiki, DrBob, Youandme, Olivier, Stevertigo, Michael Hardy, Norm, Minesweeper, Dgrant, Ahoerstemeier, TUF-KAT, Suisui, John Stewart, Smack, Tantalate, The Anomebot, Selket, DJ Clayworth, Tpbradbury, ElusiveByte, SEWilco, Omegatron, Samsara, Elwoz, David.Monniaux, PuzzletChung, Gentgeen, Robbot, Chris 73, Altenmann, Romanm, Lowellian, MaterialsScientist, Cdnc, Puckly, Andrew Levine, Hadal, Xanzzibar, Mattflaschen, Cutler, Centrx, Giftlite, Tom harrison, Ds13, Everyking, Bkonrad, No Guru, Foobar, Jurema Oliveira, Kandar, Wmahan, Espetkov, Bauke~enwiki, Chowbok, AHM, Antandrus, Beland, JoJan, Karol Langner, Rdsmith4, Icairns, Sam Hocevar, Gscshoyru, Deglr6328, Mike Rosoft, Jiy, Discospinster, Vsmith, Mani1, Edgarde, El C, RoyBoy, Aaronbrick, Bobo192, Polluks, Giraffedata, Sasquatch, 99of9, MPerel, Hagerman, Jumbuck, Danski14, Siim, Alansohn, 119, Arthena, Walkerma, Bantman, Hohum, Snowolf, Dschwen, Melaen, Velella, RainbowOfLight, Angr, Daniel Case, Knuckles, GregorB, Isnow, Laurap414, Rachack, Dysepsion, Emerson7, Graham87, Magister Mathematicae, V8rik, Saperaud~enwiki, Sjakkalle, Angusmclellan, Vary, SMC, Vegaswikian, Boccobrock, Yamamoto Ichiro, Vuong Ngan Ha, FlaBot, Major.T, Gurch, Wars, Kri, Physchim62, Chobot, DVdm, Korg, Roboto de Ajvol, Sceptre, Huw Powell, KSmrq, Hellbus, Stephenb, CambridgeBayWeather, NawlinWiki, Grafen, Jaxl, RazorICE, Irishguy, DeadEyeArrow, Botteville, 2over0, Zzuuzz, Musashi miyamoto, Closedmouth, Hurricanehink, SorryGuy, Archer7, Bluezy, Katieh5584, Kungfuadam, Junglecat, Carlosguitar, Sbyrnes321, DVD R W, CIreland, Veinor, SmackBot, UbUb, Dweller, Zsendukas, KnowledgeOfSelf, Hydrogen Iodide, Bomac, Finavon, Weiguxp, Swerdnaneb, Edgar181, TantalumTelluride, Gilliam, Miquonranger03, Complexica, Baa, Can't sleep, clown will eat me, Ajaxkroon, Lazyboi69, Yidisheryid, Dreadstar, RyGuy17, DMacks, Wizardman, Acdx, Audioiv, Springnuts, Kukini, Clicketyclack, The undertow, Mouse Nightshirt, Attys, T g7, Gobonobo, Soumyasch, Mr. Lefty, Thegreatdr, JHunterJ, Stwalkerster, George The Dragon, Optakeover, RememberMe?, Jose77, DCM~enwiki, Sonic3KMaster, Wizard191, Bonez22, R~enwiki, AGK, Az1568, Courcelles, Tawkerbot2, Pontificake, Ouishoebean, Kevin Murray, Emote, Dolphinholmer, JohnCD, Dgw, El aprendelenguas, Neelix, Kingykongy, Yaris678, Badseed, Doctormatt, Cydebot, Island Dave, Krauss, Abeg92, Vanished user vjhsduheuiui4t5hjri, Rifleman 82, Gogo Dodo, A Softer Answer, Christian75, TheCheeseManCan, Iempleh, FastLizard4, Digon3, PreRaphaelite, Ael 2, Rocket000, Thijs!bot, Epbr123, Ut Libet, Willworkforicecream, Mojo Hand, Headbomb, Куллер, John254, Tellyaddict, Grayshi, Sean William, Escarbot, Mentifisto, AntiVandalBot, Majorly, KP Botany, Dylan Lake, MECU, Lfstevens, JAnDbot, Husond, Davewho2, Gcm, MER-C, Janejellyroll, Hut 8.5, Savant13, Beaumont, Snowynight, Acroterion, Bencherlite, FaerieInGrey, Bongwarrior, VoABot II, Kaivosukeltaja, Recurring dreams, Latifshaikh20, Animum, Upholder, Daarznieks, Allstarecho, Sahil shark, Glen, DerHexer, BlackHak, B9 hummingbird hovering, MartinBot, ReikiEssentials, Arjun01, Poeloq, Rettetast,

Anaxial, Jay Litman, Tbone55, R'n'B, PrestonH, LedgendGamer, Tgeairn, Transisto, J.delanoy, Pharaoh of the Wizards, Edgeweyes, Hans Dunkelberg, Superdvd, Maurice Carbonaro, Kemiv, GeoWriter, Gzkn, Shawn in Montreal, McSly, Ultatri, Pyrospirit, AntiSpamBot, Rian-Koo, NewEnglandYankee, Knulclunk, Cherryalexandra, Doctorshim, Cmichael, Cometstyles, CardinalDan, 13creek13, Lights, VolkovBot, TreasuryTag, Meaningful Username, Macedonian, Jeff G., Indubitably, VasilievVV, Philip Trueman, TXiKiBoT, Zamphuor, MeStevo, Vipinhari, GDonato, Ann Stouter, Z.E.R.O., Ask123, Aymatth2, Devshoppe, Bmg916, Anna Lincoln, Una Smith, Jackfork, LeaveSleaves, Bob f it, Thebof, Maxim, Jmath666, Wenli, Dirkbb, Kiia15, Enviroboy, Insanity Incarnate, Legoktm, NHRHS2010, Sir Arver, Govany, Tpb, SieBot, Ttony21, Tiddly Tom, Euryalus, YKWSG, Krawi, Mbz1, Caltas, Jleela14, Tiptoety, Radon210, Oda Mari, Wilson44691, Rhanyeia, Taemyr, Crystall94, Oxymoron83, Antonio Lopez, Faradayplank, Vanzpunk0676, Lisatwo, Sarvil, Techman224, FoRgEtFuLxLuV94, Tardigrade13, Gunmetal Angel, Bob Doubles, N96, Bronzephoenix, WikiLaurent, Butane Goddess, ClueBot, Crysandlg, Ve4ernik, The Thing That Should Not Be, Matdrodes, Biggerj1, Wysprgr2005, Mr.pieface, Mild Bill Hiccup, Dengero, Trivialist, Ridge Runner, Namazu-tron, DragonBot, Awickert, Tonmad, Excirial, Jusdafax, Vanisheduser12345, Gtstricky, Lartoven, Kocoumlovesyou, MickMacNee, Jotterbot, Tagsontiago, DeltaQuad, Dancinstar, Plangent, Thehelpfulone, Aitias, Banano03, Versus22, Vanished user uih38riiw4hjlsd, Gowtham99.9, Rror, Avoided, Sergay, Perdygal90210, Tinkerbellkait, Lemchesvej, Freestyle-69, Wyatt915, Secrets.of.the.skies, Some jerk on the Internet, Johutton, Fyrael, Hempelmann, Captain-tucker, Miscibleliquids, Ronhjones, PandaSaver, TutterMouse, Fieldday-sunday, Rapidfyre, Leszek Jańczuk, Wisterlane, MrOllie, Sem-mem, CarsracBot, LAAFan, Chzz, Debresser, Favonian, 84user, Numbo3-bot, Tide rolls, BrianKnez, Luckas Blade, Teles, Білецький В.С., Tekkenmasterbrendan, Ochib, Alfie66, Reciproc08, Legobot, Luckas-bot, ZX81, Yobot, Fraggle81, TaBOT-zerem, Backslash Forwardslash, AnomieBOT, Michaeldu, 1exec1, Galoubet, Piano non troppo, AdjustShift, PuppyGirl1508, Crystal whacker, Materialscientist, Citation bot, Wafflescakes, ArthurBot, TinucherianBot II, Intelati, Pelo12354, Capricorn42, Mrba70, TheWeakWilled, Duckone, Jmundo, Jamm.T, Srich32977, -), Cnguyen123, J04n, Riotrocket8676, Call me Bubba, Logger9, Gradbrad, Raulshc, Captain-n00dle, Fortdj33, J6w5, Recognizance, Archaeodontosaurus, Alxeedo, VI, MGA73bot, Strongbadmanofme, Jon K4, HamburgerRadio, Citation bot 1, Stephane29, Bballerc28, Caitlincookegroup565, Pinethicket, I dream of horses, Calmer Waters, Outsider1994, Hoo man, BRUTE, Tinkerperson, Jauhienij, Tracyk123, Abc518, FoxBot, Ambarsande, TobeBot, Darigan, Lotje, Vrenator, January, Reaper Eternal, Canuckian89, Fastilysock, SonichuCWC, Reach Out to the Truth, Chlstal, Hornlitz, Marie Poise, Jonere, Thrind, Slon02, Cpj0788, CanadianPenguin, EmausBot, Kiernanscott, Immunize, GoingBatty, RA0808, NotAnonymous0, Tommy2010, Wikipelli, K6ka, Chiton magnificus, Cms13foreverandalways, PBS-AWB, Bongoramsey, Traxs7, Sublimehypocrisy, E557, Zap Rowsdower, Wayne Slam, Godwine1994, L Kensington, Flightx52, Kamakat, ChuispastonBot, RockMagnetist, Maxdlink, Rocksrocksrockagh, Socialservice, ResearchRave, Petrb, ClueBot NG, Dana Ashkenazi, NiallMcQueen, Chrishastoes, This lousy T-shirt, Satellizer, Virusbetax, Bcswb18, Egg Centric, Mesoderm, Asukite, Widr, Mellonhead97, 33photo, Bibcode Bot, Frogilicious, Kinaro, GeodeFinder, Davidiad, Around the world in 8 days, Mr.Kippaweda, Kimelea, HimeThy, Anbu121, Consejero89, Hansen Sebastian, Suresh Baddika, ChrisGualtieri, Besterthenyou, Khazar2, Mogism, Trollzor31, Lugia2453, Pizzalad3, Sly elf, Kevin12xd, Reatlas, Hunterk1219, Rfassbind, Everymorning, Dolly66, Rena785, DiddoMC, Juliasmexicangrill, Kramerrd, Winter355, Trackteur, Nasrullah512, Duderightnow, Lovekimchii, KasparBot and Anonymous: 878

- **Plastic crystal***Source:*https://en.wikipedia.org/wiki/Plastic_crystal?oldid=689701400*Contributors:*Frau Holle, Vegaswikian, Cydebot, Thijs!bot,VoABot II, Jcwf, Plastiskpork, Doprendek, Addbot, Yobot, PowerUserPCDude, LilHelpa, Xqbot, Tom.Reding, Marie Poise, Feeting2000 and*Anonymous: 6

- **Quasicrystal** *Source:* https://en.wikipedia.org/wiki/Quasicrystal?oldid=696624426 *Contributors:* LC~enwiki, Bryan Derksen, Olof, Stevertigo, Michael Hardy, Earth, Ahoerstemeier, Stevenj, Jpatokal, Scott, Smack, Hollgor, Tantalate, Dino, Sboehringer, Populus, AnonMoos, Romanm, Babbage, MaterialsScientist, Giftlite, Brad Bridgewater, Snowdog, Dratman, DragonflySixtyseven, Icairns, Udzu, CALR, Larrybob, D-Notice, Pavel Vozenilek, Bender235, Cyclopia, El C, Rgdboer, Diamonddavej, Enric Naval, Shenme, Timl, Crust, Lysdexia, Keenan Pepper, H2g2bob, Gene Nygaard, Ringbang, Bkkbrad, Mpatel, GregorB, Isnow, Yurik, Luh-e, Rjwilmsi, X1011, Aykroyd, YurikBot, Wavelength, Darkstar949, Hellbus, Bob0the0mighty, Bota47, Wknight94, Tetracube, 2over0, Reyk, ChristopherGautier, A13ean, SmackBot, Incnis Mrsi, Htg, Eb oesch, Jagged 85, Voulouza, Srnec, Hmains, Mirokado, Bluebot, Hibernian, George Church, LouScheffer, Bourdillona, Robofish, Zarniwoot, Uuhuuц, JMK, JRSpriggs, Thermochap, CmdrObot, Cydebot, Ntsimp, Krauss, ALittleSlow, Ael 2, Cuthbertwong, Thijs!bot, Headbomb, James Slezak, Frettloe, Coyets, Atart, Wasell, Anbaric, Father Goose, Soulbot, Sstolper, David Eppstein, Seba5618, Yannledu, Hasanisawi, BJ Axel, Hans Dunkelberg, Cdamama, Acalamari, Davecrosby uk, DrMicro, JohnBlackburne, Paul bryner, Abtinb, Quaestor23, Gilisa, Lamro, Muhassan, Atifmsiddiqui, Arcfrk, Coffee, Hertz1888, Lightmouse, RSStockdale, Videmus Omnia, Gulmammad, Qwfp, XLinkBot, Ost316, WikHead, Addbot, DOI bot, Download, Tide rolls, Zorrobot, Xenobot, Luckas-bot, Yobot, Wz33333, Amirobot, Jgmoxness, Wikihad, Cstras, AnomieBOT, Materialscientist, CoMePrAdZ, Citation bot, Mkosterv, Atumtem, FrescoBot, Sjcjoosten, Orubt, Pokyrek, Doerflermeister, Citation bot 1, RedBot, Ravn-hawk, Levochik, FoxBot, Trappist the monk, Korepin, Marie Poise, RjwilmsiBot, Skangerland, Laszlovszky András, Sreifa, Hhhippo, ZéroBot, QuentinUK, Namures, Attilabedo, Wingman4l7, Quasicrystal007, Chuispaston-Bot, ClueBot NG, Murr89, Galilsnap, Israelscitech, Peterkramer, Navasj, Helpful Pixie Bot, JohnOFL, Bibcode Bot, Ceradon, Indah blestari, Emerine, M Tuschinski, SurfChemista, ILANBLECH, Freesodas, Jimw338, Khazar2, Cjsh716, Dexbot, Deltahedron, Modeltookmodeltook, Julyancartwright, Colt browning, Anrnusna, Monkbot, Platanium, KasparBot and Anonymous: 114

- **Liquid** *Source:* https://en.wikipedia.org/wiki/Liquid?oldid=696783863 *Contributors:* Kpjas, Brion VIBBER, Bryan Derksen, Olof, Koyaanis Qatsi, William Avery, Peterlin~enwiki, Imran, Heron, Karl Palmen, Luckymama58, Olivier, Lir, Michael Hardy, Ixfd64, SebastianHelm, Ahoerstemeier, Александър, Poor Yorick, Tantalate, Jusjih, PuzzletChung, Fito, Donarreiskoffer, Gentgeen, Robbot, Swestrup, Altenmann, Lowellian, Merovingian, Blainster, Robinh, Jeremiah, Giftlite, Cfp, J heisenberg, BenFrantzDale, Ævar Arnfjörð Bjarmason, Moyogo, Bensaccount, Yekrats, Sonjaaa, Antandrus, Beland, ClockworkLunch, Vc-wp, H Padleckas, Icairns, Rgrg, Fg2, Clemwang, Frau Holle, Zondor, Mike Rosoft, Moverton, Discospinster, Cacycle, Vsmith, Kbh3rd, El C, Bobo192, Robotje, Smalljim, Mareino, Ranveig, Jumbuck, Danski14, Siim, Alansohn, Gary, Penwhale, Walkerma, Malo, KingTT, Garzo, Shoefly, Vuo, LFaraone, Agguarx, Revived, Woohookitty, Mindmatrix, Georgia guy, TigerShark, MGTom, Esben~enwiki, The Nameless, Graham87, FreplySpang, Sjakkalle, Rjwilmsi, Nneonneo, Tomtheman5, Yamamoto Ichiro, FlaBot, Mishuletz, Gurch, Tedder, Zotel, Physchim62, Chobot, DVdm, YurikBot, Wavelength, RobotE, Arado, Yyy, NawlinWiki, Brian Crawford, Moe Epsilon, Bucketsofg, Kooky, BOT-Superzerocool, Everyguy, Mike92591, Wknight94, Tetracube, Tigershrike, Closedmouth, Kevin, ArielGold, Junglecat, Hl540511, SmackBot, Chris the speller, MalafayaBot, Deli nk, Akanemoto, Zachorious, Dethme0w, Jmccabe871, Divna Jaksic, FiveRings, Paul Slocum, DRLB, B jonas, -Ozone-, Ryan Roos, DaiTengu, Acdx, Starghost, Bidabadi~enwiki, Bejnar, Chaldean, Nathanael Bar-Aur L., John, Kipala, Mbeychok, A. Parrot, JHunterJ, Jose77, Ginkgo100, CapitalR, Courcelles, Van helsing, Woudloper, MarsRover, Karenjc, ManiacalMonkey, Slazenger, Acornembryo, Gogo Dodo, Albert0, Xxanthippe, Difluoroethene, Dancter, Christian75, FastLizard4, JoshHolloway,

Omicronpersei8, Thijs!bot, Epbr123, Marek69, Nick Number, Dawnseeker2000, Mentifisto, AntiVandalBot, BokicaK, SummerPhD, MECU, Gökhan, MikeLynch, Bsmithurst, JAnDbot, Bhamv, Xeno, Hut 8.5, Steveprutz, Acroterion, Bongwarrior, VoABot II, AuburnPilot, Redaktor, Animum, Cgingold, Allstarecho, Adacus12, DerHexer, Khalid Mahmood, MartinBot, Mufo, Anaxial, CommonsDelinker, AlexiusHoratius, Tgeairn, J.delanoy, Ginsengbomb, Mintz l, Hodja Nasreddin, Rvaznyvfgxrvazny, Afluegel, Gurchzilla, Tanaats, Dhaluza, KylieTastic, Comet-styles, Tiggerjay, Useight, Omc, Dan Hickman, Black Kite, Deor, VolkovBot, CWii, Ryan032, Philip Trueman, DoorsAjar, TXiKiBoT, Yupi666, Fcb981, Sean D Martin, Psyche825, Cremepuff222, Wiae, Meters, Enviroboy, Dogah, SieBot, Mikemoral, Tresiden, Euryalus, Malcolmxl5, Laoris, Winchelsea, Da Joe, Yintan, Yankszack, Toddst1, JohnaG-A, Oxymoron83, Son111, Marquetry28, LidiaFourdraine, Kutera Genesis, Yaluen, Lehasa, SallyForth123, Loren.wilton, Sfan00 IMG, ClueBot, LAX, The Thing That Should Not Be, Remag Kee, Aria1561, DanielDeibler, Polyamorph, Boing! said Zebedee, Djr32, Excirial, Jusdafax, Cenarium, Jotterbot, Iohannes Animosus, Rash, Sdr-tirs, Aitias, Egmontaz, Crowsnest, Chhe, Tacos r friends, Matthieumarechal, Avoided, SilvonenBot, WikiDao, ZooFari, Wikiman405, Ad-dbot, Brumski, Eric Drexler, Manuel Trujillo Berges, Some jerk on the Internet, Yoenit, Fgnievinski, Richmond96, Ilovelakai, قندرمس, Debresser, Favonian, Exor674, Jasper Deng, Numbo3-bot, Erutuon, Tide rolls, MuZemike, Mozillaman425, Luckas-bot, Yobot, Zaereth, Fraggle81, Sarrus, Johnsatterfield, WizardOfOz, DemocraticLuntz, Jim1138, Memphis670, Kingpin13, Bluerasberry, Jrobinjapan, Materi-alscientist, Spirit469, The High Fin Sperm Whale, Citation bot, MauritsBot, Xqbot, JimVC3, Ryomaandres, GrouchoBot, Nayvik, Logger9, Shadowjams, WaysToEscape, LucienBOT, A little insignificant, Wireless Keyboard, Citation bot 1, Krish Dulal, Pinethicket, I dream of horses, Boulaur, Drerhymez, Hard Sin, Tom.Reding, Calmer Waters, A8UDI, Σ, Abaoabao, Robo Cop, Zorroeatsmaypo, Jauhienij, Gamewizard71, FoxBot, Vrenator, Marie Poise, DARTH SIDIOUS 2, The Utahraptor, Stj6, NerdyScienceDude, 4students, EmausBot, John of Reading, Wik-itanvirBot, Qurq, Scholar333, Racerx11, Faolin42, RA0808, RenamedUser01302013, Foodeatingperson, K6ka, Savh, Sepguilherme, JSquish, HugeGrayLover, Jman12369874, Hal3y+st3phh, Wayne Slam, Tolly4bolly, Ventilate, TyA, L Kensington, Donner60, EvenGreenerFish, Den-nisIsMe, Fowen123, Peter Karlsen, Targaryen, DASHBotAV, Rememberway, ClueBot NG, CocuBot, LOLSmaterThanYou, Satellizer, Bike CharlieCard, Widr, MerIlwBot, Helpful Pixie Bot, Bradynowinstores, Pikapika123, Bibcode Bot, 2001:db8, Mouchumi, Vagobot, 1sgfxn, Nikos 1993, Teamavolition, AwamerT, Mark Arsten, Cadiomals, Insidiae, Poophole10101, D.bolmatov, WhiteNebula, Priyamd, Ameerajan-mohamed, Physicsandshiz, Ledgermayne101, Poopyhead88, EuroCarGT, Phillip is the name, Ducknish, Dexbot, TwoTwoHello, ComfyKem, Kevin12xd, Reatlas, PhantomTech, Abc123def455, EvergreenFir, Leoesb1032, Buffbills7701, Blackbombchu, Ugog Nizdast, Aravinduser, JaconaFrere, Carlos Tao, Natty Stott, Samygemayel, Trackteur, Oisguad, CHK101, Cnbr15, Vistardhvaj, KasparBot, Quiesce, Letsdisko, The pro guy, Wikiman900909, Hermionedidallthework, Qwerty456789p and Anonymous: 459

- **Liquid crystal** *Source:* https://en.wikipedia.org/wiki/Liquid_crystal?oldid=688776259 *Contributors:* William Avery, Caltrop, Youandme, Michael Hardy, Dominus, SebastianHelm, Александър, Yoshitaka Mieda~enwiki, Glenn, Smack, Tantalate, Stone, Radiojon, Topbanana, Phil Boswell, Rogper~enwiki, ZimZalaBim, Dusik, Giftlite, BenFrantzDale, Bensaccount, Eequor, Bobblewik, Karol Langner, JohnArmagh, Gua-nabot, MuDavid, Sergei Frolov, Mykhal, Jaberwocky6669, Plugwash, Rgdboer, Femto, Tomgally, Chiral, Rumikohorie, Panjasan, Jared81, Andrewpmk, Lithian, Rwendland, RJFJR, Gene Nygaard, Nick Mks, Rvlaw, Mindmatrix, Wacko, David Haslam, Cbhiii, Jeff3000, Jwan-ders, GregorB, Eras-mus, CharlesC, V8rik, Chenxlee, Tlroche, Saperaud~enwiki, Rjwilmsi, Koavf, Eptalon, Numa, RexNL, Kebes, Wars, BjKa, Thecurran, Srleffler, Chobot, Bgwhite, DerrickOswald, YurikBot, Armistej, Epolk, Hellbus, Gaius Cornelius, Shaddack, Prime Ent-elechy, Dysmorodrepanis~enwiki, Dhollm, Misza13, SFC9394, 2over0, Equack, LeonardoRob0t, Claudiozzbo, Ilmari Karonen, SmackBot, AnOddName, M stone, Gilliam, Hugo-cs, Bluebot, Bduke, Martin Jambon, DHN-bot~enwiki, Descalzo, JustUser, Chlewbot, Waskyo, VMS Mosaic, Iamthealchemist, Akriasas, Smokefoot, DMacks, Ohconfucius, SashatoBot, DavidCooke, JorisvS, Mxreb0, Arkrishna, Kvng, Xionbox, JoeBot, Tawkerbot2, WolfgangFaber, CmdrObot, Dgw, Rifleman 82, Raul1022, Epbr123, Barticus88, HappyInGeneral, Headbomb, WVhy-brid, Sijarvis, AntiVandalBot, Guy Macon, Emmelie, Sbarnard, Markthemac, JAnDbot, Tsuji, Sangak, Magioladitis, Charlesreid1, Fallschir-mjäger, David Eppstein, LorenzoB, DerHexer, MartinBot, Byzantime, AgarwalSumeet, MrBell, Dagar1989, Pmench, 272727, Acalamari, Jcwf, IanusVentus, Liveste, Milogardner, Vanished user 39948282, Hbayat, Odysseyontario, Deor, TXiKiBoT, Showjumpersam, GroveGuy, Rei-bot, Minutemen, Laserspec, I'm with gerrit, Pitel, Rikuharts, Billygettothechopper, RJaguar3, The Unknown Hitchhiker, Harry~enwiki, Easamy, Kumioko (renamed), BlnLiCr, ClueBot, Capture88, GregVolk, Toxonomy, Gulmammad, California847, Versus22, SoxBot III, Chhe, Hotcrocodile, Wikijpp, Ngebbett, Addbot, DOI bot, TutterMouse, Chamal N, Remini2, Tassedethe, SPat, Rascallgh, Luckas-bot, Yobot, AnomieBOT, Nintend06, Prof.kulkarni, Law, Jeff Muscato, Materialscientist, 90 Auto, Citation bot, JJagomagi, MauritsBot, Brycker, Gauj-malnieks, GrouchoBot, Omnipaedista, Jezhotwells, A. di M., Dave3457, Antunbalaz, Tobby72, Sky Attacker, Pratik.mallya, Louperibot, Cita-tion bot 1, Pinethicket, Edderso, Tom.Reding, Darkwolf800, Igfmnbo, Bhxho, Σ, Pjvbra, Felix0411, Jsbeeckm, TobeBot, ItsZippy, Rnakatsuji, Callanecc, Marie Poise, NerdyScienceDude, Cogiati, Wikfr, Timeloop, RockMagnetist, ClueBot NG, Starshipenterprise, Noulin, ScottSteiner, Widr, Helpful Pixie Bot, Bibcode Bot, BattyBot, Arr4, Drjcast, Proffuwec, Z11o22, SRaemiA, Garuda0001, Mark viking, OhVerr Aiteen, Matchbox30, Ugog Nizdast, Anrnusna, Polikpolik890, Monkbot, Liquid-Crystal-Lens, Maodit, KasparBot, Namelessbj and Anonymous: 205

- **Disordered Hyperuniformity** *Source:* https://en.wikipedia.org/wiki/Disordered_Hyperuniformity?oldid=696255002 *Contributors:* Auric, Clone200 and Awesomewiki64

- **Gas** *Source:* https://en.wikipedia.org/wiki/Gas?oldid=698398049 *Contributors:* Bryan Derksen, Olof, Tarquin, Andre Engels, Youssefsan, William Avery, SimonP, Peterlin~enwiki, Ben-Zin~enwiki, Anthere, Heron, Luckymama58, Patrick, D, Michael Hardy, Nixdorf, Delirium, Looixx~enwiki, Ahoerstemeier, Stan Shebs, Mac, Jimfbleak, Александър, Glenn, Nikai, Llull, Edmilne, Malbi, Jaimeglz, Wernher, Shizhao, Bcorr, Jusjih, Johnleemk, Donarreiskoffer, Gentgeen, Robbot, Hankwang, Academic Challenger, Mervyn, LX, Dina, Tobias Bergemann, Giftlite, Tom harrison, Herbee, Theon~enwiki, Peruvianllama, Everyking, Bensaccount, Gareth Wyn, Alexf, Antandrus, OverlordQ, Scott MacLean, Karol Langner, Maximaximax, Icairns, Zfr, Iantresman, GdB, 🔲🔲, Adashiel, Grstain, D6, N328KF, Jiy, Discospinster, Qutezuce, Vsmith, Bender235, Closeapple, Brian0918, RJHall, El C, Hayabusa future, Aude, Bobo192, Fremsley, Giraffedata, Jerryseinfeld, MPerel, Haham hanuka, Jakew, Jumbuck, Danski14, Siim, Gary, Anthony Appleyard, Keenan Pepper, Katefan0, Wtshymanski, Psmither, Yuckfoo, RJFJR, Gene Nygaard, CoolMike, Noz92, Roland2~enwiki, Sylvain Mielot, Woohookitty, BillC, DrAwesome, Pdn~enwiki, Allen3, Behun, BD2412, Phillipedison1891, Rjwilmsi, Panoptical, Bruce1ee, SMC, Krash, Sango123, Yamamoto Ichiro, Gurch, Wars, Drumguy8800, Mo-gest, Jittat~enwiki, Chobot, Sharkface217, DVdm, Chachu207, Algebraist, The Rambling Man, YurikBot, Wavelength, RussBot, Jeffhoy, Backburner001, Conscious, SpuriousQ, KevinCuddeback, Stephenb, Jugander, Rsrikanth05, Wimt, NawlinWiki, Grafen, RazorICE, Lepi-doptera, Dureo, Irishguy, Dhollm, Zwobot, EEMIV, Kkmurray, Wknight94, Lt-wiki-bot, Closedmouth, Sean Whitton, Willtron, AGToth, RunOrDie, Nsevs, Junglecat, NeilN, GrinBot~enwiki, Bo Jacoby, Amberrock, DVD R W, ChemGardener, SmackBot, FocalPoint, Prodego, KnowledgeOfSelf, Unyoyega, Vald, Bomac, ScaldingHotSoup, Jrockley, Delldot, Ilikeeatingwaffles, RobotJcb, Abbeyvet, Canthusus, Nethency, Fentonrobb, Edgar181, Hmains, Carl.bunderson, Kmarinas86, Chris the speller, TimBentley, SlimJim, Quinsareth, EncMstr, MalafayaBot,

Complexica, Stevage, Akanemoto, Pencilcomics, DHN-bot~enwiki, Colonies Chris, Gracenotes, Dethme0w, Can't sleep, clown will eat me, Shalom Yechiel, DéRahier, Sephiroth BCR, Pablo9000, Amazins490, SundarBot, Jmlk17, Zrulli, Nibuod, TedE, RJN, Blake-, Astroview120mm, DMacks, Bidabadi~enwiki, Sadi Carnot, Ged UK, SashatoBot, Lambiam, BrownHairedGirl, Kuru, John, J 1982, Gobonobo, Disavian, Ishmaelblues, JorisvS, Olin, IronGargoyle, SpyMagician, Beetstra, Stizz, Doczilla, Dhp1080, Dcflyer, Ryulong, H, Etafly, Caiaffa, YipYip, Hu12, Ginkgo100, Iridescent, Wwallacee, Blehfu, CP\M, Eassin, Tawkerbot2, Stifynsemons, DangerousPanda, Dgw, Jp-hickson, McVities, MarsRover, WeggeBot, Rakwiki, Cydebot, SyntaxError55, Rifleman 82, Gogo Dodo, Flowerpotman, Corpx, Nabz~enwiki, Dumb-BOT, Narayanese, Btharper1221, Gimmetrow, Kablammo, Ucanlookitup, Mojo Hand, Headbomb, Marek69, John254, Nathaniel Zhu, Cool Blue, AgentPeppermint, CharlotteWebb, Escarbot, AntiVandalBot, M84, Majorly, Bigtimepeace, Jj137, Madbehemoth, Dylan Lake, North Shoreman, Waynesewell, Myanw, Dreaded Walrus, JAnDbot, Barek, MER-C, Hut 8.5, PhilKnight, GoodDamon, Yahel Guhan, Bongwarrior, VoABot II, AuburnPilot, Hasek is the best, Mbc362, Kaiserkarl13, Jim Douglas, Kevinmon, Avicennasis, Fabrictramp, Animum, Cgingold, Nposs, Allstarecho, Adacus12, Lukecarpenter169, Spellmaster, PrincessBrat, Glen, DerHexer, Edward321, Greenguy1090, Hdt83, MartinBot, Bullet4troubles, Chaos Wolf, Rettetast, Tholly, R'n'B, AlexiusHoratius, Johnpacklambert, Harrichr, J.delanoy, Trusilver, Sp3000, Jakesdamajorbomb, GeoWriter, Socrgrl3426, Rufous-crowned Sparrow, AtholM, Katalaveno, Abhijitsathe, Hawkmaster9, Ryan Postlethwaite, Afluegel, Mikael Häggström, Sman789, V.V zzzzz, Jonodabomb, NewEnglandYankee, SJP, Mufka, Lyctc, Uhai, Darkfrog24, Screwe, Idioma-bot, Lights, X!, VolkovBot, Cireshoe, Thedjatclubrock, TheOtherJesse, Philip Trueman, DoorsAjar, TXiKiBoT, Z.E.R.O., Woodsstock, Qxz, Redmusicjamin, Melsaran, Aycbubbles, Corvus cornix, K193, ^demonBot2, Drappel, Delbert Grady, Mannafredo, Fireman17, Isis4563, Miwanya, Venny85, Jmath666, Wenli, Greswik, Krzysfr, Sbakka, Synthebot, Falcon8765, Emily.xxo, BaByGuRRL19, Brianga, Heirloom-Gardener, AlleborgoBot, Logan, Resurgent insurgent, NHRHS2010, Comeinayeahaa, Adamboy555, Darting., SieBot, Euryalus, Nicklovesgold, Dennislee272727, Jauerback, Virtual Cowboy, Rystheguy, Gerakibot, Caltas, Cwkmail, Keilana, Flyer22 Reborn, Tiptoety, Oda Mari, Oxymoron83, Nuttycoconut, Steven Crossin, Lightmouse, Powerofgas, Dcarriker92, Alex.muller, Phoneuser, Torchwoodwho, Coldcreation, Maralia, Hooiwind, Nn123645, Denisarona, Faithlessthewonderboy, Martarius, ClueBot, Drmies, Razimantv, Mild Bill Hiccup, Counter-VandalismBot, Hschrage, Toudaiiji Neji, Puchiko, Excirial, Granvo, MorrisRob, Sun Creator, 7&6=thirteen, I luv carrots, SchreiberBike, Kpark454, I Has A Username, Katanada, Soulspick, Skier lad, DumZiBoT, Life of Riley, Clausc, Kalin1344, Poopman101, Superman1159, Bailey Pitzer, Ridhwan95, SilvonenBot, Tjmcgowan9, Noctibus, Addbot, Vejvančický, Fyrael, Ashton1983, Leszek Jańczuk, Tedmund, Ilovelakai, Skyezx, CarsracBot, Glass Sword, Mkardous, ChenzwBot, Numbo3-bot, Tide rolls, Fryed-peach, VP-bot, Luckas-bot, Yobot, Fraggle81, Azylber, DemocraticLuntz, Bjk343, Materialscientist, Spirit469, The High Fin Sperm Whale, Citation bot, ArthurBot, Obersachsebot, Xqbot, Intelati, Gilo1969, Gap9551, GrouchoBot, ChristopherKingChemist, IShadowed, GhalyBot, Schekinov Alexey Victorovich, Joaquin008, FrescoBot, LucienBOT, Thayts, Rackmount-guy, DivineAlpha, Cannolis, Krish Dulal, Pinethicket, Red banksy, Adlerbot, Tom.Reding, RedBot, Serols, Σ, IJBall, Jauhienij, Keri, Gamewizard71, FoxBot, TobeBot, Lotje, Vrenator, Begoon, 4, JamAKiska, Jeffrd10, JV Smithy, Tbhotch, DARTH SIDIOUS 2, Mean as custard, Yangosplat222, DASHBot, EmausBot, Gfoley4, 478jjjz, Abcboy151, Hudmaster, Dewritech, Racerx11, GoingBatty, RA0808, John of Lancaster, Hhhippo, JSquish, ZéroBot, The Nut, AvicAWB, Magasjukur, J1812, Ventilate, Dirtykorean, Donner60, ChuispastonBot, RockMagnetist, NTox, TYelliot, DASHBotAV, ClueBot NG, Iiii I I I, SusikMkr, Baseball Watcher, Snotbot, Corusant, Prumpa, O.Koslowski, Scottiessoulja, Kasirbot, Widr, Jalenjohnson129, Coiladam, Helpful Pixie Bot, JohnSRoberts99, Bibcode Bot, WNYY98, Lowercase sigmabot, Juro2351, Hallows AG, MusikAnimal, Nikos 1993, Amp71, Piguy101, Mark Arsten, Gorthian, Working Cat, Snow Blizzard, Hamzah145, Shawn Worthington Laser Plasma, BattyBot, RichardMills65, Cyberbot II, CarrieVS, Khazar2, MrNoSignal26, Adam8257, Murshed11, Etcyorz3t789, Saehry, Lugia2453, Frosty, Ashcool1999, Athomeinkobe, Bilalrockstar5, Tyler Daigle, Randykitty, Epicgenius, Hahabacon123, Ugog Nizdast, WikiJuggernaut, Aniapo, DudeWithAFeud, Yolokid1024, Carecat2434, AKS.9955, Joeleoj123, BethNaught, New Imam Bukhsh Gas Company, Nibgco, Pipo45, A.Minkowiski, Technick14, NQ, Plop2004, Goolongo325, Ashisbiswas560, Leocatelli09, Luistheguy, Minecrafter123456, Rahulhsp, Arnold22palmer, Kethrus, Megan Mackenzie, Siddhantsaka, Perica85, Spidermanwillruletheworld, Rylee Imhoff, KasparBot, Mynamejeff47, Benjamintylerolmstead2, Lucyiliana, Letsdisko, Ndeeper, CLCStudent and Anonymous: 602

- **Plasma (physics)** *Source:* https://en.wikipedia.org/wiki/Plasma_(physics)?oldid=698481355 *Contributors:* Trelvis, Vicki Rosenzweig, Mav, Bryan Derksen, Olof, AstroNomer~enwiki, Roadrunner, Secretsaregood, Heron, Stevertigo, Patrick, Michael Hardy, Tim Starling, Tapper of spines, Zeno Gantner, Ellywa, Stevenj, Jebba, Darkwind, Julesd, Glenn, Tantalate, Wikiborg, Reddi, David Latapie, IceKarma, Nv8200pa, Omegatron, Phoebe, Pakaran, Cdupree, April~enwiki, Donarreiskoffer, Robbot, Psychonaut, Moink, Hadal, Papadopc, David Edgar, SoLando, Dbroadwell, Wile E. Heresiarch, Giftlite, Mat-C, Art Carlson, MadmanNova, Monedula, Everyking, Jacob1207, Gracefool, Solipsist, Bobblewik, SarekOfVulcan, Beland, Mako098765, WhiteDragon, Gunnar Larsson, Karol Langner, APH, Icairns, Bk0, Nickptar, Iantresman, Kelson, Joyous!, Fermion, Jh51681, Deglr6328, Kate, PhotoBox, Spiffy sperry, Jiy, Noisy, Discospinster, Brianhe, Rich Farmbrough, Pjacobi, Vsmith, Florian Blaschke, Warpflyght, Bender235, Lou Crazy, Evice, El C, Huntster, Femto, Bobo192, Illuvatar,, Smalljim, Func, Enric Naval, Evgeny, .:Ajvol:., Elipongo, Tmh, Maurreen, I9Q79oL78KiL0QTFHgyc, Sparkgap, Photonique, Boredzo, B0at, Obradovic Goran, Yoweigh, Danski14, Alansohn, Anthony Appleyard, Ungtss, Arthena, Keenan Pepper, Craigy144, ABCD, RoySmith, PAR, Pion, SMesser, Radical Mallard, Knowledge Seeker, Cburnett, RJFJR, Sciurinæ, Mikeo, Vuo, Pauli133, DV8 2XL, Gene Nygaard, NuVanDibe, HenryLi, Pediddle, Umapathy, Brookie, Feezo, Stemonitis, Gmaxwell, Linas, Justinlebar, WadeSimMiser, Mouvement, Jwanders, Fred J, SDC, Palica, Allen3, Mr Anthem, Marudubshinki, Mandarax, Aarghdvaark, Graham87, Teknic, BD2412, RxS, Kissekatt, Grammarbot, Canderson7, Ketiltrout, Rjwilmsi, Mayumashu, Koavf, Vary, Beng341, HappyCamper, Mbutts, Frenchman113, Krash, DoubleBlue, GregAsche, Yamamoto Ichiro, Algebra, Lcolson, MinorEdit, Eyas, The ARK, Old Moonraker, JohnElder, RexNL, Crouchingturbo, Preslethe, Srleffler, Kri, Physchim62, King of Hearts, Chobot, Helios, Tone, Roboto de Ajvol, YurikBot, Borgx, Madhan49, Hairy Dude, RussBot, Arado, Conscious, Hydrargyrum, CambridgeBayWeather, Yyy, Shaddack, Giro720, Anomalocaris, Wiki alf, Bachrach44, Spike Wilbury, Grafen, NickBush24, Jaxl, JDoorjam, Nucleusboy, Brian Crawford, Coderzombie, Zirland, Figaro, DeadEyeArrow, Oliverdl, Kkmurray, Wknight94, Pr1268, Jezzabr, Tetracube, Nfm, Mütze, Tigershrike, Poppy, Phgao, 2over0, Bobryuu, Theda, Closedmouth, Jwissick, Ameyabapat, Modify, Vicarious, Peter, Willtron, Katieh5584, Kungfuadam, RG2, MAROBROS, Dkasak, Mejor Los Indios, DVD R W, Vedant lath, That Guy, From That Show!, Mhardcastle, TravisTX, Lviatour, SmackBot, Superfreaky56, Unschool, Hkhenson, Sonoma-rich, CarbonCopy, Rex the first, Prodego, KnowledgeOfSelf, Wegesrand, Sharaith, KocjoBot~enwiki, Fitch, FRS, Eskimbot, Jab843, Evanreyes, Yamaguchi⬚⬚, Brianski, Richfife, MPD01605, Bluebot, Thumperward, Miquonranger03, Mr Poo, SchfiftyThree, Complexica, Akanemoto, MIB4u, Baa, DHN-bot~enwiki, William Allen Simpson, Shamiryan, Exaudio, Can't sleep, clown will eat me, Ioscius, Berland, Rrburke, RedHillian, DavidStern, Nibuod, PointyOintment, John D. Croft, Astroview120mm, EdGl, SpiderJon, DMacks, Whoville, Mion, Bidabadi~enwiki, Tesseran, Sina2, Kuru, John, JorisvS, IronGargoyle, Ben Moore, Applejuicefool, S zillayali, Frokor, Slakr, SCOTT FISHER, Dicklyon, Waggers, Mets501, Dr.K., Arstchnca, P199, MathStuf, Dl2000, Tusenfem, Iridescent, Electrified mocha chinchilla, Laurens-af, Polymerbringer, Joseph Solis in Aus-

ibot, Oda Mari, Oxymoron83, Ioverka, Lightmouse, Chemist1828, Maderibeyza, Mygerardromance, Torkuemada, Dolphin51, Denisarona, Cjones100, SallyForth123, ClueBot, Jarihtey, Trexsandwich, The Thing That Should Not Be, Jan1nad, Freevito, Niceguyedc, Auntof6, Connor 99, AndreiDukhin, Excirial, Reethu, PixelBot, Arvz cabais, Panoramix303, Shinkolobwe, Computer97, Ottawa4ever, Cechavar, Versus22, JMatopos, XLinkBot, Spitfire, Rror, Avoided, Medomsn, Titmouse345, Fluffernutter, Chamal N, Favonian, West.andrew.g, Quercus solaris, Elen of the Roads, Freckles dog, Tide rolls, Lightbot, Teles, Zorrobot, Legobot, Luckas-bot, Yobot, KamikazeBot, AnomieBOT, 1exec1, Götz, Daniele Pugliesi, WattsJoule, Runeuser, Dinesh smita, Flewis, Materialscientist, Citation bot, Xqbot, Plumpurple, Cureden, Xadorus, Nasnema, J04n, Jhbdel, GregorTrefalt, Logger9, Greed378, Proepro, Citation bot 1, Mastereditor2000, Biker Biker, Pinethicket, Abductive, MastiBot, Reconsider the static, Trappist the monk, Foodienyc, Yunshui, Lotje, Mmnbv21, Tbhotch, Daduds 09, Hdihang, DARTH SIDIOUS 2, RjwilmsiBot, Kk lol hahaha, Midhart90, NDKilla, Danish Expert, Solarra, K6ka, Érico, Michel Awkal, Ὁ οἶστρος, Ally.b888, Erianna, Asund3r, Tomásdearg92, Donner60, Scientific29, Kyeongjae, EdoBot, ClueBot NG, Rich Smith, Ajminime97, Frietjes, Estopedist1, Widr, MrJosiahT, Helpful Pixie Bot, Bibcode Bot, Elementorix, Krenair, WikiPeterD, Geet248, Nathan2055, BrainMagnet, Karenblackhall, Snow Blizzard, Zedshort, SunKart, Anbu121, BattyBot, Tutelary, Mrt3366, Williams12357, 331dot, Lugia2453, Keamon, Reatlas, PhantomTech, Tristandilbeck, Brezhurley, Tara Zieminek, Amr94, Ramalingam1999, ACPawsey, Monkbot, Poljew, KasparBot, Sanjaysrivatsan and Anonymous: 452

- **Degenerate matter** *Source:* https://en.wikipedia.org/wiki/Degenerate_matter?oldid=696740713 *Contributors:* CYD, Bryan Derksen, Olof, AstroNomer~enwiki, Roadrunner, Michael Hardy, Alan Peakall, HarmonicSphere, Julesd, Doradus, Zoicon5, Donarreiskoffer, Robbot, Donreed, Jheise, DocWatson42, Herbee, Leperous, WhiteDragon, Superborsuk, Bender235, ZeroOne, Lycurgus, Cherlin, ChristopherWillis, Stephan Leeds, VivaEmilyDavies, RJFJR, Dirac1933, Pauli133, Gene Nygaard, Dan East, Killing Vector, Linas, Mindmatrix, Christopher Thomas, Kgbudge, Rnt20, Ashmoo, Eyu100, Wragge, Karch, YurikBot, TSO1D, Conscious, Hede2000, Ytrottier, Hellbus, Anomalocaris, Trovatore, John Newbury, Enormousdude, Modify, Mhenriday, RG2, Tim314, SmackBot, Eskimbot, Lainagier, Ohnoitsjamie, Grokmoo, Kmarinas86, Oni Ookami Alfador, Voyajer, Wikipedia brown, Percommode, Aldaron, ChowRiit, Just plain Bill, Dark Formal, Jaganath, JorisvS, Anescient, Newone, Julian.cancino, Thijs!bot, Headbomb, Yellowdesk, Myanw, JAnDbot, WolfmanSF, Dirac66, Su-no-G, Warren Dew, MartinBot, STBot, Steve98052, Sofar 2, Vranak, Holme053, Kurgus, Lechatjaune, Anna Lincoln, Mardhil, Entropy1963, Flyer22 Reborn, Anchor Link Bot, Kallog, Denisarona, WurmWoode, PipepBot, Garyzx, SuperHamster, Masterblooregard, Chief buffalo chip, Brews ohare, Coinmanj, Arjayay, 2, Tealwisp, XLinkBot, SkyLined, Addbot, Dsmith77, 84user, Phynisha 25, Luckas-bot, Captain Quirk, Hunnjazal, StrontiumDogs, Brithans, Nickkid5, Tomdo08, Gap9551, INick3, Omnipaedista, RibotBOT, Seeleschneider, FrescoBot, Mossmanj, Vhann, Fkmusgrave, Kevinpeck, Fartherred, TobeBot, Erixmix, Hhhippo, HiW-Bot, Liquidmetalrob, Ὁ οἶστρος, Quondum, MajorVariola, ClueBot NG, Bibcode Bot, BG19bot, Trevayne08, Zedshort, Shawn Worthington Laser Plasma, Sschongster, ChrisGualtieri, Bterranova, Zinganthropus, Phleg1, Jjusiopao, Volker Siegel, AKS.9955, DoisKoh, Coconutporkpie, Crito10, Vesuvius Dogg, ErikNatuurkunde and Anonymous: 113

- **Strange matter** *Source:* https://en.wikipedia.org/wiki/Strange_matter?oldid=690372782 *Contributors:* CYD, Bryan Derksen, The Anome, Frecklefoot, Oliver Pereira, SebastianHelm, Schneelocke, Vespristiano, Giftlite, Curps, Beland, EricJamesStone, Pjacobi, Sam Korn, Fwb22, Proteus71, Joriki, Scriberius, Mpatel, Squideshi, Strait, Oalsaker, Eyu100, Margosbot~enwiki, Nihiltres, Spacepotato, Bambaiah, Hellbus, Salsb, NawlinWiki, Kooky, 2over0, JDspeeder1, Algae, Mtffm, MacsBug, SmackBot, Tinz, Xaosflux, Kmarinas86, Marcus Brute, Dark Formal, Yanwen, Zzzzzzzzzzz, Dan Gluck, Siebrand, Michaelbusch, Eassin, OS2Warp, CalebNoble, Orca1 9904, Nilfanion, Zomic13, Chrislk02, Vidale, Mrph, VoABot II, CarlFeynman, Hurax, Idioma-bot, ABF, Fences and windows, Lamro, SieBot, Lethesl, ClueBot, PixelBot, Snookumz, Homocion, SkyLined, Addbot, Micke, Yobot, AnomieBOT, Xqbot, GrouchoBot, ProtectionTaggingBot, Nikto, Johann137, WikitanvirBot, Hhhippo, ZéroBot, Nobelium, Suslindisambiguator, ClueBot NG, Astrocog, Frietjes, Helpful Pixie Bot, Rasheeq1, Cavaliere1, Sebastian5059, Ajomannen, RhinoMind, Erikprantare and Anonymous: 56

- **Photonic molecule** *Source:* https://en.wikipedia.org/wiki/Photonic_molecule?oldid=697569161 *Contributors:* Velella, Bhny, Headbomb, Lfstevens, I JethroBT, Paradoctor, Yobot, AnomieBOT, Citation bot, Quondum, Dexbot, Reatlas and Anonymous: 15

- **Quantum Hall effect** *Source:* https://en.wikipedia.org/wiki/Quantum_Hall_effect?oldid=671705802 *Contributors:* CYD, The Anome, Michael Hardy, Tim Starling, Glenn, Timwi, Ozuma~enwiki, Robbot, Tonsofpcs, Xanzzibar, Saltcreek, MarkSweep, Karol Langner, Urhixidur, Bender235, Thoken, Shanes, Euyyn, CDN99, Sicherlich, Egg, Gene Nygaard, Linas, David Haslam, StradivariusTV, Josh Parris, Rjwilmsi, Chobot, Jaraalbe, YurikBot, Archelon, Shaddack, Chaiken, Sbyrnes321, SmackBot, Hkhenson, Yuyudevil, Bluebot, Complexica, DHNbot~enwiki, V1adis1av, Akriasas, Wizardman, JorisvS, NNemec, Dpb2104, Ramuman, Comech, CmdrObot, Grj23, Thijs!bot, Headbomb, Avronj, Griba2010, JAnDbot, Arch dude, Mytomi, LorenzoB, Leyo, Pyrospirit, VolkovBot, LokiClock, Dragostanasie, Ngoldman007, Lightmouse, A.C. Norman, Jilidelft~enwiki, Mild Bill Hiccup, MicroVirus, DragonBot, Addbot, Gravitophoton, DOI bot, Kahlos, Download, MrVanBot, LaaknorBot, Lightbot, SPat, Legobot, Luckas-bot, Yobot, AnomieBOT, Rubinbot, Materialscientist, Etoombs, Acpotter, Cantonsde-l'Est, Freddy78, Citation bot 1, Dm00, Pmokeefe, Trappist the monk, Spkersten, 564dude, Tpudlik, AManWithNoPlan, Tls60, Chuispastonbot, Wout Neutkens, Cphil1, JohnTsams, Bibcode Bot, BattyBot, OSU1980, SoledadKabocha, Makecat-bot, Garuda0001, Erashba, Philipphilip0001, Kfitzell29 and Anonymous: 63

- **Quantum spin Hall effect** *Source:* https://en.wikipedia.org/wiki/Quantum_spin_Hall_effect?oldid=690733469 *Contributors:* Phoebe, Michael Devore, Laurascudder, Chobot, Chris the speller, Raymond arritt, Alaibot, Dirac66, Thebeagle, Tiptoety, Quantumspinhall, Tvw4tcher, Addbot, Yobot, AnomieBOT, M&M987, Qcosmos77, Starshipenterprise, Stevesimon2, Dexbot, Feynmandiagram and Anonymous: 10

- **Rydberg matter** *Source:* https://en.wikipedia.org/wiki/Rydberg_matter?oldid=683802247 *Contributors:* William Avery, Giftlite, Dratman, Rich Farmbrough, I9Q79oL78KiL0QTFHgyc, Keenan Pepper, Ahruman, Pauli133, Woohookitty, Tabletop, Josh Parris, Rjwilmsi, Dhollm, Roques, DGaw, SmackBot, Chris the speller, Pieter Kuiper, Colonies Chris, CRGreathouse, Xxanthippe, Alaibot, JamesAM, Headbomb, R'n'B, Ojovan, Venny85, Comparat, Djr32, Gonzonoir, MystBot, Addbot, Tman1997al, MuZemike, Yobot, AnomieBOT, Materialscientist, Citation bot, Holmlid, Citation bot 1, Pastafarian32, Double sharp, ZéroBot, Parcly Taxel, Bibcode Bot, Blazespinnaker and Anonymous: 20

- **Bose–Einstein condensate** *Source:* https://en.wikipedia.org/wiki/Bose%E2%80%93Einstein_condensate?oldid=697616792 *Contributors:* Kpjas, CYD, Archibald Fitzchesterfield, Bryan Derksen, Olof, Tarquin, Gareth Owen, Josh Grosse, Hfastedge, Spiff~enwiki, Michael Hardy, Gabbe, TakuyaMurata, SebastianHelm, Alfio, Ellywa, Cyp, Stevan White, Darkwind, Glenn, Mxn, Schneelocke, Loren Rosen, Feedmecereal, Dino, Wikiborg, The Anomebot, ElusiveByte, BenRG, JorgeGG, Donarreiskoffer, Chris 73, Nurg, Robinh, GreatWhiteNortherner, Dave6, M-Falcon, Matt Gies, Giftlite, Smjg, Inter, Herbee, Dratman, Tom-, Eequor, Balenman, Chrissmith, Mooquackwooftweetmeow, Toytoy, XxPantherNovaXx, Fangz, Piotrus, Karol Langner, Brian Jackson, Spiralhighway, Sam Hocevar, Kramer, Nickptar, Vivacissamamente,

Grunt, Eep[2], NightMonkey, Lone Isle, Noisy, Discospinster, Guanabot, ThomasK, Vsmith, Aardark, Paul August, Bender235, TOR, RJHall, El C, Lycurgus, Ruyn, Laurascudder, Jpgordon, Fuxx, Directorstratton, Slicky, Sasquatch, Haham hanuka, Alansohn, Arthena, Nwinther, PAR, Pion, Kfitzgib, Cjnm, Tom12519, Snowolf, Einstein9073, BRW, KapilTagore, Pauli133, Gene Nygaard, Joriki, OwenX, Woohookitty, Linas, David Haslam, Benbest, Ruud Koot, Jeff3000, Astrophil, BlaiseFEgan, Bugman, Sjö, Rjwilmsi, Amire80, Rillian, BlueMoonlet, Salix alba, Keimzelle, Exeunt, Azure8472, FlaBot, SchuminWeb, The.valiant.paladin, Shade[2], Pete.Hurd, Srleffler, Erik4, King of Hearts, Chobot, DVdm, Sasoriza, YurikBot, Wavelength, Taurrandir, Rob T Firefly, Hairy Dude, Huw Powell, Flameviper, Michael Slone, JabberWok, David Woodward, Shell Kinney, NawlinWiki, WulfTheSaxon, Truetyper, Howcheng, Chakazul, Katrielalex, Dogcow, Grafikm fr, Zwobot, Wangi, Wknight94, FF2010, 2over0, Closedmouth, Dr.alf, Stuhacking, Otto ter Haar, Groyolo, Allium, SmackBot, Serg3d2, RossyMiles, Oxford Comma, Olegt1, Nickst, Eskimbot, Dilbert3, Gaff, Gilliam, Kmarinas86, Thumperward, Hichris, DHN-bot~enwiki, Raistuumum, Salmar, Tcb Beany, Karpita, Can't sleep, clown will eat me, Nick Levine, Kelvin Case, Neo139, Onorem, MBlume, Voyajer, Rrburke, Xyzzyplugh, George-Money, TedE, Bigmantonyd, Rich.lewis, DMacks, Xiutwel, Ligulembot, Mion, Sadi Carnot, Josellis, Tethros, Lucretius~enwiki, Lambiam, Andi47, John, MagnaMopus, Jaganath, SteveG23, Mgiganteus1, Ckatz, BillFlis, Kyoko, Dicklyon, Inquisitus, Phuzion, Brienanni, Iridescent, JMK, Clarityfiend, FelisSchrödingeris, Frank Lofaro Jr., CRGreathouse, ZICO, BeenAroundAWhile, DSachan, Orannis, Myasuda, Leakeyjee, Equendil, Stebbins, Kanags, MC10, Tashafairbairn, Mato, Gogo Dodo, JFreeman, Mattjball, Omicronpersei8, Thijs!bot, Epbr123, Wikid77, Trevheg, Fiction Alchemist, Sam Van Kooten, Headbomb, Second Quantization, Iviney, CharlotteWebb, AntiVandalBot, 17Drew, Gökhan, MSBOT, Boleslaw, Sinnerwiki, Sophosmoros, Magioladitis, WolfmanSF, VoABot II, Sushant gupta, Bakken, Ggorelik, Tonyfaull, BatteryIncluded, Dirac66, David Eppstein, LorenzoB, Refael Ackermann, Talon Artaine, Torsionalmetric, Starryharlequin, N734LQ, Anonymous 57, Sketchjoy, Custos0, J.delanoy, MITBeaverRocks, Jtw11, Bogey97, Maurice Carbonaro, AquamarineOnion, Glaux, AppleMacReporter, AntiSpamBot, Tendays, Enix150, Neil Dodgson, Idioma-bot, Austinmohr, Gnipahellir, A4bot, Qxz, Martin451, Mitchell26, Natural Philosopher, Mazarin07, Akhuettel, Spinningspark, Kapalama, Cryonic07, PaddyLeahy, Biscuittin, Awemond, WereSpielChequers, Cmossol, Matthew Yeager, Deathgleaner, Reuqr, Likebox, JD554, Reinderien, Topher385, Scorpion451, Lightmouse, Jakeng, Coldcreation, Psycherevolt, Melcombe, AllHailZeppelin, Crazz bug 5, Martarius, ClueBot, MonkeyMensch, Snigbrook, The Thing That Should Not Be, EoGuy, Emil70, Zero over zero, Razimantv, Maymay, Thegeneralguy, DrakeUnlimited, NuclearWarfare, Dboiko, Doktor Mephisto, SchreiberBike, Thingg, Jonverve, SoxBot III, Egmontaz, DumZiBoT, Ost316, Rreagan007, SilvonenBot, SkyLined, MaizeAndBlue86, Csingh23592, Addbot, DOI bot, Jojhutton, Miskaton, Friginator, Download, ChenzwBot, AtheWeatherman, 84user, Tide rolls, Lightbot, OlEnglish, Teles, SPat, Megaman en m, Ben Ben, Yobot, Wireader, VectorField, AnomieBOT, TheUfoFiles, Aaagmnr, Materialscientist, Limideen, Citation bot, MetaplecticGroup, Natural RX, Xqbot, BME-physics, Lunaintern, RibotBOT, Verbum Veritas, Nixón, HJ Mitchell, Quantum 235, Citation bot 1, Maan361, Gil987, Gaba p, I dream of horses, Coekon, RedBot, Akalabeth, Serols, Keri, Asrrin29, Senra, Canuckian89, JV Smithy, DARTH SIDIOUS 2, Obankston, Hajatvrc, Nkf31, EmausBot, John of Reading, Gfoley4, Primefac, Physics16, GoingBatty, KHamsun, Solarra, K6ka, Lent1999, H3llBot, Quondum, Timetraveler3.14, DougEFresh1122, Donner60, Fairskys, Carmichael, Jalexander-WMF, ChrisC550, ClueBot NG, Gareth Griffith-Jones, Movses-bot, All Hail Hypnotoad!, Zak.estrada, Widr, Fqr2010, Helpful Pixie Bot, HMSSolent, Jubobroff, Bibcode Bot, BG19bot, Northamerica1000, AvocatoBot, Wowwii, Rm1271, Mr.viktor.stepanov, BattyBot, Mrt3366, ChrisGualtieri, Adwaele, Baileybrooks, FlappyJenkins, Dexbot, Makecat-bot, Baldoc83, Jamesx12345, Stewwie, Avrahamleib, Sakurai23, Marcela louis, Reatlas, Epicgenius, Nonsenseferret, Aarya19991111, Ginsuloft, Aritcle, KillerKira, John Doppler, Aarjun Rampal, Happy Attack Dog, Arnaud Migres, Tusharkashyap2001, AfrikanischePost, Hans8654, Sumandark8600, Shengxingwu, Antsiepantsie, Yohoona, Mysterious Gopher, KasparBot, ProprioMe OW, Liamsmith12, Jindayat000 and Anonymous: 508

- **Fermionic condensate** *Source:* https://en.wikipedia.org/wiki/Fermionic_condensate?oldid=684258629 *Contributors:* CYD, SebastianHelm, Glenn, Cimon Avaro, Schneelocke, Charles Matthews, Fuzheado, Phys, Jeffq, Fredrik, Cedars, Xerxes314, Steuard, Gdr, Srbauer, Laurascudder, Prsephone1674, Lysdexia, Camw, Bluemoose, Nightscream, BradBeattie, YurikBot, Hellbus, Długosz, Alain r, That Guy, From That Show!, KnightRider~enwiki, SmackBot, Hmains, Kmarinas86, Kcordina, Ohconfucius, WhiteHatLurker, Usgnus, WeggeBot, Difluoroethene, Thijs!bot, Mbell, Headbomb, AntiVandalBot, Majorly, Maliz, MartinBot, Philip Trueman, ClueBot, The Help Fishy, Vanished user uih38riiw4hjlsd, MystBot, Addbot, Mjamja, Luckas-bot, AnomieBOT, Citation bot, Cowgoesmoo2, ProtectionTaggingBot, Tom.Reding, Full-date unlinking bot, EmausBot, KHamsun, ZéroBot, Alpha Quadrant (alt), Aeonx, Quondum, ClueBot NG, Yen-Tzu and Anonymous: 45

- **Superconductivity** *Source:* https://en.wikipedia.org/wiki/Superconductivity?oldid=698397737 *Contributors:* AxelBoldt, Lee Daniel Crocker, CYD, Bryan Derksen, AstroNomer~enwiki, Andre Engels, DavidLevinson, Quintanilla, Jqt, Azhyd, Waveguy, David spector, Heron, Olivier, Edward, Michael Hardy, Fred Bauder, DopefishJustin, Dominus, Karada, Tiles, Egil, Ahoerstemeier, Stevenj, Theresa knott, Snoyes, Julesd, Glenn, Cimon Avaro, GCarty, Cryoboy, Mxn, Tantalate, Reddi, Stone, Joerg Reiher~enwiki, Hao2lian, DJ Clayworth, E23~enwiki, Furrykef, Taxman, LMB, Fibonacci, Omegatron, Traroth, Topbanana, Pstudier, Pakaran, Phil Boswell, Donarreiskoffer, Robbot, Stephan Schulz, Rorro, Bkell, Hadal, UtherSRG, Robinh, Diberri, Cyberpunks~enwiki, Connelly, Giftlite, DocWatson42, MarkPNeyer, Harp, Tom harrison, Ferkelparade, Fastfission, Xerxes314, Leonard G., Foobar, Bobblewik, Wmahan, Irarum, Geni, Quadell, Spiralhighway, Icairns, Peter bertok, Gerrit, Deglr6328, Deeceevoice, Moxfyre, Reflex Reaction, Zowie, CALR, Discospinster, FT2, Rama, Vsmith, Pavel Vozenilek, Paul August, Andrejj, Kaisershatner, CanisRufus, Kwamikagami, PhilHibbs, Haxwell, Simonbp, Femto, Dalf, Bobo192, BrokenSegue, Enric Naval, Slicky, Kjkolb, Nk, Merope, PaulHanson, GiantSloth, Lightdarkness, Sligocki, Pion, Hu, Velella, Wtshymanski, Evil Monkey, RJFJR, Cmapm, Dfalkner, Gene Nygaard, Aeronautics, RHaworth, Dandv, StradivariusTV, Oliphaunt, Jeff3000, Jwanders, Alfakim, Firien, Triddle, Someone42, GregorB, Eras-mus, CharlesC, SeventyThree, Christopher Thomas, Graham87, Magister Mathematicae, Jan van Male, Josh Parris, Sjö, Sjakkalle, Rjwilmsi, Seidenstud, Fish and karate, FlaBot, PhilipSargent, Jeepo~enwiki, Gurch, Leslie Mateus, Fosnez, Goudzovski, Skierpage, Chobot, DVdm, Ahpook, Takaaki, Roboto de Ajvol, The Rambling Man, Wavelength, Mollsmolyneux, Bhny, JabberWok, Netscott, Hydrargyrum, CambridgeBayWeather, Salsb, GeeJo, Harksaw, Długosz, RyanLivingston, Ino5hiro, Mkouklis, Nineteenthly, Mccready, Dhollm, Scottfisher, Quarky2001, DeadEyeArrow, Oliverdl, Tonym88, Codell, Searchme, Light current, 2over0, DaveOinSF, Theda, Closedmouth, Bamse, Filou~enwiki, Petri Krohn, JoanneB, Alias Flood, Wylie440, Chaiken, SkerHawx, Kgf0, Children of the dragon, SmackBot, Melchoir, Gilliam, Oscarthecat, Chaojoker, Kmarinas86, Chris the speller, RevenDS, NCurse, Thumperward, Papa November, Complexica, AtmanDave, Kostmo, Dual Freq, Trekphiler, KaiserbKot, TheKMan, LouScheffer, Elendil's Heir, Toomontrangle, Pwjb, Smokefoot, Eynar, DMacks, Paulish, Simon Arnold, Lester, Nbishop, Breadbox, Kuru, John, JorisvS, Smartyllama, Manjish, IronGargoyle, Spiel496, Citicat, Majormcmuffin, Kvng, Astrobradley, JarahE, KJS77, Brienanni, Japhet, Hmtamza, Tawkerbot2, Chetvorno, CmdrObot, Van helsing, MorkaisChosen, CBM, WMSwiki, Tim1988, Lokal Profil, Phatom87, Britannic~enwiki, Cydebot, Kam42705, Neil Froschauer, Chasingsol, Myscrnnm, Lee, IComputerSaysNo, Arwen4014, Editor at Large, TrevorRC, Matwilko, Raschd, Epbr123, Kubanczyk, Dasacus, Headbomb, Dgies, Cyclonenim, Courtjester555, Mojohaza1, Casomerville, Yellowdesk, JAnDbot, Quentar~enwiki, Smartcat, Bongwarrior, VoABot II, Ginga2, SineWave,

Jjasi, Web-Crawling Stickler, Dirac66, Coolkoon, Limtohhan, Joshua Davis, Schmloof, Xantolus, CommonsDelinker, Pharaoh of the Wizards, Jtw11, Dmrmatt19, Hans Dunkelberg, Uncle Dick, Maurice Carbonaro, Nigholith, MrBell, Eliz81, Bakkouz, Rod57, Bot-Schafter, TomyDuby, Anatoly larkin, Wimox, Equazcion, Tevonic, Useight, Qaz123qaz, Bertiethecat, Idioma-bot, JeffreyRMiles, VolkovBot, TXiKiBoT, Neha simon, Calwiki, Hqb, Liquidcentre, JosephJohnCox, OlavN, Sodapopinski, Robert1947, Burntsauce, Elecwikiman, Fischer.sebastian, AlleborgoBot, Shanmugammpl, Runewiki777, Steven Weston, SieBot, Yintan, Vanished User 8a9b4725f8376, FSUlawalumni, Keilana, Hzh, Henry Delforn (old), Onopearls, Anchor Link Bot, Hamiltondaniel, Geoff Plourde, Elliott-rhodes, TubularWorld, Tegrenath, LarRan, ClueBot, Trojancowboy, Fuzzylunkinz, Ctiefel, Techdawg667, VsBot, YBCO, Niceguyedc, Rotational, Cousins.inc, CohesionBot, Jeck1335, Doctorpsi, PixelBot, Bob man801, Lartoven, Brews ohare, Neucleon, Natty sci~enwiki, Doprendek, SchreiberBike, Aitias, Subash.chandran007, SoxBot III, HumphreyW, LSTech, Tarlneustaedter, Wertuose, BodhisattvaBot, Rror, Ngebbett, Ost316, WikHead, Noctibus, ElMeBot, Addbot, Forscite, AVand, DOI bot, Melab-1, Travisoto, Flning, Jncraton, CanadianLinuxUser, Leszek Jańczuk, CarsracBot, Dr. Universe, K Eliza Coyne, Gwcdt, Lightbot, SPat, Luckas-bot, Yobot, Fraggle81, THEN WHO WAS PHONE?, CinchBug, Csmallw, MassimoAr, AnomieBOT, Cryogenics, Guff2much, Materialscientist, Citation bot, Xqbot, Eep not for fat people, Waleswatcher, NinjaDreams, Janolaf30, Dave3457, GliderMaven, FrescoBot, WikiMcGowan, Tobby72, AlanDewey, Citation bot 1, ASchwarz, Pinethicket, HRoestBot, Schrodingers rabbit, 10metreh, Tom.Reding, Gruntler, Richardc03, Mikespedia, Heller2007, Felix0411, Anoop ranjan, Aleitner, Ahsbenton, Agnel P.B., Catcamus, Akoufos, Jiyojolly, Dick Chu, Noommos, Haj33, EmausBot, John of Reading, WikitanvirBot, DonyG, JasonSaulG, Mathew10111, Pascalf, Hhhippo, ManosHacker, Medeis, A930913, Tls60, Sailsbystars, Nothingbutdreamer, ChuispastonBot, AndyTheGrump, DASHBotAV, WikiBaller, ClueBot NG, Gilderien, Hightc, Widr, Names are hard to think of, Helpful Pixie Bot, Sina.zapf, Mightyname, Nightenbelle, Jubobroff, Bibcode Bot, BG19bot, Virtualerian, Island Monkey, Ymblanter, Andol, WikiHacker187, Mark Arsten, 52 6f 62, Pong711, BattyBot, Bv.vasiliev, Chim02, MahdiBot, Jimw338, Embrittled, Adwaele, Protectionwi, Dexbot, Anandaraja, Oliver brookes, Fittold27, TwoTwoHello, Andyhowlett, Reatlas, Ruby Murray, François Robere, Rabbitflyer, Asik Ram, Monkbot, Jrafner, Laurencejwolf, Scipsycho, OzRamos, KasparBot, Superspin, Shao xc and Anonymous: 526

- **Superfluidity** *Source:* https://en.wikipedia.org/wiki/Superfluidity?oldid=696002973 *Contributors:* DragonflySixtyseven, Discospinster, Isaac Rabinovitch, Kmarinas86, Sbharris, John, Chrumps, CuriousEric, Cydebot, Lamro, Coldcreation, Addbot, Linket, PianoDan, Materialscientist, Citation bot, Abce2, Tom.Reding, Trappist the monk, LilyKitty, Nickjf22, AManWithNoPlan, David C Bailey, Vippylaman, Bibcode Bot, Ugncreative Usergname, 220 of Borg, BattyBot, Adwaele, Dexbot, GyaroMaguus, Agonbroke, Reatlas, Jonpao523, Ybidzian, JamesMoose, HamiltonFromAbove, KasparBot, Loapsodiap, Chemistry1111 and Anonymous: 16

- **Supersolid** *Source:* https://en.wikipedia.org/wiki/Supersolid?oldid=693784678 *Contributors:* Tarquin, Heron, Schneelocke, Charles Matthews, Wikiborg, Finlay McWalter, Jeffq, Merovingian, Rasmus Faber, Graeme Bartlett, Mooquackwooftweetmeow, Hellisp, Freakofnurture, Rich Farmbrough, Bender235, El C, Alamino, Lysdexia, RJFJR, Woohookitty, Yuriybrisk, MZMcBride, Wars, Bgwhite, Hairy Dude, Hellbus, GeeJo, Niccus, SmackBot, Gilliam, Kmarinas86, Chris the speller, Pionshivu, Colonies Chris, Xyzzyplugh, Daydreamer302000, Maximum bobby, Tanadeau, Docmagoo2, Samjlord, Wmullin, Em3ryguy, Arch dude, Fourchannel, Psywolf, Casmith 789, Joshua Davis, Hweimer, J.delanoy, Jtw11, Reedy Bot, Ignacio Icke, KylieTastic, VolkovBot, Pamputt, Ideal gas equation, Techdawg667, Rafaelgarcia, Connor 99, Excirial, PixelBot, Daimanius, PSimeon, Lalegria, Addbot, OlEnglish, Yobot, Amirobot, Bility, AnomieBOT, RevZoe, Materialscientist, Citation bot, Obersachsebot, RibotBOT, Saudepp, Merongb10, Tom.Reding, Uhap027, AManWithNoPlan, Calabe1992, Bibcode Bot, Trevayne08, HMman, Edursosabina, Narliu, Hangjang, Abhijitpendse and Anonymous: 64

- **Quantum spin liquid** *Source:* https://en.wikipedia.org/wiki/Quantum_spin_liquid?oldid=682215657 *Contributors:* Piranha, Melaen, V8rik, Rjwilmsi, Kgf0, CmdrObot, Headbomb, Stevvers, Buzzm, LokiClock, Addbot, SPat, Dmitry Rozhkov, AnomieBOT, Pinethicket, Tom.Reding, NortyNort, Medeis, RockMagnetist, Hazhk, Thehintsch, Bibcode Bot, BG19bot, SyntheticTwilight, BattyBot, Gjaparidze, Shaginyan, ScitDei, Ljd2, Monkbot and Anonymous: 7

- **String-net liquid** *Source:* https://en.wikipedia.org/wiki/String-net_liquid?oldid=666131821 *Contributors:* The Anome, William Avery, Rich Farmbrough, Vsmith, Kazvorpal, Gatarron, V8rik, Rjwilmsi, Smithbrenon, Stephenb, Gaius Cornelius, Dhollm, Mlouns, SmackBot, CmdrObot, Chicken Wing, Cgingold, MartinBot, Phil.e., 1ForTheMoney, SPat, Freddy78, Tom.Reding, RjwilmsiBot, Bibcode Bot, Anrnusna, Stamptrader, Monkbot and Anonymous: 13

- **Supercritical fluid** *Source:* https://en.wikipedia.org/wiki/Supercritical_fluid?oldid=688247965 *Contributors:* Bryan Derksen, Roadrunner, Ktsquare, GTBacchus, Julesd, Phoebe, Wizzy, Tom harrison, Karn, Khalid hassani, Karol Langner, Trevor MacInnis, Mike Rosoft, Brianhe, Rich Farmbrough, Vsmith, Ddlamb, Wtshymanski, RJFJR, Pauli133, Gene Nygaard, Alai, Axeman89, Woohookitty, Mindmatrix, Pol098, Tokek, V8rik, Rjwilmsi, Seidenstud, Chobot, YurikBot, Borgx, Jengelh, Dforest, Mconst, Janet13, JDoorjam, Mejor Los Indios, Veinor, SmackBot, MarcJacobs, Edgar181, Bromskloss, Yamaguchi☒☒, KennethJ, Gilliam, Colonies Chris, Cregox, Stepho-wrs, Dan Sarandon, Nojkceb, Drphilharmonic, Mytom, Hemmingsen, IronGargoyle, Utopianheaven, Fangfufu, Novangelis, Iridescent, Tjresurrection, Rifleman 82, Cuhlik, Pöllö~enwiki, Thijs!bot, Headbomb, Pjvpjv, Mentifisto, .anacondabot, Wasell, Albmont, Aznxsinn, WikiTrazom, Mythealias, Rolfs, ChemNerd, CommonsDelinker, Ash, McSly, Afluegel, Tyrerj, Bob, Albeeman, VolkovBot, Ryan032, Ask123, Greg searle, Lamro, Mattchess, AHMartin, Gbawden, WereSpielChequers, ClueBot, Jdhenry, MechEngX, Namazu-tron, Alexbot, Sun Creator, Stainless316, Rror, Avoided, Addbot, DOI bot, TStein, DebKM, Lightbot, Luckas-bot, Yobot, Materialscientist, RadioBroadcast, Sionus, Elvim, Srich32977, RHugh, Middle 8, J6w5, Citation bot 1, Jonesey95, Tom.Reding, Onel5969, Regancy42, EmausBot, Ajraddatz, ZéroBot, H3llBot, Mountainninja, FeatherPluma, ClueBot NG, Snotbot, Helpful Pixie Bot, PencilCircle, Bibcode Bot, BG19bot, Rasheeq1, Bushiki85, WikiHannibal, BattyBot, Khazar2, Shyamukumar, Dexbot, Dsokubo, Sahstar, Nick Ackerley, Qwed4404, AshtonHM, KasparBot, Top Tier Extractions, Create642 and Anonymous: 128

- **Dropleton** *Source:* https://en.wikipedia.org/wiki/Dropleton?oldid=695583850 *Contributors:* Kjkolb, Peyre, Headbomb, M.O.X, Snaily, Brandmeister, Bibcode Bot, Everymorning, Buckrogers24 and Anonymous: 1

- **Jahn–Teller effect** *Source:* https://en.wikipedia.org/wiki/Jahn%E2%80%93Teller_effect?oldid=667957975 *Contributors:* Phil Boswell, Jotomicron, GreatWhiteNortherner, Pmanderson, AliveFreeHappy, MatthewEHarbowy, Benjah-bmm27, Geraldshields11, Ericl234, Japanese Searobin, Mahanga, Linas, Chochopk, V8rik, Rjwilmsi, Rune.welsh, Cyberfunk, Random user 39849958, Sunev, Oakwood, Carabinieri, Itub, SmackBot, Unyoyega, Srnec, Chris the speller, Cadmium, Smokefoot, DMacks, Vina-iwbot~enwiki, Ravenous75, Dicklyon, Rifleman 82, Alaibot, Savager, Brichcja, JAnDbot, Leyo, P.wormer, Kyle the bot, Djkrajnik, AlleborgoBot, Jdaloner, Chousze, Sun Creator, Hess88, Good

Olfactory, Addbot, Out of Phase User, DOI bot, LaaknorBot, Yobot, Crystal whacker, Bci2, Citation bot 1, Vacalm, Double sharp, HiW-Bot, Ready, Helpful Pixie Bot, Bibcode Bot, MickeyDonald, Serioso95, L.Sochava, Project Osprey, Andrea.zitolo, Anrnusna, KasparBot and Anonymous: 38

- **Quark–gluon plasma** *Source:* https://en.wikipedia.org/wiki/Quark%E2%80%93gluon_plasma?oldid=693347500 *Contributors:* Taw, Michael Hardy, Cyde, Karada, SebastianHelm, Charles Matthews, David Newton, Grendelkhan, Phys, Dmytro, David Edgar, JerryFriedman, Art Carlson, Herbee, Rick Block, HorsePunchKid, Mako098765, Deglr6328, Squash, MuDavid, Ylai, Bender235, CheekyMonkey, Haxwell, Bradkittenbrink, Enric Naval, Cmdrjameson, Supercrisis, Jag123, Sam Korn, Fwb22, Anthony Appleyard, Axl, Hu, Knowledge Seeker, Cal 1234, Vuo, Joriki, Firsfron, Mpatel, GregorB, SDC, Palica, RichardWeiss, Ashmoo, Yuriybrisk, Maros, Ae77, Bubba73, Nihiltres, Goudzovski, Silversmith, YurikBot, Wavelength, Mushin, Bambaiah, Hairy Dude, Hellbus, Salsb, NawlinWiki, CecilWard, E2mb0t~enwiki, Curpsbot-unicodify, Ilmari Karonen, Ybbor, KasugaHuang, Neier, SmackBot, Stepa, PeterSymonds, Skizzik, Kmarinas86, Chris the speller, Silly rabbit, Jbergquist, Khazar, Dark Formal, Vampus, Fangfufu, JayHenry, Petr Matas, CmdrObot, Foice, Van helsing, Ruslik0, Michael C Price, Thijs!bot, Headbomb, Nick Number, Eb.eric, JAnDbot, Xeno, Yill577, Savant13, 28421u2232nfenfcenc, Ethron, MartinBot, Pagw, CommonsDelinker, J.delanoy, Maurice Carbonaro, Jeepday, DorganBot, 1812ahill, Momo Hemo, Fences and windows, 0nlyth3truth, BotKung, Pamputt, Ptrslv72, AlleborgoBot, Logan, SieBot, BotMultichill, Triwbe, Maelgwnbot, ClueBot, Flaming, Thunderhippo, Brews ohare, Mstrickl, Healyhatman, DumZiBoT, XLinkBot, Oldnoah, SkyLined, Truthnlove, Stormcloud51090, Addbot, Mjamja, Qmark42, Tide rolls, Lightbot, OlEnglish, מלמד ר, Luckas-bot, Yobot, Amirobot, 4th-otaku, AnomieBOT, Essin, Citation bot, ArthurBot, ProtectionTaggingBot, False vacuum, Davdde, Ciceronibus, FrescoBot, Citation bot 1, Naxuesen, Tom.Reding, RedBot, Johann137, IVAN3MAN, Meier99, Puzl bustr, EmausBot, Mnkyman, Naznin farhah, ZéroBot, SalGiandinoto, Arbnos, Yiosie2356, SporkBot, Jesanj, Rangoon11, ClueBot NG, Jack Greenmaven, Raktimabir, Theopolisme, Helpful Pixie Bot, Bibcode Bot, 2001:db8, Shawn Worthington Laser Plasma, BattyBot, Kalmiopsiskid, Chemya, Saehry, Epicgenius, Prokaryotes, Polytope24, Pcharito, Vieque, Sofia Koutsouveli, KH-1, Crystallizedcarbon, Isambard Kingdom, Srednuas Lenoroc, Qulos, Louis de Brogile and Anonymous: 120

- **Strongly symmetric matter** *Source:* https://en.wikipedia.org/wiki/Strongly_symmetric_matter?oldid=613154560 *Contributors:* Cmdrjameson, I9Q79oL78KiL0QTFHgyc, BRW, JIP, SmackBot, Headbomb, Goldenrowley, Cgingold, Addbot, Yobot, Daniele Pugliesi, Erik9bot, BG19bot, ChessBOT and Anonymous: 6

- **Glass** *Source:* https://en.wikipedia.org/wiki/Glass?oldid=698413266 *Contributors:* MichaelTinkler, Derek Ross, Bryan Derksen, Olof, The Anome, Koyaanis Qatsi, DanKeshet, Mark, Rmhermen, Novalis, PierreAbbat, Roadrunner, SimonP, Jtoomim, Ben-Zin~enwiki, Anthere, DrBob, Heron, Mintguy, Montrealais, Stevertigo, Edward, Patrick, JohnOwens, Ken Arromdee, Michael Hardy, Tim Starling, Booyabazooka, JakeVortex, Llywrch, Pstreck, Ixfd64, Chmouel, Karada, Gbleem, Arpingstone, Shimmin, Tregoweth, Ahoerstemeier, Mac, Jimfbleak, Theresa knott, Snoyes, Angela, Darkwind, Aarchiba, Александър, Julesd, Lupinoid, Glenn, Susurrus, Evercat, Pizza Puzzle, Feedmecereal, The Tom, RodC, Timwi, Ww, Stone, Choster, JCarriker, Visorstuff, Doradus, Markhurd, Tpbradbury, Taxman, Paul-L~enwiki, Omegatron, Ed g2s, Samsara, Thue, Bevo, Carax, Raul654, Dpbsmith, Dysfunktion, Wetman, Pakaran, Jerzy, Jusjih, PuzzletChung, Robbot, Ke4roh, Josh Cherry, Noldoaran, Pigsonthewing, Kristof vt, Altenmann, Yosri, Meelar, DHN, Bkell, Moink, Mervyn, Hadal, Robinh, Xanzzibar, Iain.mcclatchie, Alan Liefting, Mor~enwiki, Marc Venot, Parasite, Giftlite, DocWatson42, Christopher Parham, Elf, BenFrantzDale, Lethe, Tom harrison, Meursault2004, Lupin, Ferkelparade, Hagedis, Ahltorp, Wwoods, Everyking, Bkonrad, Michael Devore, Stevei, Guanaco, Yekrats, Eequor, Solipsist, Foobar, Dan Gardner, Darrien, SWAdair, Tneidt, Bobblewik, Tagishsimon, Edcolins, Wmahan, Utcursh, Uranographer, Bact, Knutux, Slowking Man, LucasVB, Quadell, Antandrus, DragonflySixtyseven, Cynix, Peter bertok, Ukexpat, DanBlackham, Sonett72, Memento-Vivere, Deglr6328, ErikvDijk, Trevor MacInnis, Squash, Grunt, Freakofnurture, Poccil, Imroy, DanielCD, Zaphod-Swe, Discospinster, Rich Farmbrough, Kevinb, Vsmith, Ardonik, Ericamick, LindsayH, Xezbeth, Ponder, Mani1, Pavel Vozenilek, Martpol, Night Gyr, ESkog, Mashford, JoeSmack, Z2trillion, Sgeo, MisterSheik, CanisRufus, Adrianward, Sfahey, El C, Bletch, Shanes, Susvolans, Dennis Brown, Raverdrew, Femto, Kompas, Bobo192, Bakerq, Fir0002, HiddenInPlainSight, Dreish, Duk, Filiocht, Vortexrealm, Apyule, Savvo, ParticleMan, Harvestgalaxy, Homerjay, Vystrix Nexoth, Aquillion, Jojit fb, Bert Hickman, Chuckstar, Pschemp, Nsaa, Stephen Bain, Jumbuck, Wendell, Alansohn, Duffman~enwiki, Mo0, Melromero, Andrewpmk, AzaToth, SeanLegassick, Mailer diablo, Walkerma, Dark Shikari, CJ, Bart133, Svartalf, Ayeroxor, Melaen, ClockworkSoul, TaintedMustard, XB-70, BRW, Wtshymanski, Evil Monkey, Grenavitar, Pytheas, Bsadowski1, DV8 2XL, Gene Nygaard, Goulo, LukeSurl, HenryLi, Bookandcoffee, Ceyockey, Sheynhertz-Unbayg, Crosbiesmith, Roland2~enwiki, The JPS, Simetrical, Woohookitty, 2004-12-29T22:45Z, LOL, LogicX, Lunar Jesters, StradivariusTV, Uncle G, Benbest, Polyparadigm, WadeSimMiser, MONGO, Rtdrury, Tabletop, Oldie~enwiki, GregorB, SCEhardt, Macaddct1984, Eyreland, Wdanwatts, Zzyzx11, Prashanthns, Sweetfreek, Dysepsion, Graham87, Marskell, Sparkit, V8rik, BD2412, FreplySpang, Csnewton, Haikupoet, Yurik, Josh Parris, Sjakkalle, Rjwilmsi, Pdelong, CristianChirita, Valentinejoesmith, DeadlyAssassin, Darguz Parsilvan, SpNeo, SMC, DonSiano, SeanMack, Brighterorange, The wub, Matt Deres, Lotu, Rachelcutie, Yamamoto Ichiro, Algebra, SystemBuilder, FlaBot, RobertG, Intersofia, Nihiltres, Nivix, Rune.welsh, RexNL, Gurch, Subversive, Sperxios, Tysto, Srleffler, Smithbrenon, King of Hearts, Nicholasink, Chobot, Sbrools, DVdm, 334a, Adoniscik, Gwernol, Straker, FrankTobia, Roboto de Ajvol, YurikBot, Spacepotato, Mare, RobotE, Sceptre, Hairy Dude, Jimp, Kafziel, JarrahTree, RussBot, BruceDLimber, Muchness, Witan, SpuriousQ, IByte, ClareyF, M0RHI, RadioFan, Hydrargyrum, Stephenb, Gaius Cornelius, Shaddack, Rsrikanth05, Wimt, Anomalocaris, Rat144, Royalbroil, NawlinWiki, Wiki alf, Wiktionary4Prez!, Shayden, Spike Wilbury, Semolo75, Rieger, Grafen, Cquan, ImGz, Taco325i, MacGyver07, Caladein, Dmoss, Matticus78, Moe Epsilon, Scs, Emersoni, EEMIV, DeadEyeArrow, Bota47, AgentLewis, Wknight94, Breakfastchief, Jwissick, Sotakeit, Pb30, JoanneB, Femmina, LeonardoRob0t, RenamedUser jaskldjslak904, David Biddulph, RunOrDie, Argo Navis, NeilN, Maxamegalon2000, Mejor Los Indios, DVD R W, CIreland, Kf4bdy, Tobyk777, AndrewWTaylor, ChemGardener, Yvwv, Veinor, Yakudza, SmackBot, Formativ, Amcbride, Errarel, Madking, Kendai, Slashme, KnowledgeOfSelf, Hydrogen Iodide, Zerida, McGeddon, Ufundo, Vald, Flyer 13, Kilo-Lima, Yuyudevil, KocjoBot~enwiki, Jagged 85, RedSpruce, Jrockley, Eskimbot, Darkain, Kintetsubuffalo, Edgar181, Yamaguchi⬜⬜, Gilliam, Ohnoitsjamie, Folajimi, Skizzik, JMiall, Anwar saadat, Chris the speller, Beineix, Ccshan, Jprg1966, Thumperward, Skomae, Anchoress, Fluri, LaggedOnUser, SchfiftyThree, Hibernian, Moshe Constantine Hassan Al-Silverburg, Sbharris, Darth Panda, Lightspeedchick, Nichetas, TheNate, Dethme0w, Can't sleep, clown will eat me, MJCdetroit, Rrburke, Fredfly, Addshore, Midnightcomm, Stevenmitchell, Nahum Reduta, COMPFUNK2, Theonlyedge, Nakon, Caniago, OranL, MichaelBillington, RandomP, Fods12, DMacks, Springyard, Acdx, Daniel.Cardenas, Mion, Springnuts, LaoAn, Bejnar, Kukini, Cookie90, Sarfa, Ohconfucius, SashatoBot, Lester, Harryboyles, Rodrigo.siqueira, John, Scientizzle, Mugsywwiii, Kipala, Shadowlynk, JorisvS, Mr.K., Peterlewis, Jim.belk, Majorclanger, Fig wright, Mr. Lefty, SpyMagician, Mathel, Stratadrake, A. Parrot, Dr Smith, Beetstra, Noah Salzman, Laogeodritt, Sandb, Jon186, Waggers, AdultSwim, Peter Horn, Jose77, Peyre, Ahhwhereami, Hu12, OnBeyondZebrax, Nehrams2020, Iridescent, Lord Anubis, Iepeulas, Laurens-af, Shoeofdeath, Mclowes, ScottHolden, StephenBuxton, Az1568, Yskyflyer, Tawkerbot2, George100, MightyWarrior, Sky-

Walker, Dia^, Xcentaur, JForget, RSido, James pic, Mikiemike, CmdrObot, Bridesmill, Ale jrb, Whiskers165, ShelfSkewed, Joelholdsworth, WeggeBot, MeekMark, Grj23, Aihtdikh, IrishJew, Karenjc, Danward, IMattUK, Peinwod, AndrewHowse, Cydebot, Kupirijo, Mblumber, WoodenFeet, Glassguy, Rifleman 82, Gogo Dodo, JFreeman, Jon Stockton, Llort, Chasingsol, Baudoin, Dusty relic, Luckyherb, Amandajm, Sadharan, Tawkerbot4, Herorev, Poodleboy, Mike1942f, Omicronpersei8, Pustelnik, Mtijn, Epbr123, Barticus88, Pajz, N5iln, Mdawg728, Dtgriscom, Marek69, A3RO, Electron9, Ufwuct, James086, Sturm55, Nick Number, Elert, Escarbot, Sean.barton, Mentifisto, Zack Howes, Cyclonenim, AntiVandalBot, Majorly, Luna Santin, Widefox, Seaphoto, Opelio, Just Chilling, Dcstreit, Jayron32, Julia Rossi, Sukh17, Kevin Hughes, Jj137, Tmopkisn, Postlewaight, Danny lost, Darklilac, Farosdaughter, Gdo01, Evan Roberts, Jasonblizzard, Altamel, Dougher, PresN, JAnDbot, Barek, Omeganian, MER-C, Plantsurfer, Aderksen, ZZninepluralZalpha, Miltopia, Hello32020, Panarjedde, 100110100, Hku04, SiobhanHansa, Acroterion, RadicalPi, Easchiff, Secret Squïrrel, Jaysweet, Bongwarrior, VoABot II, Dekimasu, JamesBWatson, Kinston eagle, Scipio Carthage, Swpb, Doug Coldwell, Voloshinov, Steven Walling, Nyttend, Midgrid, Leventozler, Theroadislong, Indon, SHCGRA Max, Allstarecho, LorenzoB, Cpl Syx, DerHexer, Floria L, Lawrie.skinner, Gun Powder Ma, Wikianon, Black Stripe, Jemijohn, MartinBot, Phantomsnake, General Jazza, Arjun01, Quanticle, Ravichandar84, TheEgyptian, Bus stop, R'n'B, Robertmiklus, CommonsDelinker, AlexiusHoratius, Redshoe2, Nono64, Lilac Soul, Noplacethatfar00, Tgeairn, Tadpole9, J.delanoy, Trusilver, Rlsheehan, StewE17, Uncle Dick, Gem-fanat, Jreferee, Extransit, Cdamama, Thaurisil, GeoWriter, Gzkn, Johnbod, Smeira, Teeler7, CMoG, Ignatzmice, Koven.rm, Samtheboy, Afluegel, Jon Ascton, AntiSpamBot, Eastcheap, WHeimbigner, Johnhardcastl, Belovedfreak, Trilobitealive, Losmilzo, KylieTastic, Juliancolton, Plindenbaum, RB972, Ojovan, Tradsud, Gtg204y, Michael Angelkovich, Iceddragons, Ja 62, TheNewPhobia, SoCalSuperEagle, Squids and Chips, Nitpikr, Idioma-bot, X!, VolkovBot, CWii, Alexm7, Barefoot Steve, Jeff G., Brando130, Jmrowland, Lear's Fool, Soliloquial, Dougie monty, Capsot, Philip Trueman, Madmikeuk, TXiKiBoT, Oshwah, Mcavenue, Zidonuke, Dawidbernard, WANGXD, Spallen, Miranda, Walor, Anonymous Dissident, Sankalpdravid, Someguy1221, Anna Lincoln, Fghhytwehg, Sancasa, Martin451, Broadbot, Jackfork, LeaveSleaves, K Watson1984, Piro-san, RiverStyx23, Pinkxmastree, Eubulides, Tri400, Andy Dingley, Applesmeagol, Y, Synthebot, Vintage lu, Tonyvan67, Df747jet, Richtom80, Joshuachen1816, Necris, Amber48, S326021, Bylandl, Why Not A Duck, Locke9k, Brianga, Mike4ty4, Hs12345, Calumite, Pjoef, AlleborgoBot, Nagy, Rahxephon001, Steven Weston, AdRock, Jannopian, MissMJ, SieBot, StAnselm, Iamemery, Tiddly Tom, Scarian, Laoris, NB-NB, Jauerback, Dawn Bard, Titands64, Matthew Yeager, Eleanor Blakelock, Triwbe, Yintan, GerardusS, Keilana, Happysailor, Flyer22 Reborn, Blackfiredaemon, Loustar99, Arknascar44, Bananastalktome, Oxymoron83, Antonio Lopez, Inderkpriyan, Nuttycoconut, Steven Crossin, Lightmouse, RW Marloe, Polbot, JackSchmidt, Alex.muller, Quisquillian, OKBot, Thomasmc14, La Parka Your Car, Moeng, Jongleur100, HighInBC, Denisarona, Davidz07, Escape Orbit, Crazz bug 5, Martarius, ClueBot, GorillaWarfare, Bob1960evens, Snigbrook, IceUnshattered, Hurleydog3, Jaded-view, Windria2009, Mild Bill Hiccup, PJFitzpatrick, Polyamorph, J8079s, SuperHamster, Hafspajen, Smart racist, Niceguyedc, MARKELLOS, Rotational, Trivialist, Bensci54, Say2anniyan, Mspraveen, Sv1xv, Excirial, -Midorihana-, Alexbot, Jusdafax, Art-top, CrazyChemGuy, Kedaung, PixelBot, Eeekster, Oam123, Blabla123456789, Lartoven, MickMacNee, Qqwwwwwwww, JamieS93, Dekisugi, Njsnjsnjs, Bubblesnstuff, The Red, SchreiberBike, La Pianista, Ryan Burke, Addhoc 6, جمزۃ جامع, Horselover Frost, Versus22, Chloenora, PotentialDanger, Paradoksikal, Robdb2, V09s82, SoxBot III, Kirby dew, JKeck, Eriryoutan, Skunkboy74, Bobbob123456, XLinkBot, Reallysmartgrammarwiz, Rror, Duncan, Lynn Biggs, Jgamleus, Dthomsen8, Zombaid, Avoided, Badgernet, Jbeans, Ruth Fillery-Travis, Dayosh, Tylnsnbl, The Rationalist, Airplaneman, Ssupercrazyy, Cewvero, Gasscg94wa, Screwup 10, Domainpubber, Addbot, Mr0t1633, Avandalen, Taaanique, Rhiltons, Element16, Landon1980, AnnaJGrant, Ronhjones, Kptanna1, FlabberGASTED7777777, CanadianLinuxUser, Fluffernutter, Ronkonkaman, Ptdreamer, Skyezx, Download, Paul o'grady, RTG, Bloodkith, Bong.dreamer, Glane23, Atler smth, Jorghex, Glass Sword, AndersBot, Roux, Favonian, SpBot, 5 albert square, Numbo3-bot, Ehrenkater, Tide rolls, Lightbot, Krano, Teles, Gail, WikiDreamer Bot, Jarble, Bermicourt, StAuNcH ChArAcTeR, Angrysockhop, Legobot, Charlottie1994, Luckas-bot, Yobot, Zaereth, Senator Palpatine, Fraggle81, Tinnalach, Mauler90, Jagclarke, THEN WHO WAS PHONE?, Sweerek, Offlobby, Eric-Wester, AnomieBOT, XL2D, Shootbamboo, Atomes, Jim1138, IRP, Galoubet, JackieBot, Piano non troppo, Det går an, Kingpin13, Ulric1313, Bobthej, Materialscientist, RobertEves92, Citation bot, Jasion982, Bob Burkhardt, DynamoDegsy, Kevinwil, Neurolysis, ArthurBot, Kiwi20004, Xqbot, The Banner, Capricorn42, Wayne Roberson, Austin, Texas, Renaissancee, Dunott, DSisyphBot, Locos epraix, The Evil IP address, Forton, K-22-22, Jashudson, Almabot, Vanroc, J04n, Pmlineditor, GrouchoBot, Awesomekk, Pipy09, Riotrocket8676, ProtectionTaggingBot, Nardisoero, Neonbabyblue, Chicoisadog, Amaury, Energybender, Peterdanielfazakas, Mailsewer, Sarinspill, Glassblock, Logger9, Wicca800, GhalyBot, Natural Cut, Spot Lips, Justin320, Shadowjams, E0steven, Erik9, SD5, Pauswa, Photnart, Dan larsen, YusrSehl, NinetyTwo, FrescoBot, Tangent747, Fortdj33, LucienBOT, Z8, Riventree, Jason asselin, RoyGoldsmith, Lenawayt, ReedH, Madababest, Freebirds, Bison05, Citation bot 1, DrilBot, Pinethicket, Vicenarian, Jucool0405, Me63636363, Abductive, Jomuma94, Triplestop, V.narsikar, Jschnur, RedBot, Ezhuttukari, Serols, WP addict 0, Jauhienij, Incubusman, Cmdahler, TobeBot, Vrenator, TBloemink, Clarkcj12, Bubbles77777, Foxterdexter, Jeffrd10, Tbhotch, Minimac, Sideways713, Marie Poise, Duckyplox, Pi zza314159, RjwilmsiBot, Sugar of love, Jackehammond, Slon02, AmmarAhmad007, Enauspeaker, DASHBot, Mr. Anon515, Orphan Wiki, WikitanvirBot, Immunize, Never give in, Boundarylayer, Dewritech, Britannic124, RA0808, JacobParker100, Minimac's Clone, Ebe123, Kavo456, Mr.PHP, Tibfulv, Gangsta420, Falconerd, Wikipelli, The Sisser, AquaThane, JackClint, Fæ, Bollyjeff, Shuipzv3, Jenks24, Dken, EdEColbert, H3llBot, Rizzoent, Jackman10, SporkBot, GianniG46, Pussybandit, Erianna, Dfdffddfdffddffdffdfd, Sepehrjamali, L Kensington, Donner60, Inka 888, Hrgriffey, Equilux, Guardianofhistory, TYelliot, DASHBotAV, Ebehn, Solelis, Petrb, ClueBot NG, BobBarker7, Jackl69007, A520, Amybethlive, Zeta0134, Bped1985, PaleCloudedWhite, Kewliopimpstuff, JETZ000, Pen glass, Tideflat, Bigjhoninelcasa28, O.Koslowski, ScottSteiner, Adwiii, Widr, Antiqueight, Ryan Vesey, Kharma9287, Helpful Pixie Bot, Bibcode Bot, Kevin451~enwiki, DBigXray, Swimmer633, Lowercase sigmabot, BG19bot, Betty Noire, Krenair, Sefuhotman3, Ngtihgiterughwet4bg4tu, Glasseswiki, The Mark of the Beast, Wiki13, AvocatoBot, Mark Arsten, TriviaMonth2011, Robodude2000, Mesconsing, Mmovchin, Reporter shrimpy, Adamchan94, Snow Blizzard, Mycosys, Glacialfox, Klilidiplomus, Achowat, M.barreda, Hdgveuy, Cebrin, IrlSmith, BattyBot, Shubhrose, Bbarazi, Mahmud Alimi Wardag, DBestofknowledge, Sameersbkrgupta, Cyberbot II, Lewis8967, ChrisGualtieri, EuroCarGT, Jesseakabob, MadGuy7023, SanctusofRajput, Dexbot, Hmainsbot1, Lugia2453, WilliamDigiCol, Wikieditor26, Sly elf, Memberofthephamily, Liam123fat, Kevin12xd, 900mill, Qzykcc, Avivgy, Reatlas, Seiharu85, Faizan, KBRATTON5, Epicgenius, Timend, TJROCKS000, Sampleez, Sevınti faıv, Greenworldinfo, Felipe Marra, Damon.cross, PhantomTech, Poonsmasher, Extabulis, Pokechu22, ElHef, DavidLeighEllis, CensoredScribe, Connorrandall, Cahhhh, 296.x, 101kittykat, Tonyman3, Dinosarelife, XLEG1TxGAMINGx, RJW2013, Jianhui67, AtticTapestry, Glassman90, The Gragster, SwagmasterE 4, Devinpooop, M3owUsername1, Dave1634, Narry hemmo7, Sqyv, Moutmat, Dumbarse12321, Trackteur, Retard1114, Jackhhay, Garfield Garfield, I will destroy you 420, Kaitlyn153, DSCrowned, Abulasar, Davebignob, TheMagikCow, Julietdeltalima, Iwilsonp, NoBodyKnowZ, Rissalynn13, TeaLover1996, Msanitam, Cubicalgauardian, Glass is a solid, Mediavalia, Edgelessroom, David4114, Zortwort, Elfman1616, Merkcid, Joe898989, Schwabyboy108, TessaDiamond, KasparBot, BaronVonVeon, Flance master11, IanMcBian, Hax hacks hax, Majora, Magdaphanaflmao, Kenschutz, Zombie-crusher72, Bobbypope101, Monstercookie72, TrollBroXD, 360mlgnoscoper420, Zippyzam, Moist As

Usual and Anonymous: 1493

- **Superglass** *Source:* https://en.wikipedia.org/wiki/Superglass?oldid=696633044 *Contributors:* Velella, Dhollm, Headbomb, Steveprutz, Afluegel, Nubiatech, Happysailor, KoshVorlon, Wolfch, Addbot, Tcncv, Luckas-bot, IRP, Tom.Reding, ZéroBot, ClueBot NG, HazelAB, Bibcode Bot, Wwaters1 and Anonymous: 7

33.4.2 Images

- **File:1-cooling-crystallizer-schladen.JPG** *Source:* https://upload.wikimedia.org/wikipedia/commons/e/ee/1-cooling-crystallizer-schladen. JPG *License:* CC BY-SA 3.0 *Contributors:* Own work *Original artist:* Elmschrat Coaching38

- **File:1D_normal_modes_(280_kB).gif** *Source:* https://upload.wikimedia.org/wikipedia/commons/9/9b/1D_normal_modes_%28280_kB% 29.gif *License:* CC-BY-SA-3.0 *Contributors:* This is a compressed version of the Image:1D normal modes.gif phonon animation on Wikipedia Commons that was originally created by Régis Lachaume and freely licensed. The original was 6,039,343 bytes and required long-duration downloads for any article which included it. This version is 4.7% the size of the original and loads *much* faster. This version also has an interframe delay of 40 ms (v.s. the original's 100 ms). Including processing time for each frame, this version runs at a frame rate of about 20–22.5 Hz on a typical computer, which yields a more fluid motion. Greg L 00:41, 4 October 2006 (UTC). (from http://en.wikipedia.org/wiki/Image: 1D_normal_modes_%28280_kB%29.gif) *Original artist:* Original Uploader was Greg L (talk) at 00:41, 4 October 2006.

- **File:2006-01-14_Surface_waves.jpg** *Source:* https://upload.wikimedia.org/wikipedia/commons/4/43/2006-01-14_Surface_waves.jpg *License:* CC-BY-SA-3.0 *Contributors:* picture taken by Roger McLassus (improved by DemonDeLuxe, Sep 2006) *Original artist:* Roger McLassus

- **File:3D_model_hydrogen_bonds_in_water.svg** *Source:* https://upload.wikimedia.org/wikipedia/commons/c/c6/3D_model_hydrogen_bonds _in_water.svg *License:* CC BY-SA 3.0 *Contributors:* File:3D model hydrogen bonds in water.jpg *Original artist:* User Qwerter at Czech wikipedia:Qwerter. Transferred from cs.wikipedia; Transfer was stated to be made by User:sevela.p. Translated to english by by Michal Maňas (User:snek01). Vectorized by Magasjukur2

- **File:6Cube-QuasiCrystal.jpg** *Source:* https://upload.wikimedia.org/wikipedia/commons/9/95/6Cube-QuasiCrystal.jpg *License:* CC BY-SA 3.0 *Contributors:* Own work *Original artist:* Jgmoxness

- **File:Al71Ni24Fe5_TEM.jpg** *Source:* https://upload.wikimedia.org/wikipedia/commons/2/26/Al71Ni24Fe5_TEM.jpg *License:* CC BY 4.0 *Contributors:* http://www.nature.com/srep/2015/150313/srep09111/full/srep09111.html *Original artist:* Paul J. Steinhardt et al.

- **File:Ambox_important.svg** *Source:* https://upload.wikimedia.org/wikipedia/commons/b/b4/Ambox_important.svg *License:* Public domain *Contributors:* Own work, based off of Image:Ambox scales.svg *Original artist:* Dsmurat (talk · contribs)

- **File:Amethystemadagascar2.jpg** *Source:* https://upload.wikimedia.org/wikipedia/commons/7/71/Amethystemadagascar2.jpg *License:* CC BY 3.0 *Contributors:* Own work *Original artist:* Didier Descouens

- **File:Blacksmoker_in_Atlantic_Ocean.jpg** *Source:* https://upload.wikimedia.org/wikipedia/commons/6/6f/Blacksmoker_in_Atlantic_Ocean. jpg *License:* Public domain *Contributors:* NOAA Photo Library *Original artist:* P. Rona

- **File:Bose-Einstein_Condensation.ogv** *Source:* https://upload.wikimedia.org/wikipedia/commons/d/d9/Bose-Einstein_Condensation.ogv *License:* CC BY-SA 3.0 *Contributors:* Own work *Original artist:* Juboroff Juboroff J.Bobroff and full list in credits

- **File:Bose_Einstein_condensate.png** *Source:* https://upload.wikimedia.org/wikipedia/commons/a/af/Bose_Einstein_condensate.png *License:* Public domain *Contributors:* NIST Image *Original artist:* NIST/JILA/CU-Boulder

- **File:Boyle_air_pump.jpg** *Source:* https://upload.wikimedia.org/wikipedia/commons/3/31/Boyle_air_pump.jpg *License:* Public domain *Contributors: New Experiments ... Touching the Spring of the Air ... Original artist:* Robert Boyle

- **File:Bulk_Metallic_Glass_Sample.jpg** *Source:* https://upload.wikimedia.org/wikipedia/commons/e/e6/Bulk_Metallic_Glass_Sample.jpg *License:* Public domain *Contributors:* Photo taken in the lab *Original artist:* George Stobbart

- **File:CD-RW_bottom.jpg** *Source:* https://upload.wikimedia.org/wikipedia/commons/c/cf/CD-RW_bottom.jpg *License:* CC-BY-SA-3.0 *Contributors:* No machine-readable source provided. Own work assumed (based on copyright claims). *Original artist:* No machine-readable author provided. Invalid username 18786~commonswiki assumed (based on copyright claims).

- **File:CERN-cables-p1030764.jpg** *Source:* https://upload.wikimedia.org/wikipedia/commons/c/cc/CERN-cables-p1030764.jpg *License:* CC BY-SA 2.0 fr *Contributors:* Own work *Original artist:* Rama

- **File:CFD_Shuttle.jpg** *Source:* https://upload.wikimedia.org/wikipedia/commons/b/ba/CFD_Shuttle.jpg *License:* Public domain *Contributors:* http://www.nasa.gov/multimedia/imagegallery/image_feature_431.html *Original artist:* NASA

- **File:COT_rad_anion_Frost_orbs.png** *Source:* https://upload.wikimedia.org/wikipedia/commons/d/d7/COT_rad_anion_Frost_orbs.png *License:* Public domain *Contributors:* Transferred from en.wikipedia to Commons. *Original artist:* Cyberfunk at English Wikipedia

- **File:CalciteEchinosphaerites.jpg** *Source:* https://upload.wikimedia.org/wikipedia/commons/8/87/CalciteEchinosphaerites.jpg *License:* Public domain *Contributors:* Own work *Original artist:* Mark A. Wilson (Department of Geology, The College of Wooster).[1]

- **File:Carbon_dioxide_density-pressure_phase_diagram.jpg** *Source:* https://upload.wikimedia.org/wikipedia/commons/3/37/Carbon_dioxide _density-pressure_phase_diagram.jpg *License:* Public domain *Contributors:* M.A.Jacobs,Measurement and modeling of thermodynamic properties for the processing of polymers in supercriticalfluids,PhD thesis,Eindhoven University of Techynology,Eindhoven,2005
  based on data from:

  - S. Angus, B. Armstrong and K.M. de Reuck, International thermodynamic tables of the fluid state. Carbon dioxide, Pergamon Press, Oxford, 1976.

- R. Span and W. Wagner, J. Phys. Chem. Ref. Data 25 (1996) 1509.

Original artist: Dr. Marc A. Jacobs

- **File:Carbon_dioxide_pressure-temperature_phase_diagram.svg** *Source:* https://upload.wikimedia.org/wikipedia/commons/1/13/Carbon_dioxide_pressure-temperature_phase_diagram.svg *License:* CC0 *Contributors:* Commons, Image:Carbon dioxide pressure-temperature phase diagram.jpg *Original artist:*

  - Ben Finney
  - Mark Jacobs

- **File:Cascade-process-of-ionization.svg** *Source:* https://upload.wikimedia.org/wikipedia/commons/c/ce/Cascade-process-of-ionization.svg *License:* CC BY-SA 3.0 *Contributors:* Cascade process of ionization.png *Original artist:* Original png by Rudolfensis; svg version by Angelito7

- **File:Chem_template.svg** *Source:* https://upload.wikimedia.org/wikipedia/commons/a/ac/Chem_template.svg *License:* Public domain *Contributors:* own work inspired by <a href='//commons.wikimedia.org/wiki/File:Chem_template.png' class='image'><img alt='Chem template.png' src='https://upload.wikimedia.org/wikipedia/commons/6/64/Chem_template.png' width='48' height='48' class='thumbborder' data-file-width='48' data-file-height='48' /></a><a href='//commons.wikimedia.org/wiki/File:Chem_template.svg' class='image'><img alt='Chem template.svg' src='https://upload.wikimedia.org/wikipedia/commons/thumb/a/ac/Chem_template.svg/46px-Chem_template.svg.png' width='46' height='46' class='thumbborder' srcset='https://upload.wikimedia.org/wikipedia/commons/thumb/a/ac/Chem_template.svg/69px-Chem_template.svg.png 1.5x, https://upload.wikimedia.org/wikipedia/commons/thumb/a/ac/Chem_template.svg/92px-Chem_template.svg.png 2x' data-file-width='400' data-file-height='400' /></a> *Original artist:* Amada44

- **File:Cholesterinisch.png** *Source:* https://upload.wikimedia.org/wikipedia/commons/d/d2/Cholesterinisch.png *License:* CC-BY-SA-3.0 *Contributors:* retirved form german Wikipedia originaly uploaded on 00:55, 19. Jan 2005 *Original artist:* de:Benutzer:Heimoponnath

- **File:Cholesteryl_benzoate.png** *Source:* https://upload.wikimedia.org/wikipedia/commons/9/9a/Cholesteryl_benzoate.png *License:* Public domain *Contributors:* ? *Original artist:* ?

- **File:Chrysler_Building_detail.jpg** *Source:* https://upload.wikimedia.org/wikipedia/commons/b/b3/Chrysler_Building_detail.jpg *License:* CC-BY-SA-3.0 *Contributors:* Digital photo by en:User:Postdlf. *Original artist:* User Postdlf on en.wikipedia

- **File:ColloidalStability.png** *Source:* https://upload.wikimedia.org/wikipedia/commons/c/cc/ColloidalStability.png *License:* CC BY 3.0 *Contributors:* SunKart (talk) (Uploads) *Original artist:* SunKart at en.wikipedia

- **File:Commons-logo.svg** *Source:* https://upload.wikimedia.org/wikipedia/en/4/4a/Commons-logo.svg *License:* ? *Contributors:* ? *Original artist:* ?

- **File:ComparisonStericStab-ShearThinningFluids2.png** *Source:* https://upload.wikimedia.org/wikipedia/en/e/ef/ComparisonStericStab-png *License:* CC-BY-3.0 *Contributors:*
  SunKart (talk) (Uploads) *Original artist:*
  SunKart (talk) (Uploads)

- **File:Compressibility_Factor_of_Air_75-200_K.png** *Source:* https://upload.wikimedia.org/wikipedia/commons/8/84/Compressibility_Factor_of_Air_75-200_K.png *License:* Public domain *Contributors:* Own work *Original artist:* Power.corrupts

- **File:Crystal_facet_formation.svg** *Source:* https://upload.wikimedia.org/wikipedia/commons/9/95/Crystal_facet_formation.svg *License:* CC0 *Contributors:* Own work *Original artist:* Sbyrnes321

- **File:Crystalline_polycrystalline_amorphous2.svg** *Source:* https://upload.wikimedia.org/wikipedia/commons/0/05/Crystalline_polycrysta amorphous2.svg *License:* CC BY-SA 3.0 *Contributors:*

- Cristal ou amorphe.svg *Original artist:* Cristal_ou_amorphe.svg: Cdang

- **File:Cu_water.png** *Source:* https://upload.wikimedia.org/wikipedia/commons/a/a7/Cu_water.png *License:* CC BY-SA 3.0 *Contributors:* Own work *Original artist:* Andrea.zitolo

- **File:Cvandrhovst.png** *Source:* https://upload.wikimedia.org/wikipedia/commons/0/08/Cvandrhovst.png *License:* CC-BY-SA-3.0 *Contributors:* ? *Original artist:* ?

- **File:Daltons_symbols.gif** *Source:* https://upload.wikimedia.org/wikipedia/commons/3/39/Daltons_symbols.gif *License:* Public domain *Contributors:* ? *Original artist:* ?

- **File:Decay_schematic.jpg** *Source:* https://upload.wikimedia.org/wikipedia/en/5/52/Decay_schematic.jpg *License:* GFDL *Contributors:* ? *Original artist:* ?

- **File:Different_minerals.jpg** *Source:* https://upload.wikimedia.org/wikipedia/commons/a/af/Different_minerals.jpg *License:* CC BY-SA 3.0 *Contributors:* Own work *Original artist:* Brocken Inaglory

- **File:Diffusion_animation.gif** *Source:* https://upload.wikimedia.org/wikipedia/commons/f/fe/Diffusion_animation.gif *License:* Public domain *Contributors:* http://classwww.gsfc.nasa.gov/CAGESite/pages/demo3.htm *Original artist:* ?

- **File:Edit-clear.svg** *Source:* https://upload.wikimedia.org/wikipedia/en/f/f2/Edit-clear.svg *License:* Public domain *Contributors:* The *Tango! Desktop Project. Original artist:*
  The people from the Tango! project. And according to the meta-data in the file, specifically: "Andreas Nilsson, and Jakub Steiner (although minimally)."

- **File:Ehrenfest_Lorentz_Bohr_Kamerlingh_Onnes.jpg** *Source:* https://upload.wikimedia.org/wikipedia/commons/9/99/Ehrenfest_Lorentz_Bohr_Kamerlingh_Onnes.jpg *License:* Public domain *Contributors:* http://www.luf.nl/site/start.asp?hoofdcategorieID=2&paginaID=365 *Original artist:* unbekannt

33.4.3 Content license